ADHESION PROMOTION TECHNIQUES

MATERIALS ENGINEERING

Additional Volumes in Preparation

ADHESION PROMOTION TECHNIQUES

Technological Applications

edited by

K. L. MITTAL
Hopewell Junction, New York

A. PIZZI
ENSTIB
University of Nancy I
Epinal, France

CRC Press
Taylor & Francis Group
Boca Raton London New York

CRC Press is an imprint of the
Taylor & Francis Group, an **informa** business

CRC Press
Taylor & Francis Group
6000 Broken Sound Parkway NW, Suite 300
Boca Raton, FL 33487-2742

First issued in paperback 2019

© 2002 by Taylor Francis Group, LLC
CRC Press is an imprint of Taylor & Francis Group, an Informa business

No claim to original U.S. Government works

ISBN-13: 978-0-8247-0239-7 (hbk)
ISBN-13: 978-0-367-39996-2 (pbk)

**Visit the Taylor & Francis Web site at
http://www.taylorandfrancis.com**

**and the CRC Press Web site at
http://www.crcpress.com**

Preface

Adhesion plays an important role in a legion of technological areas ranging from aerospace to biomedical applications to microelectronics. When two materials are brought in contact, the proper or adequate adhesion between them is of great importance, so it is necessary to devise ways to attain the requisite adhesion strength between different materials. Even a cursory look at the literature shows that there has been brisk activity in ameliorating the existing techniques or in devising new and improved methods to modify different materials to enhance their adhesion. Polymer materials innately are poorly adhesionable because of their relative inertness and low surface energy, and there has been tremendous R&D activity in the arena of polymer surface modification to render them adhesionable.

This book was conceived with the idea of providing a summary of state-of-the-art anent adhesion promotion techniques for different materials. It contains 12 chapters written by internationally renowned authors. The first chapter by Schultz and Nardin expatiates upon the many theories or mechanisms of adhesion, which provide the requisite basis to create the apropos chemistry or morphology on the material surfaces to enhance their adhesion. The next chapter by Chehimi provides an exegesis on acid-base interactions and how to harness these to improve adhesion. Incidentally, the acid-base mechanism of adhesion is in vogue today and the number of examples where the acid-base concept is operative or useful has been increasing over the last decade. The chapter by Pizzi concentrates on the fundamental approach to adhesion by addressing the relevance/importance of molecular mechanics/dynamics modeling in adhesion. Atomic/molecular understanding of adhesion should be extremely beneficial in selecting or creating the appropriate materials to attain the desired adhesion strength. The next chapter by Hayes and Ralston deals with the application of atomic force microscopy (AFM) in fundamental adhesion studies. There has been brisk interest and activity in the last few years in utilizing AFM in the field of adhesion. Such AFM studies are unraveling the reasons why different materials show different adhesion behavior.

The next three chapters provide an exposition on different ways to treat polymer surfaces to improve their adhesion to different materials. The chapter by Wertheimer and colleagues concentrates on plasma treatment of polymers to improve their adhesion. Such treatment is commonly used these days and has been shown to be very useful in improving adhesion of polymeric materials. Flame treatment of polymers to improve adhesion is the subject of the chapter by Brewis and Mathieson. It should be mentioned here that flame treatment is used commercially on large polymer parts. Uehara discusses the corona treatment of polymers to improve adhesion, a treatment that has also found much commercial application. The chapter by Buchman and Dodiuk-Kenig summarizes the recent developments in laser surface treatment of various materials to improve adhesion. Such treatment is a relatively new addition to the arsenal of surface modification techniques but it is followed these days at an accelerated pace. The chapter by Pedraza discusses the adhesion enhancement of metallic films to ceramic substrates using lasers and low energy ions. Metal/ceramic systems are of paramount importance in many advanced technologies. The chapter by Kang and colleagues provides a review of surface graft copolymerization and grafting of polymers for adhesion improvement. Such approaches offer the opportunity to tailor the surface chemistry of polymers (e.g., from the acid-base viewpoint) and thus enhance adhesion. The next chapter by Pisanova summarizes the microbial treatment of polymer surfaces to improve adhesion. Such treatment is relatively new but offers exciting possibilities. As we learn more about the interactions of microorganisms with polymers, new vistas will emerge. The book concludes with the chapter by Matisons and colleagues which discusses silanes on glass fibers as adhesion promoters for composite applications. In reinforced composites, proper adhesion between the reinforcement and the matrix is crucial, and today composites constitute a very important class of materials.

We certainly hope that anyone with core or tangential interest in adhesion will find this book a fountain of information. Also it should serve as a Baedeker to the novice and a commentary on the current activity on adhesion promotion techniques for the veteran researcher. The book is profusely referenced and amply illustrated. Furthermore, it should be recorded here that this book is the first one dedicated exclusively to the topic of adhesion promotion techniques.

Now it is our pleasure to express our sincere thanks to the authors for their interest and contributions without which this book would not have seen the light of day.

K. L. Mittal
A. Pizzi

Contents

Contributors

Derek McHardy Brewis Department of Physics, Loughborough University, Loughborough, Leicestershire, United Kingdom

Leanne Britcher Ian Wark Research Institute, University of South Australia, Mawson Lakes, South Australia, Australia

Alisa Buchman Materials and Processes, RAFAEL, Haifa, Israel

Mohamed Mehdi Chehimi Institut de Topologie et de Dynamique des Systèmes, Université Paris 7—Denis Diderot, Associé au CNRS (UPRESA 7086), Paris, France

Gregory Czeremuszkin Polyplasma, Inc., Montreal, Quebec, Canada

Hanna Dodiuk-Kenig Materials and Processes, RAFAEL, Haifa, Israel

Robert A. Hayes Ian Wark Research Institute, University of South Australia, Mawson Lakes, South Australia, Australia

E. T. Kang Department of Chemical Engineering, National University of Singapore, Kent Ridge, Singapore

Scott Kempson Ian Wark Research Institute, University of South Australia, Mawson Lakes, South Australia, Australia

Jolanta E. Klemberg-Sapieha   Department of Engineering Physics and Materials Engineering, Ecole Polytechnique, Montreal, Quebec, Canada

Der-Jang Liaw   Department of Chemical Engineering, National Taiwan University of Science and Technology, Taipei, Taiwan, Republic of China

Ludvik Martinu   Department of Engineering Physics and Materials Engineering, Ecole Polytechnique, Montreal, Quebec, Canada

Isla Mathieson   Department of Physics, Loughborough University, Loughborough, Leicestershire, United Kingdom

Janis Matisons   Ian Wark Research Institute, University of South Australia, Mawson Lakes, South Australia, Australia

M. Nardin   Institut de Chimie des Surfaces et Interfaces (CNRS), Mulhouse, France

Koon Gee Neoh   Department of Chemical Engineering, National University of Singapore, Kent Ridge, Singapore

A. J. Pedraza   Department of Materials Science and Engineering, The University of Tennessee, Knoxville, Tennessee

Elena V. Pisanova   Metal-Polymer Research Institute, Academy of Sciences of Belarus, Gomel, Republic of Belarus

A. Pizzi   ENSTIB, University of Nancy I, Epinal, France

John Ralston   Ian Wark Research Institute, University of South Australia, Mawson Lakes, South Australia, Australia

J. Schultz   Institut de Chimie des Surfaces et Interfaces (CNRS), Mulhouse, France

Kuang Lee Tan   Department of Physics, National University of Singapore, Kent Ridge, Singapore

Tohru Uehara Faculty of Science and Engineering, Shimane University, Shimane, Japan

Michael R. Wertheimer Department of Engineering Physics and Materials Engineering, Ecole Polytechnique, Montreal, Quebec, Canada

1
Theories and Mechanisms of Adhesion

J. Schultz and M. Nardin
Institut de Chimie des Surfaces et Interfaces (CNRS), Mulhouse, France

I. INTRODUCTION

Adhesion phenomena are relevant to many scientific and technological areas and in recent years have become a very important field of study.

The main application of adhesion is bonding by adhesives, which is replacing, at least partially, more classical mechanical attachment techniques such as bolting or riveting. It is considered to be competitive primarily because it saves weight, ensures better stress distribution, and offers better aesthetics because the glue line is practically invisible. Applications of bonding by adhesives can be found in many industries, particularly in advanced technological domains such as the aeronautical and space industries, automobile manufacture, and electronics. Adhesives have also been introduced in areas such as dentistry and surgery.

However, adhesive joints are not the only application of adhesion. Adhesion is concerned whenever solids are brought into contact, for instance, in coatings, paints, and varnishes; multilayered sandwiches; polymer blends; filled polymers; and composite materials. Because the final performance or use properties of these multicomponent materials depend significantly on the quality of the interface that is formed between the solids, it is understandable that a better knowledge of adhesion phenomena is required for practical applications.

The field of adhesion began to create real interest in scientific circles only about 50 years ago. Thus, adhesion became a scientific subject in its own right, but it is still a subject in which empiricism and technology are slightly ahead of science, although the gap between theory and practice has been narrowed considerably.

In fact, the term *adhesion* covers a wide variety of concepts or ideas, depending on whether the subject is broached from a molecular, microscopic, or macroscopic point of view or whether one talks about the formation of the interface or the failure of the bonded system. The term adhesion is, therefore, ambiguous, meaning both the establishment of interfacial bonds as well as the mechanical load required to break an assembly. As a matter of fact, one of the main difficulties in the study of adhesion mechanisms is that the subject is at the boundary of several scientific fields, including macromolecular science, physical chemistry of surfaces and interfaces, materials science, mechanics and micromechanics of fracture, and rheology. Consequently, the study of adhesion uses various concepts, depending very much on one's special field of expertise, and therefore treatment of the phenomena observed can be considerably different. This variety of approaches is emphasized by the fact that many theories of adhesion have been proposed, which together are both complementary and contradictory:

> Mechanical interlocking
> Electronic theory (also known as electrical double layer, or electrostatic, or parallel plate capacitor theory)
> Theory of boundary layers and interphases (also known as weak boundary layer theory)
> Adsorption (thermodynamic) theory (also referred to as wettability and acid-base theory)
> Diffusion theory
> Chemical bonding theory

Among these models, one usually distinguishes rather arbitrarily between mechanical and specific adhesion, the latter being based on the different types of bonds (electrostatic, secondary, acid-base, chemical) that can develop between two solids. Actually, each of these theories is valid to some extent, depending on the nature of the solids in contact and the conditions of formation of the bonded system. Therefore, they do not negate each other and their respective importance depends largely on the system considered.

II. MECHANISMS OF ADHESION

A. Mechanical Interlocking

The mechanical interlocking model, proposed by McBain and Hopkins as early as 1925 [1], considers that mechanical keying, or interlocking, of the adhesive into the cavities, pores, and asperities of the solid surface is the major factor in determining adhesion strength. One of the most consistent examples illustrating the contribution of mechanical anchoring was given many years ago by Borroff and Wake [2], who measured the adhesion between rubber and textile fabrics. These authors clearly showed that penetration of the protruding fiber ends into the rubber was the most important parameter in such bonded systems. However, the possibility of establishing good adhesion between smooth surfaces leads to the conclusion that the theory of mechanical keying cannot be considered universal. To overcome this difficulty, following the approach suggested primarily by Gent and Schultz [3,4], Wake [5] proposed that the effects of both mechanical interlocking and thermodynamic interfacial interactions could be taken into account as multiplying factors for estimating the joint strength *G*:

$$G = (\text{constant}) \times (\text{mechanical keying component})$$
$$\times (\text{interfacial interactions component})$$

Therefore, according to the foregoing equation, a high level of adhesion should be achieved by improving both the surface morphology and physicochemical surface properties of both the substrate and the adhesive. However, in most cases, the enhancement of adhesion by mechanical keying can be attributed simply to the increase of interfacial area due to surface roughness, insofar as the wetting conditions are fulfilled to permit penetration of the adhesive into pores and cavities. However, total penetration of the adhesive into the cavities is not always possible, due to the existence of a back pressure stemming from air entrapped in these cavities. The final depth and the rate of penetration are strongly dependent on the shape of the pores (cylindrical, conical) as shown schematically in Fig. 1.

The work by Packham and coworkers [6–9] has further stressed the notable role played by the surface texture of substrates in determining the magnitude of the adhesion strength. In particular, they found [6] high values of peel strength of polyethylene on metallic substrates when a rough and fibrous-type oxide surface was formed on the substrate. Ward and collaborators [10–12] have empha-

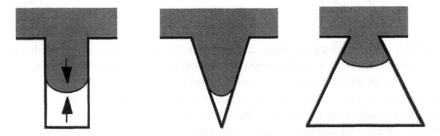

Figure 1   Schematical representation of the effect of pore shape on the penetration of the adhesive. (Arrows are related, respectively, to the wetting forces in the liquid and to the back pressure due to air entrapped in the cavity.)

sized the improvement in adhesion, measured by means of a pull-out test, between plasma-treated polyethylene fibers and epoxy resin. In that case, long-time plasma treatments created a pronounced pitted structure on the polyethylene surface, which was easily filled by the epoxy resin as a result of good wetting.

One of the most important criticisms of the mechanical interlocking theory, as suggested in different studies [9,13,14], is that improvement of adhesion does not necessarily result from a mechanical keying mechanism but that the surface roughness can increase the energy dissipated viscoelastically or plastically around the crack tip and in the bulk of the materials during joint failure. Effectively, it is now well known that this energy loss is often the major component of adhesion strength.

B. Electronic Theory (Also Known as Electrical Double Layer, or Electrostatic, or Parallel Plate Capacitor Theory)

The electronic theory of adhesion was primarily proposed by Deryaguin and co-workers in 1948 [15–19]. These authors suggested that an electron transfer mechanism between the substrate and the adhesive, having different electronic band structures, could occur in order to equalize the Fermi levels. This phenomenon induced the formation of a double electrical layer at the interface, and Deryaguin et al. proposed that the resulting electrostatic forces could contribute significantly to the adhesion strength. Therefore, the adhesive-substrate junction can be analyzed as a capacitor. During interfacial failure of this system, separation of the two plates of the capacitor leads to an increasing potential difference until a discharge occurs. Consequently, it is considered that adhesion strength results from the attractive electrostatic forces across the electrical double layer. The energy of separation of the interface, G_e, is therefore related to the discharge potential V_e as follows:

$$G_e = \frac{h\varepsilon_d}{8\pi}\left(\frac{\partial V_e}{\partial h}\right)^2 \tag{1}$$

where h is the discharge distance and ε_d the dielectric constant of the medium. Moreover, according to such an approach, adhesion would vary with the pressure of the gas in which the measurement was performed. Hence, Deryaguin et al. measured, by means of a peel test, the peel energy at different polymer-substrate interfaces, such as poly(viny chloride)/glass, natural rubber/glass, or steel systems, in argon and air environments at various gas pressures. A significant variation of peel energy versus gas pressure was evidenced and very good agreement between the theoretical values, calculated from Eq. (1), and the measured values of G_e was obtained, regardless of the nature of the gas used. However, several other studies [5,20] did not confirm these results and thus seem to indicate that the good agreement previously obtained was rather casual. According to Deryaguin's approach, the adhesion depends on the magnitude of the potential barrier at the adhesive-substrate interface. Although this potential barrier does exist in many cases (Refs. 21 and 22, for example), no clear correlation between the electronic interfacial parameters and the measured adhesion strength is usually found. Moreover, for systems constituted of glass substrate coated with a vacuum-deposited layer of gold, silver, or copper, von Harrach and Chapman [23] showed that the electrostatic contribution to peel strength, estimated from the measurement of charge densities, could be considered negligible.

Furthermore, as already mentioned, the energy dissipated viscoelastically or plastically during fracture experiments plays a major role in the measured adhesion strength but is not included conceptually in the electronic theory of adhesion.

Finally, it could be concluded that the electrical phenomena often observed during failure processes are the consequence rather than the cause of high bond strength.

C. Theory of Weak Boundary Layers—Concept of Interphase

It is now well known that alterations and modifications of the adhesive and/or adherend can be found in the vicinity of the interface, leading to the formation of an interfacial zone exhibiting properties (or property gradients) that differ from those of the bulk materials.

The first approach to this problem is due to Bikerman [24], who stated that the cohesive strength of a weak boundary layer (wbl) could always be considered the main factor in determining the level of adhesion, even when the failure appears to be interfacial. According to this assumption, the measured adhesion energy G is always equal to the cohesive energy G_c(wbl) of the weaker interfacial layer. This theory is based mainly on probability considerations showing that the fracture should never propagate only along the adhesive-substrate interface for pure statistical reasons and that cohesive failure within the weaker material near the interface is a more favorable event. Therefore, Bikerman has proposed several types of wbl's, such as those resulting from the presence at the interface of impurities, small molecules, or short polymer chains.

Nevertheless, two main criticisms of the wbl argument can be invoked. First, there is much experimental evidence that purely interfacial failure does occur for many different systems. Second, although the failure is cohesive in the vicinity of the interface in at least one of the materials in contact, this cannot necessarily be attributed to the existence of a wbl. According to several authors [25,26], the stress distribution in the materials and the stress concentration near the crack tip certainly imply that the failure must propagate very close to the interface, but not exactly at the interface.

However, the creation of interfacial layers has received much attention and has led to the concept of "thick interface" or "interphase" widely used in adhesion science [27]. Such interphases are formed irrespective of the nature of both adhesive and substrate; their thickness ranges between the molecular level (a few angstroms or nanometers) and the microscopic scale (a few micrometers or more). Many physical, physicochemical, and chemical phenomena are responsible for the formation of such interphases, as shown from examples taken from our own work [28]:

> The orientation of chemical groups or the overconcentration of chain ends to minimize the free energy of the interface [29], as well as special conformation of macromolecules at the interface [30,31]
> The migration toward the interface of additives or low-molecular-weight fraction [32–34]
> The growth of a transcrystalline structure, for example, when the substrate acts as a nucleating agent [35]
> The formation of a pseudovitrous zone resulting from a reduction in chain mobility through strong interactions with the substrate [36]
> The modification of the thermodynamics and/or kinetics of the polymerization or cross-linking reaction at the interface through preferential adsorption of reaction species or catalytic effects [37–40]

It is clear that the presence of such interphases can strongly alter the strength of multicomponent materials and that the properties of these layers must not be ignored in the analysis of adhesion measurement data. A complete understanding of adhesion allowing performance prediction must take into account the potential formation of these boundary layers.

D. Adsorption (or Thermodynamic) Theory (Also Referred to as Wettability or Acid-Base Theory)

The thermodynamic model of adhesion, generally attributed to Sharpe and Schonhorn [41], is certainly the most widely used approach in adhesion science at present. This theory considers that the adhesive will adhere to the substrate because of interatomic and intermolecular forces established at the interface, provided that intimate contact is achieved. The most common interfacial forces result from van der Waals and Lewis acid-base interactions, as described below. The magnitude of these forces can generally be related to fundamental thermodynamic quantities, such as surface free energies of both adhesive and adherend. Generally, the formation of an assembly goes through a liquid-solid contact step and, therefore, criteria for good adhesion become essentially criteria for good wetting, although this is a necessary but not sufficient condition.

In the first part of this section, wetting criteria as well as surface and interface free energies are defined quantitatively. The estimation of a reversible work of adhesion W from the surface properties of materials in contact is therefore considered. In the second part, different models relating the measured adhesion strength G to the free energy of adhesion W are examined.

1. Wetting Criteria, Surface and Interface Free Energies, and Work of Adhesion

In a solid-liquid system, wetting equilibrium is defined from the profile of a sessile drop on a planar solid surface. Young's equation [42], relating the surface tension γ of materials at the three-phase contact point to the equilibrium contact angle θ (Fig. 2), is written as

$$\gamma_{SV} = \gamma_{SL} + \gamma_{LV} \cos \theta \qquad (2)$$

Subscripts S, L, and V refer, respectively, to solid, liquid, and vapor phases such that a combination of two of these subscripts corresponds to the given interface (for example, SV corresponds to the solid-vapor interface).

The term γ_{SV} represents the surface free energy of the substrate after equilibrium adsorption of vapor from the liquid and is sometimes lower than the surface free energy γ_S of the solid in vacuum. This decrease is defined as the spreading pressure π ($\pi = \gamma_S - \gamma_{SV}$) of the vapor on the solid surface. In most cases, in particular when dealing with polymeric materials, π could be neglected and, to a first approximation, γ_S is used in place of γ_{SV} in wetting analyses. When the contact angle has a finite value ($\theta > 0°$) the liquid does not spread on the solid surface. On the contrary, when $\theta = 0°$ the liquid totally wets the solid and spreads

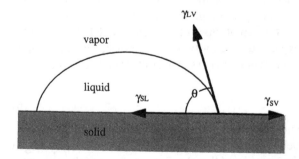

Figure 2 Drop of a liquid on a solid surface at equilibrium.

over the surface spontaneously. Hence, neglecting π, a condition for spontaneous wetting to occur is

$$\gamma_S \geq \gamma_{SL} + \gamma_{LV} \tag{3}$$

or

$$S = \gamma_S - \gamma_{SL} - \gamma_{LV} \geq 0 \tag{4}$$

the quantity S being called the spreading coefficient. Consequently, Eq. (4) constitutes a wetting criterion. It is worth noting that geometric aspects or processing conditions, such as the surface roughness of the solid (e.g., see Fig. 1) and applied external pressure, can restrict the applicability of this criterion.

However, a more fundamental approach leading to the definition of other wetting criteria is based on the analysis of the nature of forces involved at the interface and allows the calculation of the free energy of interactions between two materials to be bonded.

For low-surface-energy solids, such as polymers, many authors have estimated the thermodynamic surface free energy from contact angle measurements. The first approach was an empirical one developed by Zisman and coworkers [43–45]. They established that a linear relationship often existed between the cosine of contact angle, cos θ, of several liquids and their surface tension, γ_{LV}. Zisman introduced the concept of critical surface tension, γ_c, which corresponds to the value of the surface energy of an actual or hypothetical liquid that will just spread on the solid surface, giving a zero contact angle. However, there is no general agreement about the meaning of γ_c and Zisman himself always emphasized that γ_c was not the surface free energy of the solid but only a closely related empirical parameter.

For solid-liquid systems, according to Dupré's relationship [46], the adhesion energy W_{SL} is defined as

$$W_{SL} = \gamma_S + \gamma_{LV} - \gamma_{SL} = \gamma_{LV}(1 + \cos \theta) \tag{5}$$

in agreement with Eq. (2) and neglecting the spreading pressure. Fowkes [47] proposed that the surface free energy γ of a given entity could be represented by the sum of the contributions of different types of interactions. Schultz et al. [48] suggested that γ could be expressed by only two terms, a dispersive component (London interactions) and a polar component (superscripts D and P, respectively), as follows:

$$\gamma = \gamma^D + \gamma^P \tag{6}$$

The last term in the right-hand side of this equation corresponds to all the nondispersion forces, including Debye and Keesom interactions, as well as hydrogen bonding. Fowkes [49] has also considered that the dispersive part of these interactions between solids 1 and 2 can be well quantified as twice the geometric mean of the dispersive components of the surface energy of the two entities. Therefore, in the case of interactions involving only dispersion forces, the adhesion energy W_{12} is given by

$$W_{12} = 2(\gamma_1^D\gamma_2^D)^{1/2} \tag{7}$$

By analogy with the work of Fowkes, Owens and Wendt [50] and then Kaelble and Uy [51] have suggested that the nondispersive part of interactions between two materials can be expressed as the geometric mean of the nondispersive components of their surface energies, although there is no theoretical reason to represent all the nondispersive interactions by this type of expression. Hence, the work of adhesion W_{12} becomes

$$W_{12} = 2(\gamma_1^D\gamma_2^D)^{1/2} + 2(\gamma_1^P\gamma_2^P)^{1/2} \tag{8}$$

For solid-liquid equilibrium, a direct relationship between the contact angle θ of the drop of a liquid on a solid surface and the surface properties of both liquid and solid is obtained from Eqs. (5) and (8). By contact angle measurements of droplets of different liquids of known surface properties, the components γ_S^D and γ_S^P of the surface energy of the substrate can then be determined.

A major difficulty in using such surface energy components, especially in the case of polar polymers in contact with solids, stems from the ability of macromolecules to take particular orientations and form particular arrangements or packing near interfaces, due to the force field created by the solid materials. Therefore, according to the orientation and the location of polar groups in the vicinity of the interface, the nondispersive component of the surface energy of the polymer can be strongly modified. Such an observation has led to the new and more "dynamic" concept of *potential* surface energy of polymers [29]. This concept can be illustrated, for example, by the influence of the nature of the mold on the surface properties of ethylene copolymers [52], i.e., poly(ethylene-co-vinylacetate) and poly(ethylene-co-acrylic acid). It has been shown that after molding in contact with a nonpolar polymer, such as poly(tetrafluoroethylene) (PTFE; $\gamma_S^P \approx 0$), the nondispersive component of the surface energy of a statistical copolymer of ethylene and acrylic acid (20%) is close to zero. On the contrary, when molding is performed in contact with a polar polymer, such as poly(ethylene terephthalate) (PET; $\gamma_S^P \approx 7 \text{ mJ/m}^2$), the nondispersive component of the surface energy of this copolymer becomes significantly larger as determined by contact angle measurements of water droplets. Such an example clearly shows that the surface energy of a polymer is not an intrinsic property of the polymer and that a real difficulty does exist in estimating accurately the reversible work of adhesion.

However, there is no doubt that the notion of dispersive and nondispersive components of the surface energy constitutes a significant advance compared with the very empirical one of γ_c, but it is still controversial from a theoretical point of view. That is why other approaches were investigated and, particularly, the generalized Lewis acid-base concept. This concept describes the acid-base interactions in terms of molecular frontier orbitals and makes it possible to analyze all the different degrees of electronic sharing between two entities. For this approach, a base and an acid are defined according to the energy level of molecular orbitals, the well-known highest occupied molecular orbital (HOMO) and lowest unoccupied molecular orbital (LUMO), respectively [53]. The magnitude of acid-base interactions is directly related to the difference in energy between HOMO and LUMO levels of the interacting entities. Finally, the equilibrium intermolecular or interatomic distance between these entities decreases when the energy of interactions increases. Effectively, short intermolecular distances, i.e., 0.18 to 0.2 nm, close to those corresponding to covalent or ionic bonds, correspond to the strongest acid-base interactions (hydrogen bonds for water molecule association, for example), whereas interatomic distances of about 0.5 nm, similar to distances currently observed for van der Waals interactions, are related to the weakest acid-base interactions [54].

Fowkes and coworkers [55–58] have shown that these electron acceptor and donor interactions constitute a major type of interfacial force between the adhesive and the substrate. Such an approach is able to take into account hydrogen bonds, which are often involved in adhesive joints. Moreover, Fowkes and Mostafa [56] have shown that the contribution of the polar (dipole-dipole) interactions to the thermodynamic work of adhesion could generally be neglected compared with both dispersive and acid-base contributions. They have also considered that the acid-base component W^{ab} of the adhesion energy can be related to the variation of enthalpy, $-\Delta H^{ab}$, corresponding to the establishment of acid-base interactions at the interface, as follows:

$$W^{ab} = f(-\Delta H^{ab})n^{ab} \qquad (9)$$

where f is a factor that converts enthalpy into free energy and is taken equal to unity and n^{ab} is the number of acid-base bonds per unit interfacial area, close to about 6 μmol/m^2. Therefore, from Eqs. (7) and (9), the total adhesion energy W_{12} becomes [59]

$$W_{12} = 2(\gamma_1^D\gamma_2^D)^{1/2} + f(-\Delta H^{ab})n^{ab} \qquad (10)$$

The experimental values of the enthalpy $(-\Delta H^{ab})$ can be estimated from the work of Drago et al. [60,61], who proposed the following relationship:

$$-\Delta H^{ab} = C^A C^B + E^A E^B \qquad (11)$$

where C^A and E^A are two quantities that characterize the acidic material at the interface, whereas, similarly, C^B and E^B characterize the basic material. The validity of Eq. (11) was clearly evidenced for polymer adsorption on various substrates [58]. Another estimation of $(-\Delta H^{ab})$ can be carried out by the semiempirical approach defined by Gutmann [62], who proposed that liquids might be characterized by two constants: an electron acceptor number AN and an electron donor number DN. For solid surfaces, electron acceptor and donor numbers, K_A and K_D, respectively, have also been defined and measured by inverse gas chromatography [63–65]. In this approach, the enthalpy of formation $(-\Delta H^{ab})$ of acid-base interactions at the interface between two solids 1 and 2 is now given by [62,63]

$$-\Delta H^{ab} = K_{A1}K_{D2} + K_{A2}K_{D1} \tag{12}$$

This expression was successfully applied to describe fiber-matrix adhesion in the field of composite materials by Schultz et al. [65–67]. In these studies, the practical adhesion, defined as a normalized interfacial shear strength, was linearly related to the work of adhesion W between the fiber and the matrix (Fig. 3), estimated from Eqs. (10) and (12).

It must be mentioned that other semiempirical approaches have been proposed to quantify acid-base interactions, in particular that of van Oss et al. [68–71], who define surface energy components γ^+ and γ^- to describe, respectively, the electron acceptor and electron donor characters of liquids and solid surfaces. Finally, acid-base interactions can also be analyzed in terms of Pearson's hard-soft acid-base (HSAB) principle [72,73], the application of which to solid-solid interactions and thus to adhesion is under investigation.

At present, it is worth mentioning that a considerable amount of work and an extensive debate inside the scientific community are devoted to the acid-base concept. This community is clearly divided into two main approaches: the first one, maintained by physicists, who state the continuity of interactions in terms of magnitude and intermolecular equilibrium distance, and the second one, by physical chemists, who propose to classify these interactions according to their

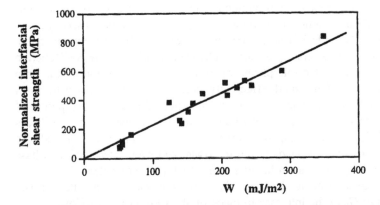

Figure 3 Linear relationship between fiber-matrix interfacial shear strength (practical adhesion) in model composites, measured by a fragmentation test, and reversible work of adhesion W, calculated from Eqs. (10) and (12). (From Ref. 66.)

nature. Nevertheless, it is now well established that the acid-base adduct resulting from the acid-base interactions has to be considered as a new entity with its own properties and, therefore, it is difficult a priori to attribute acid-base characters to a single species, a liquid or a solid surface.

2. Models Relating the Measured Adhesion Strength *G* to the Adhesion Energy *W*

Although the present subsection deals with secondary forces, these models can also be applied to other types of interfacial interactions.

One of the most important models in adhesion science, usually called the *rheological model* or *model of multiplying factors*, was proposed primarily by Gent and Schultz [3,4] and then re-examined using a fracture mechanics approach by Andrews and Kinloch [74] and Maugis [75]. In this model, the measured peel strength is simply equal to the product of the reversible work of adhesion *W* and a loss function Φ, which corresponds to the energy irreversibly dissipated in viscoelastic or plastic deformations in the bulk materials and at the crack tip and depends on both peel rate *v* and temperature *T*:

$$G = W\Phi(v, T) \tag{13}$$

As already mentioned, the value of Φ is usually far higher than that of *W* and the energy dissipated can then be considered as the major contribution to the measured adhesion strength *G*. In the case of assemblies involving elastomers, it has been clearly shown in different studies [3,4,74,76–78] that the viscoelastic losses during peel experiments and, consequently, the function Φ follow a time-temperature equivalence law such as that of Williams, Landel and Ferry (WLF) [79]. Relation (13) is illustrated in Fig. 4, where *G*, measured in air and in the presence of a liquid (methanol), is plotted against the peel rate for styrene-butadiene-rubber (SBR)/aluminum assemblies [80,81]. For such assemblies, only physical interactions (van der Waals) are involved at the interface. On the one hand, the linear increase of *G* versus peel rate results from the effect of the loss function Φ. On the other hand, the role played by *W* is clearly pointed out by the drop in *G* in the presence of the liquid. Effectively, because this liquid reduces the level of van der Waals interfacial interactions and does not affect the bulk properties of the two materials in contact, the relative decrease of the peel energy, Δ*G*/*G*, is equal to Δ*W*/*W*, as shown in Fig. 4. A lower relative variation of *G* would indicate the existence of covalent bonds at the interface, and even interfacial acid-base adducts, insofar as they cannot be altered by the liquid molecules.

From a general point of view, it is more convenient to use the intrinsic fracture energy G_0 of the interface in place of *W* in Eq. (13), as follows:

$$G = G_0\Phi(v, T) \tag{14}$$

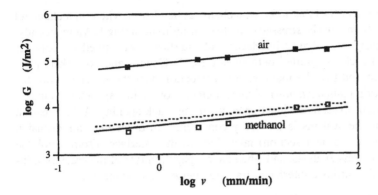

Figure 4 Influence of peel rate (v) and presence of methanol(—, theoretical curve) on the measured peel energy G of SBR/aluminum assemblies. (From Ref. 80.)

Effectively, when viscoelastic losses are negligible, i.e., when performing experiments at a very low peel rate or high temperature, $\Phi \rightarrow 1$ and G must tend toward W. However, the resulting threshold value G_0 is generally 100 to 1000 times higher than the thermodynamic work of adhesion, W.

From a famous fracture analysis of weakly cross-linked rubbers, called the *trumpet* model, de Gennes [82] has derived an expression similar to Eq. (14), when the crack propagation rate v is sufficiently high. He distinguished three different regions along the trumpet starting from the crack tip: a hard, a viscous, and, finally, a soft zone. The length of the hard region is equal to $v\tau$, where τ is the relaxation time, and then the viscous region extends to a distance $\lambda v\tau$. Factor λ is the ratio of the high-frequency viscous modulus to the zero-frequency viscous modulus of the material and obviously represents the viscoelastic behavior of the rubber. Hence, according to this approach, it is shown that the maximum adhesion strength is given by the following expression, similar to Eq. (14):

$$G \cong G_0\lambda \tag{15}$$

where G_0 is the intrinsic fracture energy for low velocities, i.e., when the polymer near the crack behaves as a soft material.

Carré and Schultz [83] have re-examined the significance of G_0 for cross-linked elastomer-aluminum assemblies and proposed that it can be expressed as

$$G_0 = Wg(M_c) \tag{16}$$

where g is a function of the molecular weight M_c between cross-link nodes and corresponds to a molecular dissipation. Such an approach is based on Lake and

Thomas' argument [84] that to break a chemical bond somewhere in a chain, all bonds in the chain must be stressed close to their ultimate strength. More recently, concerning polymer-polymer interfaces and interfaces constituted by polymer and connecting chains grafted onto substrates, de Gennes and coworkers [85–87] have proposed that the main energy dissipation near the interface was due to the extraction (called *suction*) of short segments of chains, as well as connecting chains themselves, in the junction zone during crack opening. At least partial miscibility of the systems is implicitly considered, in order that interdiffusion phenomena (see the next section) take place at the interface. From a volume balance and a stress analysis, and when the propagation rate of the crack is sufficiently low, the intrinsic interfacial fracture energy G_0 is given by

$$G_0 \cong W(1 + \sigma N) \approx W\sigma N \tag{17}$$

where σ and N are, respectively, the interfacial areal density and the number of monomers constituting the chains or the chain segments sucked out during the crack propagation. Obviously, this analysis holds only for values of N less than N_e, i.e., the critical number of monomers from which physical entanglements between macromolecular chains just occur in the polymer bulk. The case in which $N > N_e$ implies at least a disentanglement process but more likely a process of chain scission, which will be analyzed below. It is also obvious that the crack propagation rate plays a major role, because at high velocities scission of chains is more likely to occur.

More precise analyses of the suction process, the influence of crack velocity, and more complex cases (many-stitch problem, etc.) are performed in several studies [88–97]. However, it is worth mentioning some results concerning the effect of the interfacial areal density σ and molecular weight M (or N) of connecting chains on the fracture energy between a polymer and solid surfaces. The adhesion energy between polydimethyl siloxane (PDMS) elastomeric networks and flat surfaces (silicon wafer), modified by irreversible adsorption and end grafting of PDMS chains, was particularly studied [98,99]. This energy was measured, either by peel test or by Johnson, Kendall, and Roberts (JKR) [100] experiments (contact between a flat surface and a hemisphere), at very low crack opening rates (a few tens of nm at most), so it can be considered that the experimental values of G are close to those of the intrinsic interfacial fracture energy G_0. PDMS networks exhibiting different values of molecular weight between cross-links M_c (or N_c) were used. The most important result of these studies is that, in each case, G_0 passes through a maximum when the surface density σ of grafted chains increases (Fig. 5). Such a new result can be understood if the following fact is taken into account: when a contact is established between a cross-linked elastomer and a layer of surface-anchored chains, the penetration of these chains into the network (*interdigitation*) is always accompanied by swelling of the latter. Some energy is required for this swelling, and this may limit the interdigitation

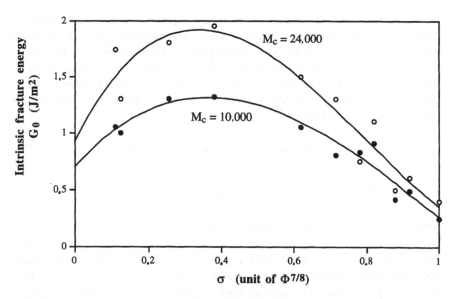

Figure 5 Typical evolution of intrinsic fracture energy G_0, measured by 90° peel experiments at a velocity of 50 nm/s, as a function of the surface density σ of the grafted chains in the surface layer (σ is proportional to $\Phi^{7/8}$, where Φ is the polymer volume fraction in the bath used to graft the chains). (From Ref. 98.)

process. This problem was investigated for end-grafted brushes by Brochard-Wyart et al. [101], who predicted the existence of a critical interfacial density σ^* above which total penetration of the chains over their full length into the elastomeric network could no longer be achieved. Clearly, this threshold depends on both the molecular weight between cross-links of the elastomer and the molecular weight of the surface chains, its theoretical expression being $\sigma^* \approx N_c/N^{3/2}$.

Therefore, in terms of adhesion energy, when σ is low ($\sigma < \sigma^*$), interdigitation can occur, leading to an increase of G in agreement with Eq. (17). Above σ^*, the network starts to reject more and more the surface chains, which can then only partly penetrate into the elastomer. If it is considered that this partial penetration is obtained with only a fraction of the grafted chains, the other chains are therefore reflected at the interface and form a "passive" layer between the substrate and the network, which tends to decrease strongly the level of adhesion. At very high values of σ, when a dense surface layer is obtained, no interdigitation can be considered and the fracture energy tends toward the Dupré work of adhesion W between two elastomers in contact.

Finally, it is worth examining the analyses concerning tack, in other words, the instantaneous adhesion when a substrate and an adhesive are placed in contact for a short time t (of the order of 1 s) under a given pressure. This tack phenomenon is of great importance for processes involving hot-melt or pressure-sensitive adhesives. First, it has clearly been shown [102] that the viscoelastic characteristics of the adhesive, and in particular its viscous modulus, plays a major role in the separation energy. De Gennes [103] has suggested that the measured tack could be related to both the adhesion energy W and the rheological properties of the bulk adhesive as follows:

$$G_{tack} \cong W \frac{\mu_\infty}{\mu_0} \quad \text{for weakly cross-linked elastomers} \tag{18}$$

$$G_{tack} \cong W \frac{t}{\tau} \quad \text{for uncross-linked elastomers} \tag{19}$$

where μ_∞ and μ_0 are the high-frequency and zero-frequency moduli of the adhesive, respectively, and τ is the reptation time of the macromolecular chains (see the next section). Equation (19) holds for the time t much larger than this reptation time. However, experimental investigations [104–107] show clearly that G_{tack} varies as a power law of the contact time t, that is, $G_{tack} \approx t^\alpha$, with an exponent α in the range of about ca. 0.1 to 0.5, rather than unity as stated by de Gennes. When rough materials are in contact, such variations can be explained by the new approach of Creton and Leibler [104], who proposed that tack should be proportional to the real contact area. This contact area, initially small due to roughness, increases as a function of time in relation with the time-dependent stress relaxation modulus of the elastomer. Rather good agreement is obtained between theoretical and experimental values of the exponent α [105,106]. However, when smooth but polar materials are considered, such variations of G_{tack} versus t could also be interpreted in terms of the evolution, as a function of time, of the level of interfacial interactions due to molecular reorientation and reorganization at the interface [107]. Such phenomena tend to increase the reversible work of adhesion, therefore increasing tack energy.

E. Diffusion Theory

The diffusion theory of adhesion is based on the assumption that the adhesion strength of polymers to themselves (autohesion), or to one another, is due to mutual diffusion (interdiffusion) of macromolecules across the interface, thus creating an interphase. Such a mechanism, mainly supported by Voyutskii [108], implies that the macromolecular chains or chain segments are sufficiently mobile and mutually soluble. This is of great importance for many adhesion problems, such as healing and welding processes. Therefore, if interdiffusion phenomena are involved, the adhesion strength should depend on different factors such as contact time, temperature, and the nature and molecular weight of polymers. Actually, such dependences are experimentally observed for many polymer-polymer junctions. Vasenin [109] has developed, from Fick's first law, a quantitative model for the diffusion theory that correlates the amount of material w diffusing in a given x direction across a plane of unit area to the concentration gradient $\partial c/\partial x$ and the time t:

$$\partial w = -D_f \partial t \frac{\partial c}{\partial x} \tag{20}$$

where D_f is the diffusion coefficient. To estimate the depth of penetration of the molecules that interdiffused into the junction region during the time of contact t_c, Vasenin assumed that the variation of the diffusion coefficient with time was of the form $D_d t_c^{-\beta}$, where D_d is a constant characterizing the mobility of the polymer chains and β is of the order of 0.5. Therefore, it is possible to deduce the depth of penetration l_p as well as the number N_c of chains crossing the interface, as

$$l_p \approx k(\pi D_d t_c^{1/2})^{1/2} \tag{21}$$

$$N_c = \left(\frac{2N\rho}{M}\right)^{2/3} \tag{22}$$

where k is a constant, N is Avogadro's number, and ρ and M are, respectively, the density and the molecular weight of the polymer. Finally, Vasenin assumed that the measured peel energy G was proportional to both the depth of penetration and the number of chains crossing the interface between the two polymers. From Eqs. (21) and (22), G becomes

$$G \cong K\left(\frac{2N\rho}{M}\right)^{2/3} D_d^{1/2} t_c^{1/4} \tag{23}$$

where K is a constant that depends on molecular characteristics of the polymers in contact. Experimental results and theoretical predictions from Eq. (23) were found [109] to be in very good agreement in the case of junctions between polyisobutylenes of different molecular weights. In particular, the dependence of G on $t_c^{1/4}$ and $M^{-2/3}$ was clearly evidenced.

One important criticism of the model proposed by Vasenin is that the energy dissipated viscoelastically or plastically during peel measurements does not appear in Eq. (23). Nevertheless, in his work, the values of coefficients K and D_d are not theoretically quantified but determined only by fitting. Therefore, it can be considered that the contribution of hysteretic losses to the peel energy is implicitly included in these constants.

In fact, the major scientific aspect of interdiffusion phenomena is concerned with the dynamics of polymer chains in the interfacial region. The fundamental understanding of the molecular dynamics of entangled polymers has advanced significantly during the past two decades due to the theoretical approach proposed by de Gennes [110], extended later by Doi and Edwards [111] and Graessley [112]. This approach stems from the idea that the chains cannot pass through each other in a concentrated polymer solution, a melt, or a solid polymer. Therefore, a chain with a random coil conformation is trapped in an environment of fixed obstacles. This constraint confines each chain inside a tube. De Gennes has analyzed the motion, limited mainly to effective one-dimensional diffusion along a given path, of a polymer chain subjected to such confinement. He described this type of motion as wormlike and gave it the name *reptation*. The reptation relaxation time τ, associated with the movement of the center of gravity of the entire chain through the polymer, was found to vary with the molecular weight M as M^3. Moreover, the diffusion coefficient D, which defines the diffusion of the center of mass of the chain, takes the form $D \sim M^{-2}$.

One of the most important and useful applications of the reptation concept concerns crack healing, which is primarily the result of the diffusion of macromolecules across the interface. The healing process was particularly studied by Kausch and coworkers [113]. The problem of healing is to correlate the macroscopic strength measurements with a microscopic description of motion. The difference between self-diffusion phenomena in the bulk polymer and healing is that the polymer chains in the former case move over distances many times larger than their gyration radii, whereas in the latter case, healing can be considered as efficient in terms of adhesion strength in the time that a macromolecule initially close to the interface needs to move about halfway across this interface. This problem has been analyzed by several authors, who have considered that the healing process is controlled by different factors, such as (1) the number of bridges across the interface (de Gennes [114]); (2) the crossing density of molecular contacts or bridges (Prager and Tirrell [115]); (3) the center-of-mass Fickian interdiffusion distance (Jud et al. [113]); and (4) the monomer segment interpenetration distance (Kim and Wool [116]). The resulting scaling laws for the fracture energy G versus time t during healing are the following:

$$G \cong t^{1/2}M^{-3/2} \quad \text{for (1) and (2)} \tag{24}$$

$$G \cong t^{1/2}M^{-1} \quad \text{for (3)} \tag{24'}$$

$$G \cong t^{1/2}M^{-1/2} \quad \text{for (4)} \tag{24''}$$

Even if there are some differences in the value of the exponent of the molecular weight in these expressions, all the approaches agree with the dependence of G on the square root of healing time; such a dependence has been clearly evidenced experimentally for poly(methyl methacrylate) polymer, for example [113], in contradiction of Vasenin's model.

Finally, it can be concluded that diffusion phenomena do contribute greatly to the adhesion strength in many cases involving polymer-polymer junctions. Nevertheless, the interdiffusion of macromolecular chains requires both polymers to be sufficiently soluble and the chains to have sufficient mobility. These conditions are obviously fulfilled for autohesion, healing, or welding of identical polymers. However, diffusion can become a most unlikely mechanism if the polymers are not or only slightly soluble, are highly cross-linked or crystalline, or are placed in contact at temperatures far below their glass transition temperature. Nevertheless, in the case of junctions between two immiscible polymers, the interface could be strengthened by the presence of a diblock copolymer, in which each molecule consists of a block of the first polymer bonded to a block of the second polymer, or each of the two blocks is miscible with one of the polymers. The copolymer molecules generally concentrate at the interface, and each block diffuses or "dissolves" into the corresponding polymer. Therefore, the improvement in joint strength can also be related to an interdiffusion process. When the molecular weight M of each block of the copolymer is less than the critical entanglement weight M_e for which entanglements of chains just occur in the polymer, the adhesion strength could be interpreted in terms of a suction mechanism as described in the previous section. On the contrary, when $M > M_e$, the failure of the joint generally requires rupture of the copolymer chains. This latter phenomenon, i.e., chain scission or more precisely rupture of chemical bonds, is analyzed in the next section.

F. Chemical Bonding Theory

It is easily understandable that chemical bonds formed across the adhesive-substrate interface can greatly enhance the level of adhesion between the two materials. These bonds are generally considered as primary bonds in comparison with physical interactions, such as van der Waals, which are called *secondary force interactions*. The terms *primary* and *secondary* stem from the relative strength or bond energy of each type of interaction. The typical strength of a covalent bond, for example, is of the order of 100 to 1000 kJ/mol, whereas those of van der Waals interactions and hydrogen bonds do not exceed 50 kJ/mol. It is clear that the formation of chemical bonds depends on the reactivity of both the adhesive and the substrate. Different types of primary bonds, such as ionic and covalent bonds, at various interfaces have been reported in the literature. The most famous example concerns the bonding to brass of rubber cured with sulfur, adhesion resulting from the creation of polysulfide bonds [117]. One of the most important adhesion areas involving interfacial chemical bonds is the use of adhesion promoter molecules, generally called *coupling agents*, to improve the joint strength between adhesive and substrate. These species are able to react chemically at both ends, with the substrate on one side and the polymer on the other side, thus creating a chemical bridge at the interface. Coupling agents based on silane molecules are the most common type of adhesion promoters [118,119]. They are widely employed in systems involving glass or silica substrates, particularly in the case of polymer-based composites reinforced by glass fibers. In addition to the improvement in joint strength, a significant enhancement of the environmental resistance of the interface, in particular to moisture, can be achieved in the presence of such coupling agents.

The influence of chemical bonds on the joint strength G, and more precisely on the intrinsic adhesion fracture energy G_0, defined in a previous section, has been analyzed in several studies. The most relevant and elegant work in this area was performed by Gent and Ahagon [120], who examined the effect on the adhesion of polybutadiene to glass of chemical bonds established at the interface by using silane coupling agents.

In these experiments the surface density of interfacial covalent bonds between the glass substrate and the cross-linked elastomer was varied by treating the glass plates with different mixtures of vinyl- and ethyl-terminated silanes. Obviously, both species form siloxane bonds on the glass surface. Moreover, it was assumed that the vinylsilane could react chemically with the polybutadiene during the cross-linking treatment of this rubber, in which a radical reaction was involved. On the contrary, a chemical reaction between the ethyl group of the latter silane and the elastomer is unlikely. Therefore, Gent and Ahagon [120] have shown that the intrinsic fracture energy G_0 increases linearly (Fig. 6) with the surface concentration of vinylsilane, in good agreement with their assumptions, and thus proved the important effect of primary bonds on adhesion strength.

Other experimental evidence for the chemical bond effect on the interfacial strength is related to the adhesion between two sheets of cross-linked polyethyl-

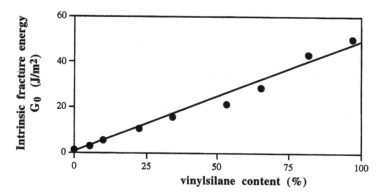

Figure 6   Intrinsic fracture energy G_0 of polybutadiene-glass assemblies as a function of the concentration of vinylsilane in the mixture of vinyl- and ethylsilanes on the glass surface. (From Ref. 120.)

ene [121]. In order to control the number of chemical bonds at the interface, the assemblies were prepared as follows. First, polyethylene containing 2% by weight of dicumylperoxide (DCP) was molded into sheets at a rather low temperature (120°C) to prevent the decomposition of DCP. Second, partial pre-cross-linking of the two separate polymer sheets was performed at 140°C for a given time. As the decomposition kinetics of DCP is known at this temperature, the degree of cross-linking can be varied as a function of time. Finally, assemblies of the two resulting sheets were obtained under pressure by heating at 180°C to ensure the total decomposition of DCP. Hence, this technique leads to complete cross-linking in the bulk of the assembly, whose mechanical properties therefore remain constant, whereas the surface density of interfacial bonding can be varied. In agreement with previous results obtained by Gent and Ahagon [120], a linear relationship has been established between the peel energy G and the number of bonds σ per unit interfacial area, insofar as σ does not exceed 1×10^{13} bonds/cm^2.

Analyzing again the effect of connecting chains grafted onto solid surfaces, but now considering that chain scission is the major phenomenon in controlling the adhesion strength instead of a suction process, de Gennes and coworkers [86,87] proposed that the intrinsic interfacial fracture energy G_0 should vary as follows:

$$G_0 \cong W + W_r N\sigma \approx W_r N\sigma \tag{25}$$

where W, N, and σ have the same significance as in Eq. (17) and W_r is the energy required to break the network of covalent bonds formed by the connecting chains (a few J/m^2). Equation (25) indicates clearly that if W_r can be considered as a constant, the intrinsic fracture energy increases linearly with the surface density of interfacial chemical bonds σ, in good agreement with experimental observations previously described. Moreover, for the results obtained by Gent and Ahagon [120], it is possible to consider that the treatment of the glass surface by pure vinylsilane may correspond to the rupture of chemical bonds [Eq. (25)], whereas the treatment with pure ethylsilane may be related to the suction phenomenon [Eq. (17)]. Therefore, the ratio of G_0 obtained in the former case to that in the latter case should be equal to the ratio W_r/W. A theoretical value of 35 has been estimated [87] for this ratio, in rather good agreement with the results presented in Fig. 6.

In a series of papers [122–124], Brown and colleagues analyzed the improvement of adhesion between two immiscible polymers, poly(methyl methacrylate) (PMMA) and poly(phenylene oxide) (PPO), by the presence of polystyrene-PMMA diblock copolymers. Since one member of the block is PMMA and the other is polystyrene (PS), which is totally miscible with PPO, it was reasonably expected that the copolymer organized at the interface, because each block dissolved in the respective homopolymer. The molecular weight of these blocks is always higher than the critical molecular weight M_e, for which entanglements of chains occur in the homopolymers. Experimentally, Brown et al. [123] employed partially or fully deuterated copolymers in order to be able to determine the deuterium on the fracture surface after separation by secondary ion mass spectrometry and forward-recoil spectroscopy. A scission of the copolymer chains near the junction point of the two blocks was observed, indicating that the diblock copolymers were well organized at the interface, whatever their molecular weights, with their junction accurately located at the PMMA-PPO interface. Moreover, Brown [124] has proposed a molecular interpretation of the toughness of glassy polymers, which can also be applied to the failure of interfaces between immiscible polymers. This approach stems from the idea that the cross-tie fibrils, which exist between primary fibrils in all crazes, can transfer mechanical stress between the broken and unbroken fibrils and thus strongly affect the failure mechanics of a craze. It is based on a simple model of crack tip stress concentration. Finally, assuming that all the effectively entangled chains in the material are drawn into the fibril, the fracture energy G of a polymer is found to be directly related to the square of both the areal density σ of entangled chains and the force f required to break a polymer chain:

$$G \cong \sigma^2 f^2 \frac{D}{S} \qquad (26)$$

where D is the fibril diameter and S is the stress at the craze-bulk interface, which is assumed to be constant. Brown [124] has considered that diblock copolymer coupled interfaces between PMMA and PPO are ideal experimental systems for testing the validity of his model. Indeed, a linear dependence of the interfacial fracture energy G on the diblock copolymer surface density σ, on logarithmic scales, is observed for copolymers of different molecular weights. A slope of 1.9 $\pm$ 0.2 was found for the master straight line, in good agreement with Eq. (26). Nevertheless, it is worth noting that Brown's results involving chain scission at the interface and leading to a dependence of G on σ^2 contradict previous examples in which linear relationships between G and σ were established. More precise analyses of the effect on the adhesion strength of crack velocity, molecular weight and chemical structure (di- and triblocks, alternated structure, etc.) of the copolymers, and so on have been made [90,92,93,95,96,125–130].

III. CONCLUDING REMARKS

Adhesion is a very complex field beyond the reach of any single model or theory. Given the number of phenomena involved in adhesion, the variety of materials to be bonded, and the diversity of bonding conditions, the search for a unique, universal theory capable of explaining all the experimental facts is useless.

In practice, several adhesion mechanisms can be involved simultaneously. However, it is generally considered that the adsorption or thermodynamic theory offers the main mechanism with the widest applicability. It describes the achievement of intimate contact and the development of physical forces at the interface. This is a necessary step for interlocking, interdiffusion, and chemical bonding mechanisms to occur subsequently, further increasing the adhesion strength.

Finally, one can consider that the measured adhesion strength of an assembly could be expressed as a function of three terms related, respectively, to (1) the interfacial molecular interactions, (2) the mechanical and rheological properties of bulk materials, and (3) the characteristics of the interphase. The two first terms have received a great deal of attention in studies in the field of physical chemistry of surfaces and fracture mechanics. The third term constitutes the challenge for a real and complete understanding of adhesion.

REFERENCES

1. JW McBain, DG Hopkins. J Phys Chem 29:88, 1925.
2. EM Borroff, WC Wake. Trans Inst Rubber Ind 25:190,199, 210, 1949.
3. AN Gent, J Schultz. Proceedings Symposium on Recent Advances in Adhesion, 162nd ACS Meeting, Vol 31, No 2, 1971, p. 113.
4. AN Gent, J Schultz. J Adhesion 3:281, 1972.
5. WC Wake. Adhesion and the Formulation of Adhesives. London: Applied Science, 1982.
6. JR Evans, DE Packham. In: KW Allen, ed. Adhesion 1. London: Applied Science, 1977, p. 297.
7. DE Packham. In: AJ Kinloch, ed. Developments in Adhesives—2. London: Applied Science, 1981, p 315.
8. DE Packham. In: KL Mittal, ed. Adhesion Aspects of Polymeric Coatings. New York: Plenum, 1983, p 19.
9. PJ Hine, S El Muddarris, DE Packham. J Adhesion 17:207, 1984.
10. NH Ladizesky, IM Ward. J Mater Sci 18:533, 1983.
11. M Nardin, IM Ward. Mater Sci Technol 3:814, 1987.
12. NH Ladizesky, IM Ward. J Mater Sci 24:3763, 1989.
13. HE Bair, S Matsuoka, RG Vadimsky, TT Wang. J Adhesion 3:89, 1971.
14. TT Wang, HN Vazirani. J Adhesion 4:353, 1972.
15. BV Deryaguin, NA Krotova. Dokl Akad Nauk SSSR 61:843, 1948.
16. BV Deryaguin. Research 8:70, 1955.
17. BV Deryaguin, NA Krotova, VV Karassev, YM Kirillova, IN Aleinikova. Proceedings 2nd International Congress on Surface Activity, Vol III. London: Butterworths, 1957, p 417.

18. BV Deryaguin, VP Smilga. In: Adhesion: Fundamentals and Practice. London: McLaren and Son, 1969, p 152.
19. BV Deryaguin, NA Krotova, VP Smilga. Adhesion of Solids. New York: Consultants Bureau (Plenum), 1978.
20. CL Weidner. Adhesives Age 6(7):30, 1963.
21. J Krupp, W Schnabel. J Adhesion 5:296, 1973.
22. L Lavielle, JL Prévot, J Schultz. Angew Makromol Chem 169:159, 1989.
23. G von Harrach, BN Chapman. Thin Solid Films 13:157, 1972.
24. JJ Bikerman. The Science of Adhesive Joints. New York: Academic Press, 1961.
25. WD Bascom, CO Timmons, RL Jones. J Mater Sci 19:1037, 1975.
26. RJ Good. J Adhesion 4:133, 1972.
27. LH Sharpe. Proceedings Symposium on Recent Advances in Adhesion, 162nd ACS Meeting. Vol 31, No 2. 1971, p 201.
28. J Schultz. J Adhesion 37:73, 1992.
29. J Schultz, A Carré, C Mazeau. Int J Adhesion Adhesives 4:163, 1984.
30. M Brogly, Y Grohens, C Labbe, J Schultz. Int J Adhesion Adhesives 17:257, 1997.
31. Y Grohens, M Brogly, C Labbe, J Schultz. Polymer 38:5913, 1997.
32. J Schultz, L Lavielle, A Carré, P Comien. J Mater Sci 24:4363, 1989.
33. M Brogly, M Nardin, J Schultz. Macromol Symp 119:89, 1997.
34. M Brogly, M Nardin, J Schultz. Polymer 39:2185, 1998.
35. M Nardin, EM Asloun, F Muller, J Schultz. Polym Adv Technol 2:161, 1991.
36. M Nardin, A El Maliki, J Schultz. J Adhesion 40:93, 1993.
37. J. Schultz, A Carré. J Appl Polym Sci Appl Polym Symp 39:103, 1984.
38. V Péchereaux, J Schultz, X Duteurtre, JM Gombert. Proceedings 2nd International Conference on Adhesion and Adhesives, Bordeaux, France, Adhecom 89. Vol 1. 1989, p 103.
39. X Dirand B Hilaire, E Lafontaine, B Mortaigne, M Nardin. Composites 25:645, 1994.
40. X Dirand, B Hilaire, J-J Hunsinger, M Nardin. J Mater Sci 32:5753, 1997.
41. LH Sharpe, H Schonhorn. Chem Eng News 15:67, 1963.
42. T Young. Philos Trans R Soc 95:65, 1805.
43. HW Fox, WA Zisman. J Colloid Sci. 5:514, 1950; ibid. 7:109, 428, 1952.
44. EG Shafrin, WA Zisman. J Am Ceram Soc 50:478, 1967.
45. WA Zisman. In: Contact Angle, Wettability and Adhesion. Adv Chem Ser No 43. Washington, DC: American Chemical Society, 1964, p 1.
46. A Dupré. Théorie Mécanique de la Chaleur. Paris: Gauthier-Villars, 1869, p 369.
47. FM Fowkes. J Phys Chem 67:2538, 1963.
48. J Schultz, K Tsutsumi, JB Donnet. J Colloid Interface Sci 59:277, 1977.
49. FM Fowkes. Ind Eng Chem 56(12):40, 1964.
50. DK Owens, RC Wendt. J Appl Polym Sci 13:1740, 1969.
51. DH Kaelble, KC Uy. J Adhesion 2:50, 1970.
52. T Chihani, P Bergmark, P Flodin. J Adhesion Sci Technol 7:327, 1993.
53. LH Lee, ed. Fundamentals of Adhesion. New York: Plenum, 1991.
54. JN Israelachvili. Intermolecular and Surface Forces. 2nd ed. London: Academic Press, 1991.
55. FM Fowkes, S Maruchi. Org Coatings Plast Chem 37:605, 1977.

56. FM Fowkes, MA Mostafa. Ind Eng Chem Prod Res Dev 17:3, 1978.
57. FM Fowkes. Rubber Chem Technol 57:328, 1984.
58. FM Fowkes. J Adhesion Sci Technol 1:7, 1987.
59. KL Mittal, HR Anderson Jr, eds. Acid-Base Interactions: Relevance to Adhesion Science and Technology. Utrecht: VSP, 1991.
60. RS Drago, GC Vogel, TE Needham. J Am Chem Soc 93:6014, 1971.
61. RS Drago, LB Parr, CS Chamberlain. J Am Chem Soc 99:3203, 1977.
62. V Gutmann. The Donor-Acceptor Approach to Molecular Interactions. New York: Plenum, 1978.
63. C Saint-Flour, E Papirer. Ind Eng Chem Prod Res Dev 21:337, 666, 1982.
64. E Papirer, H Balard, A Vidal. Eur Polym J 24:783, 1988.
65. J Schultz, L Lavielle, C Martin. J Adhesion 23:45, 1987.
66. M Nardin, J Schultz. Composite Interfaces 1:177, 1993.
67. M Nardin, J Schultz. Langmuir 12:4238, 1996.
68. CJ van Oss, MK Chaudhury, RJ Good. Adv Colloid Interface Sci 28:35, 1987.
69. CJ van Oss, MK Chaudhury, RJ Good. Chem Rev 88:927, 1988.
70. RJ Good, MK Chaudhury, CJ van Oss. In: LH Lee, ed. Fundamentals of Adhesion. New York: Plenum, 1991, p 153.
71. RJ Good, CJ van Oss. In: ME Schrader, GI Loeb, eds. Modern Approaches to Wettability—Theory and Applications. New York: Plenum, 1992, p 1.
72. RG Pearson. J Am Chem Soc 85:3533 1963.
73. RG Pearson. J Chem Educ 64:561, 1987.
74. EH Andrews, AJ Kinloch. Proc R Soc Lond A332:385, 401, 1973.
75. D Maugis. J Mater Sci 20:3041, 1985.
76. AN Gent, RP Petrich. Proc R Soc Lond A310:433, 1969.
77. GR Hamed. In: RL Patrick, ed. Treatise on Adhesion and Adhesives. Vol 6. New York: Marcel Dekker, 1989, p 33.
78. MF Vallat, J Schultz. Proceedings of International Rubber Conference, Paris, II (16), 1982, p 1.
79. ML Williams, RF Landel, JD Ferry. J Am Chem Soc 77:3701, 1955.
80. A Carré, J Schultz. J Adhesion 18:171, 1984.
81. A Carré, J Schultz. J Adhesion 18:207, 1985.
82. PG de Gennes. C R Acad Sci Paris Ser II 307: 1949, 1988.
83. A Carré, J Schultz. J Adhesion 17:135, 1984.
84. GJ Lake, AG Thomas. Proc R Soc Lond A300:103, 1967.
85. PG de Gennes. J Phys (Paris) 50:2551, 1989.
86. E. Raphaël, PG de Gennes. J Phys Chem 96:4002, 1992.
87. F Brochard-Wyart, PG de Gennes. J Adhesion 57:21, 1996.
88. HR Brown. Macromolecules 26:1666, 1993.
89. Hong Ji, PG de Gennes. Macromolecules 26:520, 1993.
90. J Washiyama, EJ Kramer, C-Y Hui. Macromolecules 26:2928, 1993.
91. KP O'Connor, TCB McLeish. Macromolecules 26:7322, 1993.
92. HR Brown, C-Y Hui, E Raphaël. Macromolecules 27:608, 1994.
93. J Washiyama, EJ Kramer, C Creton, C-Y Hui. Macromolecules 27:2019, 1994.
94. HR Brown. Science 263:1411, 1994.
95. L Kogan, C-Y Hui, A Ruina. Macromolecules 29:4090, 1996.

96. L Kogan, C-Y Hui, A Ruina. Macromolecules 29:4101, 1996.
97. GT Pickett, AC Balazs, D Jasnow. Trends Polym Sci 5:128, 1997.
98. M Deruelle, M Tirrell, Y Marciano, H Hervet, L Léger. Faraday Discuss 98:55, 1994.
99. M Deruelle, L Léger, M Tirrell. Macromolecules 28:7419, 1995.
100. KL Johnson, K Kendall, AD Roberts. Proc R Soc Lond A324:301, 1971.
101. F Brochard-Wyart, PG de Gennes, L Léger, Y Marciano. J Phys Chem 98:9405, 1994.
102. MF Tse. J Adhesion Sci Technol 3:551, 1989.
103. PG de Gennes. C R Acad Sci Paris Ser II 312:1415, 1991.
104. C Creton, L Leibler. J Polym Sci Polym Phys 34:545, 1996.
105. A Zosel. Proceedings 3rd European Adhesion Conference, EURADH'96, Cambridge (UK), 3-6 September 1996. London: Institute of Materials, 1996, p 47.
106. A Zosel. J Adhesion Sci Technol 11:1447, 1997.
107. M Nardin, E Cheret-Bitterlin, J Schultz. Proceedings 20th Annual Meeting Adhesion Society, Hilton Head Island, February 23-26, 1997, p 137.
108. SS Voyutskii. Autohesion and Adhesion of High Polymers. New York: Wiley Interscience, 1963.
109. RM Vasenin. In: Adhesion: Fundamentals and Practice. London: McLaren and Son, 1969, p 29.
110. PG de Gennes. J Chem Phys 55:572, 1971.
111. M Doi, SF Edwards. J Chem Phys Faraday Trans 74:1789, 1802, 1818, 1978.
112. WW Graessley. Adv Polym Sci 47:76, 1982.
113. K Jud, HH Kausch, JG Williams. J Mater Sci 16:204, 1981.
114. PG de Gennes. C R Acad Sci Paris Ser B 291:219, 1980; ibid. 292:1505, 1981.
115. S Prager, M Tirrell. J Chem Phys 75:5194, 1981.
116. YM Kim, RP Wool. Macromolecules 16:1115, 1983.
117. S Buchan, WD Rae. Trans Inst Rubber Ind 20:205, 1946.
118. EP Plueddemann. Silane Coupling Agents. New York: Plenum, 1982.
119. KL Mittal, ed. Silanes and Other Coupling Agents. Utrecht: VSP, 1992.
120. AN Gent, A Ahagon. J Polym Sci Polym Phys Ed 13:1285, 1975.
121. P Delescluse, J Schultz MER Shanahan. In: KW Allen, ed. Adhesion 8. London: Elsevier Applied Science, 1984, p 79.
122. HR Brown. Macromolecules 22:2859, 1989.
123. HR Brown, VR Deline, PF Green. Nature 341:221, 1989.
124. HR Brown. Macromolecules 24:2752, 1991.
125. C Creton, EJ Kramer, C-Y Hui, HR Brown. Macromolecules 25:3075, 1992.
126. WF Reichert, HR Brown. Polymer 34:2289, 1993.
127. E Boucher, JP Folkers, H Hervet, L Léger. Macromolecules 29:774, 1996.
128. R Kulasekere, H Kaiser, JF Ankner, TP Russell, HR Brown, CJ Hawker, AM Mayes. Macromolecules 29:5493, 1996.
129. TP Russell. Curr Opin Colloid Interface Sci 1:107, 1996.
130. C-A Dai, KD Jandt, DR Iyengar, NL Slack, KH Dai, WB Davidson, EJ Kramer, C-Y Hui. Macromolecules 30:549, 1997.

2
Harnessing Acid-Base Interactions to Improve Adhesion

Mohamed Mehdi Chehimi
Université Paris 7—Denis Diderot, Associé au CNRS (UPRESA 7086), Paris, France

I. INTRODUCTION

The reversible work of adhesion (W) is the free energy change per unit area in creating an interface between two bodies (Fig. 1). The work W is related to the intermolecular forces that operate at the interface between two materials, e.g., an adhesive and an adherend. However, in practice, the reversible work of adhesion may be obscured by other factors (e.g., mechanical interlocking, interdiffusion) because it is always a few orders of magnitude lower than the measured adhesive joint strength [1,2]. One important contribution to practical joint strength is the energy loss due to irreversible deformation processes within the adhesive. Nevertheless, Gent and Schultz [3] showed using peel strength measurements that viscoelastic losses were proportional to the reversible work of adhesion. For this reason, one of the important tasks is to determine the nature of interfacial chemical and physical forces and to understand how they control the reversible work of adhesion.

In 1964, Fowkes [4] proposed that both the reversible work of adhesion (W) and the surface tension (γ) had additive components:

$$W = W^d + W^p + W^h + W^m \cdots$$

and

$$\gamma = \gamma^d + \gamma^p + \gamma^h + \gamma^m \cdots$$

since the intermolecular attractions at interfaces result from independent phenom-

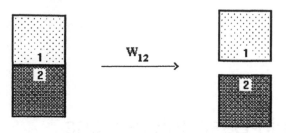

Figure 1 Definition of the work of adhesion.

ena such as dispersion forces (d), dipole interactions (p) and hydrogen bonding (h), a subset of Lewis acid-base interactions, and metallic bonds (m). For convenience, these intermolecular interactions were split into additive dispersive and nondispersive forces, the latter being unfortunately attributed to polar interactions including the hydrogen bond or acid-base interactions. However, as early as 1960, Pimentel and McClellan [5] demonstrated that the heat of hydrogen bonding between two distinct molecules was related to the acid strength of the proton donor (or electron acceptor) and to the base strength of the proton acceptor (or electron donor) and was completely unrelated to their dipolar moments. This led Fowkes [6] to propose that the so-called polar term in the reversible work of adhesion was due to Lewis acid-base interactions (including hydrogen bonding), whereas the true contribution of permanent dipole-dipole interactions and dipole-induced dipole interactions could rather be lumped with the dispersive interactions term, since it is negligible in the condensed phase (about 1%) [7]. Distinguishing between acid-base interactions and "polar" interactions is thus fundamentally important and also has a practical implication, since Fowkes demonstrated for complex systems that the former but not the latter led to a substantial improvement of adhesion. It is also important to point out that acid-acid and base-base interactions do not improve adhesion for they are of the van der Waals type only [8–10]. This is illustrated by the determination of the acid-base contribution to the work of adhesion (W_{SL}^{AB}) of liquids to poly (ethylene-co-acrylic acid), P(E-AA), with varying percentages of acrylic acid [8]. Figure 2 shows for dimethylformamide (DMF), dimethyl sulfoxide (DMSO) and 0.1 N NaOH solution (all three test liquids are basic) that the W_{SL}^{AB} increases with the percentage of acrylic acid. By contrast, W_{SL}^{AB} for the phenol solution (Lewis acid) in tricresyl phosphate is zero and independent of the acrylic acid content of the copolymer.

Another important example of the role of acid-base interactions concerns polymer adsorption. Fowkes and Mostafa [11] demonstrated that the amount of polyl(methyl methacrylate) (PMMA) (electron donor or Lewis base) adsorbed onto silica (electron acceptor or Lewis acid) was much higher than that of ad-

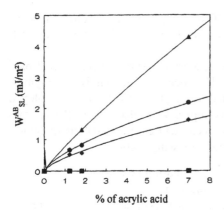

Figure 2   Acid-base contribution to the work of adhesion (W^{AB}_{SL}) determined by contact angle measurements for various liquid–acrylic acid copolymer pairs versus the acrylic acid content. (▲) 0.1 N NaOH; (●) DMSO; (◆) DMF; (■) phenol, 35% in Tcp. (From Ref. 8.)

sorbed chlorinated poly(vinyl chloride) (CPVC, Lewis acid). When $CaCO_3$ (Lewis base) was used as the substrate, CPVC adsorbed in a greater amount than PMMA. In the case of the PMMA-silica system, it was demonstrated that the acid-base properties of the solvent were of significant importance because the solvent can interact via specific acid-base forces with the polymer (chloroform-PMMA interaction) or preferentially adsorb onto the substrate (tetrahydrofuran-silica interaction). Both phenomena result in hindering polymer adsorption. By contrast, in a noncompeting solvent such as CCl_4, the authors obtained a much higher amount of PMMA adsorbed on silica because in this case the polymer-substrate acid-base interactions were maximized.

The pioneering developments by Professor Frederick M. Fowkes of the acid-base theory in adhesion have attracted the attention of several laboratories to this subject. A festschrift in his honor on the occasion of his 75th birthday was published in four issues of the Journal of Adhesion and Technology [12] and in the first monograph, entitled Acid-Base Interactions: Relevance to Adhesion Science and Technology [13]. These publications constitute, without a shadow of a doubt, an important step in the history of acid-base chemistry in general and adhesion science in particular.

There is now a massive number of fundamental and applied examples in the literature showing that acid-base interactions are effective in improving adhesion, adsorption, dispersibility, solubility, and mixing of polymers and other materials [12–19]. They culminated with a recent interfacial force microscopy (IFM) study of adhesion forces at the molecular scale between various mercaptan-functionalized gold probes and substrates [20]. The maximal work of adhesion of the probe to the substrate was obtained for the acid-base ($-NH_2$ versus —COOH) combinations (see Sec. V.A).

The aim of the present contribution is to review some aspects of acid-base interactions in adsorption and adhesion. The focus will be mainly on polymers (see Appendix), given the very large variety of polymer-based systems employed in coatings, packaging, electronics, etc. Examples were selected from the author's research work and from a survey of the recent acid-base interactions literature.

This chapter is organized in the four following sections:

Scope, properties, and evaluation of acid-base interactions
Theory of acid-base interactions in adhesion
Experimental assessment of acid-base properties of polymers and other materials
Some practical applications of acid-base interactions

II. SCOPE, PROPERTIES, AND EVALUATION OF ACID-BASE INTERACTIONS

Acid-base interactions including hydrogen bonds are specific and not ubiquitous like the London dispersive interactions. They occur when a base (electron donor or a proton acceptor) and an acid (electron acceptor or proton donor) are brought close together. This can be described by the general equation

$$A \; + \; :B \; \rightarrow \qquad A:B \tag{1}$$
$$\text{acid} \quad \text{base} \quad \text{acid-base complex}$$

Table 1 indicates the three possible types of acids and bases and examples of corresponding molecules. These types of acids and bases lead to nine possible acid-base adducts. Five of these combinations;namely n-n, n-σ*, n-π*, π-σ*, and π-π*, yield addition-type complexes, whereas the other four combinations lead to adducts with displacement [21]. For example, the interaction of PMMA in chloroform results in the formation of n-σ* acid-base adducts. PMMA is a Lewis

Table 1   Types of Lewis Acids and Bases

Electrons donors (bases)		Electron acceptors (acids)	
n	Pyridine, EtAc	n	BF_3, $AlCl_3$
σ	Alkanes	σ*	I_2, $HCCl_3$
π	Benzene	π*	C_6H_5—NO_2

base due to the nonbonding electron doublets from the oxygen in the $C{=}O$ group, whereas the acceptor site in chloroform is its C—H antibonding σ^* orbital.

Amphoters are species that bear both acidic and basic sites and can thus interact specifically with either pure acids or bases. In the terminology of van Oss et al. [22], pure acids and bases are called "monopolar" whereas amphoters are called "bipolar." This is a rather unfortunate terminology because in modern approaches to interfacial phenomena acid-base interactions are distinguished from "polar" interactions. For this reason, Berg [17] preferred the terms "monofunctional" for pure acids and pure bases and "bifunctional" for amphoteric ones.

Water-water hydrogen bonding is, for example, responsible for the anomalously high boiling point of water and contributes to 70% of the surface tension of this liquid at ambient temperature. It is also well known that the hydrogen bonds between the complementary base pairs thymine-adenine (two bonds) and cytosine-guanine (three bonds) are the key to the double helical structure of DNA. Finally, it has long been recognized that acid-base interactions have a drastic effect on polymer macroscopic properties such as glass transition temperature [23,24], polymer miscibility [25,26], solubility in common solvents [15,27], swelling [27], adsorption [11], and adhesion [6].

Hydrogen or acid-base bonds are exothermic and their energy ranges from 8 to 50 kJ/mol [28,29]. This compares with the energy of London dispersion forces but exceeds that of dipole-dipole (Keesom) and dipole–induced dipole (Debye) interactions. With a large and negative value, the heat of an acid-base interaction can overcome the positive or the negligibly small negative entropic term $-T\Delta S$, so that adhesion and mixing can be substantially improved. The high energy associated with acid-base interactions is due to Coulombic forces acting at intermolecular distances of about 0.2–0.3 nm. Acid-base interactions are thus of the short-range type by comparison with the long-range London dispersive interactions, which can operate at distances exceeding 10 nm. For example, Nardin and Schultz [29] have demonstrated for a series of single-fiber composites that the maximal work of adhesion (W) was obtained for fiber-matrix systems interacting via both dispersive and acid-base interactions on the one hand and for the smallest intermolecular distance λ of about 0.2 nm on the other hand (see Fig. 3). For such short distance, the highest heat of acid-base interaction (50 J/mol) between the fiber and the matrix was obtained.

The importance of acid-base interactions in various fields of chemistry led to extensive research in the 1960s to obtain acid-base scales. This resulted in the hard and soft acids and bases (HSAB) scales of Pearson [30], Drago's E and C constants [31] and Gutmann's donor and acceptor numbers [32]. It is also important to recall the approach of Bolger and Michaels [33] for predicting the adhesion of Brönsted organic and inorganic species.

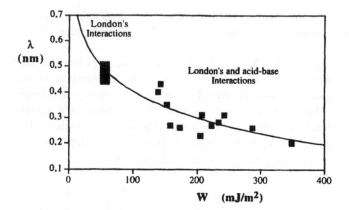

Figure 3 Variation of the intermolecular distance λ at the interface versus the reversible work of adhesion W. (From Ref. 29.)

A. Hard and Soft Acids and Bases

Pearson [30] proposed qualitative scales of acidity and basicity based on the numerical values of equilibrium constants for nucleophilic substitution reactions. Pearson noted that the stability of the acid-base adducts depended on the size and the charge of the adjacent acids and bases. Pearson identified hard and soft types of acids and hard and soft types of bases.

> Hard acids (or electrophiles) have a positive charge, are hard to reduce due to their high-energy low unoccupied molecular orbital (LUMO), and have a small size (e.g., H^+).
> Soft acids have a low-energy LUMO and are thus easy to reduce, do not necessarily have a positive charge, and have a large size (e.g., I_2, metals).
> Hard bases (or nucleophiles) are difficult to oxidize for they have a low-energy high occupied molecular orbital (HOMO), are usually negatively charged, have a small size, and have a high pK_a (e.g. O^{2-}, ketones).
> Soft bases are easy to oxidize because of their high-energy HOMO, do not necessarily have a negative charge, have a large size, and have a small pK_a (e.g., amines).

Pearson proposed an expression to rationalize his HSAB concept:

$$\log K = S_A S_B + \sigma_A \sigma_B \tag{2}$$

where S is a hardness factor, σ is a softness parameter, and A and B stand for acid and base, respectively. Implicitly, Eq. (2) indicates that like species form stable adducts. In other words, "hard acids prefer to bind to hard bases, and soft acids prefer to bind to soft bases." Unfortunately, the HSAB theory remained of very limited utility because it failed to predict quantitatively the stability of the adducts. Drago [34] pointed out that in the HSAB literature, results are explained after the answer is known. Nevertheless, Lee [35] has related chemical hardness to the average energy gap of a solid and has proposed the following classification of solids:

> Metals: soft and mostly acidic.
> Semimetals: soft
> Semiconductors: rather soft and mostly basic
> Most insulators including polymers: hard

Lee reported that a metal-metal interaction could be viewed as an acid-base interaction. This is, for example, the case of the chemical interaction at the Cr-Cu interface, which has been modeled as an acid-base interaction where Cr is a Lewis acid and Cu a Lewis base because it has more filled than empty orbitals [35].

The work of Lee has contributed considerably to the extension of the HSAB principles, established for liquid solutions, to solid-solid interactions.

B. Drago's *E* and *C* Parameters

Drago [31] proposed a four-parameter equation to predict the heat of acid-base adduct formation:

$$-\Delta H^{AB} = (E_A E_B + C_A C_B) \tag{3}$$

where E and C are the susceptibilities of the acid (A) and the base (B) to undergo an electrostatic interaction (E) and a covalent bond (C), respectively. Drago showed that his equation estimated ΔH^{AB} for almost 1600 adducts with an accuracy of 0.1 to 0.2 kcal/mol (0.4–0.8 kJ/mol). Stable adducts are obtained when both the acid and the base have large E or C constants. Fowkes suggested determining E and C parameters for polymers and other materials using a set of reference acids and bases of known Drago parameters. However, this is best achieved by choosing a set of reference species of widely differing C/E ratios (where C/E ratio can be considered to be a measure of relative softness). Table 2 displays Drago's parameters for some frequently used acidic and basic probes.

Fowkes and coworkers have used test acids (e.g., phenol and chloroform) and bases (e.g., pyridine and ethyl acetate) to determine Drago's parameters for polymers and metal oxides using essentially calorimetric heats and infrared measurements [14,27,37,38]. The approach of Fowkes was applied in combination with inverse gas chromatography (IGC) to determine E and C parameters for conventional polymers [39], conducting polymers [40,41], and untreated and silane-treated glass beads [42]. There are also some great expectations for the use

Table 2 Drago's Parameters for Some Commonly Used Acids and Bases

Acids	C_A	E_A	C_A/E_A
Iodine	1.00	1.00	1.0
$SbCl_5$	5.13	7.38	0.695
t-BuOH	0.30	2.04	0.147
Pyrrole	0.30	2.54	0.116
$CF_3CH(OH)CF_3$	0.62	5.93	0.105
Phenol	0.44	4.33	0.102
Chloroform	0.16	3.02	0.053
H_2O	0.26	2.61	0.010

Bases	C_B	E_B	C_B/E_B
TCHP$=$S[a,b]	9.67	0.61	15.8
Triethylamine	11.1	0.99	11.19
Pyridine	6.40	1.17	5.47
THF	4.27	0.98	4.37
Diethyl ether	3.25	0.96	3.38
1,4-Dioxane	2.38	1.09	2.18
Acetone	2.33	0.98	2.36
Ethyl acetate	1.74	0.98	1.79
$Et_3P=O$[a,c]	2.70	1.64	1.65

E and C in (kcal/mol)$^{1/2}$ from Ref. 31, except:
[a] Ref. 36.
[b] Tricyclohexyl phosphine oxide.
[c] Triethyl phosphine oxide.

of contact angles of diiodomethane solutions of specific probes such as phenol to determine E and C parameters of polymer surfaces. The determination of ΔH^{AB} for probe-surface systems was suggested by Fowkes et al. [43] on the basis of temperature-dependent contact angles and substitution of the Young equation into the Gibbs equation for solute adsorption from diiodomethane onto the surface under investigation. This novel approach was applied to PMMA [43] and chemically modified Teflon surfaces [44]. Perhaps, the combination of ΔH^{AB} data determined via contact angles for two or more specific probes will lead to E and C parameters. It is also worth noting the potential use of nuclear magnetic resonance (NMR) [14] and X-ray photoelectron spectroscopy (XPS) [16,45] for the assessment of E and C parameters.

Table 3 shows E and C parameters for various polymers and metal oxides obtained by a variety of techniques.

Table 3 Drago's E and C Parameters ($kcal^{1/2}$ $mol^{-1/2}$) for Some Polymers and Fillers

	C_A	E_A	C_B	E_B	Method[a]	Reference
Polymers						
CPVC	—	3			IR	27
	0.36	2.70			IR	38
PVB	—	4			IR	27
Phenoxy resin	0.24	1.53			IR	46
Epoxy resin	0.29	1.72			NMR	14
PVdF	0.7	1.8			IR	47
PMMA			1.18	0.59	IR	46
			0.96	0.68	IR	27
PEO			5.64	0.77	IR	46
PPyCl	0.27	4.17	0.45	1.09	IGC	40
PPyTS	0.27	4.35	24.5	−0.36	IGC	41
PPO			9.5	≈0	XPS/IR	45
Fillers						
SiO_2	1.14	4.39			MC/IR	37
TiO_2	1.02	5.67			MC	37
α-Fe_2O_3	0.8	4.50			MC	48
	1.1	0.50–1.0			MC	49
γ-Fe_2O_3	0.79	5.4			MC/IR	38
E-glass	0.02	0.15	0.39	0.2	MC	37
Glass beads	0.70	6.0			IGC	42
APS-treated glass	1.60	0.62			IGC	42

[a] MC, microcalorimetry; APS, aminopropyltriethoxysilane.

C. Gutmann's Donor and Acceptor Numbers

Gutmann [32] proposed a two-parameter equation for the estimation of ΔH^{AB}:

$$-\Delta H^{AB}(kcal/mol) = (AN \times DN)/100 \qquad (4)$$

where AN is the acceptor number of the acidic species and DN the donor number of the basic species. DN was defined as the negative of the enthalpy of formation of the acid-base adduct between the base under investigation and a reference Lewis acid, antimony pentachloride ($SbCl_5$) in 1,2-dichloroethane (inert solvent):

$$DN = -\Delta H(SbCl_5: base) \qquad (5)$$

However, Lim and Drago [50] showed that solvation effects contributed to most of the DN values by 0.7 unit. More important, it should be noted that since $SbCl_5$ is a soft acid, the DN scale is a classification of softness for bases.

The AN, the acceptor number of Lewis acids, was defined as the relative ^{31}P NMR shift obtained when triethylphosphine oxide (Et_3PO) was dissolved in the candidate acid. The scale was normalized by assigning an AN value of 0 to the NMR shift obtained with hexane and 100 to that obtained from the $SbCl_5$: Et_3PO interaction in dilute 1,2-dichloroethane solution. However, the total shift of ^{31}P NMR in two-component systems has an appreciable contribution of van der Waals interactions that must be accounted for in correlating spectral shifts with heats of acid-base interactions. A mathematical remedy based on the dispersive contribution to the surface tension of solvents (derived from experimental measurements of surface tensions and interfacial tensions versus squalane) was then proposed by Riddle and Fowkes [7] to correct ^{31}P NMR shifts for van der Waals interactions leading to a new scale of acceptor numbers. The authors also proposed the following equation in order to convert the corrected AN values ($AN-AN^d$) in ppm to AN^* in kcal/mol units:

$$AN^* = 0.228(AN - AN^d) \text{ in kcal/mol} \tag{6}$$

where AN^d is the dispersive component of the original AN values published by Gutmann. Besides the van der Waals contribution to the original AN values, this scale has also been criticized because it is viewed as a scale of hardness for acids since Et_3PO is a hard reference base. Despite these criticisms, the merit of the Gutmann approach lies in the fact that his scales provide both acidic and basic parameters for amphoteric species, which is not the case with Drago's E and C classifications.

D. Bolger's Δ_A and Δ_B Interaction Parameters

In the case of interaction of organic-inorganic materials (for example, polymer–metal oxide), Bolger and Michaels [33] suggested a model based on Brönsted acid-base chemistry to account for the strength of the interaction. They defined a Δ parameter for organic acids and bases:

$$\Delta_A = \text{IEPS}(B) - pK_a(A) \tag{7a}$$

and

$$\Delta_B = pK_a(B) - \text{IEPS}(A) \tag{7b}$$

where K_a is the dissociation constant of the organic species and IEPS the isoelectric point of the metal oxide. Bolger and Michaels identified three regimes of acid-base interactions:

Table 4 Bolger's Δ Parameters for Selected
Organic–Metal Oxide Pairs

	SiO$_2$ (2)[a]	Al$_2$O$_3$ (8)	MgO (12)
CH$_3$COOH	−2.7	3.3	7.3
CH$_3$NH$_2$	8.6	2.6	−1.4

[a] Numbers in parentheses correspond to the IEPS values.

1. $\Delta \ll 0$: negligibly weak acid-base interactions
2. $\Delta \approx 0$: acid-base interactions comparable to those due to dispersive interactions
3. $\Delta > 0$: strong acid-base interactions perhaps resulting in chemical attack or (metal) corrosion

Table 4 shows Δ parameters for acetic acid [$pK_a(A) = 4.7$] and methyl amine [$pK_a(B) = 10.6$] interacting with SiO$_2$, Al$_2$O$_3$, and MgO, whose IEPS values are 2, 8, and 12, respectively. Maximal positive values of Δ are obtained for the amine-SiO$_2$ and carboxylic acid–MgO interactions, thus for acid-base adducts, whereas the Δ parameters predict no such interactions for the carboxylic acid–silica and amine-MgO adducts, for these are of the acid-acid and base-base types, respectively.

Bolger's concept has been used successfully to interpret the failure mechanisms of polyimide-MgO joints [51]. Similarly, a Δ_A value of 6.5 was estimated for the interaction of PMDA-ODA PAA [pyromellitic dianhydride oxydianiline poly(amic acid), $pK_a(A) = 3$] with copper (IEPS of copper oxide = 9.5), a too strong predicted interaction that parallels the migration of copper in the polymer film [52].

III. THEORY OF ACID-BASE INTERACTIONS IN ADHESION

A. The Thermodynamic Work of Adhesion

The thermodynamic work of adhesion W is by definition the free energy change per unit area required to separate to infinity two surfaces initially in contact with the result of creating two new surfaces (see Fig. 1). In the absence of chemisorption and interdiffusion, W is the sum of the various intermolecular forces involved and can be related to the surface free energies (Dupré's equation):

$$W = \gamma_1 + \gamma_2 - \gamma_{12} \tag{8}$$

where γ_1 and γ_2 are the surface free energies of components 1 and 2 and γ_{12} is the interfacial free energy. For two materials interacting via London dispersive forces only across their interface, Fowkes [4] suggested that W be described by

$$W = W^d = 2(\gamma_1^d \gamma_2^d)^{1/2} \tag{9}$$

where W^d is the dispersive contribution to the work of adhesion and γ_i^d the dispersive contribution to the surface energy γ_i.

In the case in which both materials have ''polar'' interfacial sites, W can be described by

$$W = W^d + W^p \tag{10}$$

where W^p is the polar contribution to the reversible work of adhesion. The contribution W^p was described by [53]

$$W^p = 2(\gamma^p_1\gamma^p_2)^{1/2} \tag{11}$$

where γ_i^p is the polar contribution to the surface energy of the ith species. This is known as the extended Fowkes' equation. However, Fowkes et al. [14,54] have demonstrated that Eq. (11) is incorrect and cannot predict the magnitude of the nondispersive interactions. The main problem with the extended Fowkes' equation is the wrong assumption that the nondispersive contribution to W of two polar materials is a function of their polar properties. Indeed, when Eq. (8)–(11) are applied to a liquid-liquid system, such as water-ethanol, they cannot predict their miscibility or immiscibility. Although the γ^p value for ethanol is only 1.1 mJ/m^2, this liquid is very hydrophilic and miscible in water in all proportions. Fowkes has shown that the use of a geometric mean expression for estimating the work of adhesion and interfacial tension between water and ethanol predicts that these two liquids are completely immiscible with an interfacial tension of 37.7 mJ/m^2, which, of course, is absurd. For this reason, γ^p is usually a very inadequate measure of polarity or hydrophilicity [14].

In modern approaches to interfacial interactions, the nondispersive contribution to the work of adhesion is attributed to Lewis acid-base interactions, and W^{AB}, the acid-base component of W, can be evaluated from

$$W^{AB} = W - 2(\gamma_1^d\gamma_2^d)^{1/2} \tag{12}$$

Two methods can be recommended for determining W^{AB}; the first one was suggested by Fowkes and Mostafa in 1978 [11] and the second approach was introduced by van Oss et al. [22] in 1988.

B. The 1978 Method of Fowkes and Mostafa

This method makes use of ΔH^{AB} to assess W^{AB}:

$$W^{AB} = -fn_{AB}\Delta H^{AB} \tag{13}$$

where f is a conversion factor from free energy to enthalpy and n_{AB} the number of acid-base contacts per unit area. The ΔH^{AB} can be evaluated experimentally, e.g., by microcalorimetry [14], infrared spectroscopy [14,16,55] and contact angle measurements [43,44], or evaluated with the Drago four-parameter equation [31]. Equation (10) can thus be rewritten

$$W = 2(\gamma^d_A \gamma^d_B)^{1/2} + f n_{AB}(E_A E_B + C_A C_B) \tag{14}$$

Equations (3) and (12) were applied to the benzene-water interface using Drago's E_A and C_A constants for water and E_B and C_B for benzene. Drago's equation predicts a ΔH^{AB} value of -5.0 kJ/mol. The cross-sectional area of benzene (0.50 nm^2) leads to $n_{AB} = 3.3$ μmol/m^2. Applying Eq. (13) yields $W_{12}{}^{AB}/f = 16.5$ mJ/m^2, which compares well with the value determined at 20°C using Eqs. (8) and (12):

$$\begin{aligned}
W_{12}^{AB} &= \gamma_1 + \gamma_2 - \gamma_{12} - 2(\gamma_1^d \gamma_2^d)^{1/2} \\
&= 72.8 + 28.9 - 35 - 2\,(22 \times 28.9)^{1/2} = 16.3 \text{ mJ/m}^2
\end{aligned} \tag{15}$$

This implies that $f = 1$. However, f cannot be set equal to unity as found by Vrbanac and Berg [56] in their study of various neutral, acidic, and basic polymer surfaces. They checked Eq. (13) using a combination of wettability to determine W, conductimetric titrations for the assessment of n_{AB}, flow calorimetry to determine ΔH^{AB}, and temperature-dependent determination of surface tension and contact angle to estimate f. It was concluded that f (temperature dependent) was significantly below unity in most cases and that even including this effect, the preceding equation was still not verified quantitatively when the terms were measured independently. Indeed, for a DMSO-poly(ethylene-co-acrylic acid 5%) system, $W^{AB} = 1.3$ and 3 mJ/m^2 from wetting measurements and independent measurements of the various parameters in Eq. (13), respectively. As this procedure was applied to a single system, one cannot claim or conclusively deny the quantitative (dis)agreement between the two ways of estimating W^{AB}.

C. The 1988 Method of van Oss, Good, and Chaudhury

van Oss et al. [22] introduced the notion of acidic and basic components of the surface energy (γ^+ and γ^-, respectively) to characterize the acid-base properties of materials and predict W^{AB}:

$$W^{AB} = 2(\gamma_1^+ \gamma_2^-)^{1/2} + 2(\gamma_1^- \gamma_2^+)^{1/2} \tag{16}$$

The γ^+ and γ^- for a solid can be determined by contact angle measurements using three reference liquids of known γ_L^d, γ_L^+, and γ_L^-. The γ_L^+ and γ_L^- for test liquids were established with model surfaces and test liquids on the arbitrary assumption that for water $\gamma_L^- = \gamma_L^+ = 25.5$ mJ/m^2 and

$$\gamma_L^{AB} = 2(\gamma_L^+ \gamma_L^-)^{1/2} \tag{17}$$

Application of Eq. (17) to water (w) yields

$$\gamma_w^{AB} = 2(25.5 \times 25.5)^{1/2} = 51 \text{ mJ/m}^2 = \gamma^{nd}$$

where γ^{nd} is the nondispersive or the so-called polar component of the surface energy of water.

Once γ^+ and γ^- are determined for a given material, the overall acid-base ("polar") contribution to its γ can be evaluated as for water in Eq. (17). Table 5 shows γ^+ and γ^- values for liquids, polymers, and other materials.

van Oss and Good [15] have shown that their approach can predict the solubility of poly(ethylene oxide) (PEO) and dextran in water but not the "γ^p approach," which leads to a positive interfacial free energy for the polymer-

Table 5 Surface Energy Components for Commonly Used Test Liquids and for Polymers and Other Materials

	γ	γ^d	γ^{AB}	γ^+	γ^-
Liquids					
Water	72.8	21.8	51	25.5	25.5
Glycerol	64	34	30	3.92	57.4
Formamide	58	39	19	2.28	39.6
DMSO	44	36	8	0.5	32
Ethylene glycol[a]	48	29	19	3	30.1
α-Bromonaphthalene	44.4	43.5	—	—	—
CH_2I_2	50.8	50.8	—	—	—
Polymers					
PEO 6000[b]	43	43	0	0	64
Dextran 10000[b]	61.2	47.4	13.8	1.0	47.4
PMMA	39–43	39–43	0	0	9.5–22.4
PVC	43.7	43	0.7	0.04	3.5
PS	42.0	42	0	0	1.1
PE[d]	33	33	0	0	0.1
PE[e]	57.9–62.5	42	15.9–20.5	2.1	30–50
PAI[c,f]	52.6	42.8	9.8	1.04	23.15
Cellulose acetate	40.2	35	5.2	0.3	22.7
Cellulose nitrate	45	45	0	0	16
Agarose	44.1	41	3.1	0.1	24
Gelatin	38	38	0	0	19
Human serum albumin	44.3	41	3.3	0.15	18

All values in mJ/m² from Ref. 10 except [a]Ref. 57, [b]Ref. 15, and [c]Ref. 58. [d] Based on advancing angles. [e] Based on receding angles. [f] PAI, iodinated polyacetylene.

water systems (5.8 and 6 mJ/m² for PEO and dextran, respectively) and thus to the insolubility of these polymers in water.

Although the novel approaches to interfacial interactions just described were shown to be more valid than the "polar interactions" approach in predicting adhesion, solubility, and miscibility phenomena [11,14–16], they must be viewed with caution as controversial publications have shown their shortcomings [14,56]. Nevertheless the qualitative and/or quantitative determination of acid-base properties for polymers and other materials is an important step toward the understanding and the prediction of their interfacial interactions. This is a very delicate task because it requires the determination of the heat or the free energy changes of acid-base interactions of reference acidic and basic chemical species (of known acid-base properties) with the material. The choice of reference test acids and bases is also crucial and usually depends on the nature of the material under investigation and the experimental conditions associated with the technique used for the assessment of acid-base properties. As pointed out by Dwight [19], "strongly interacting probes are useless in IGC because they will not desorb, but are necessary in XPS so that they will not be removed in the ultra high vacuum."

IV. EXPERIMENTAL ASSESSMENT OF ACID-BASE PROPERTIES OF POLYMERS AND OTHER MATERIALS

The arsenal of techniques available for assessing materials' acid-base properties include contact angle measurements [44,59], microcalorimetry [13], inverse gas chromatography [18,41,60], Fourier transform infrared (FTIR) [27], NMR [14], and XPS [16,45,61–68]. All these methods and others were reviewed on several occasions [6,14,16–19,38,45,68]. However, for the purpose of this contribution, the focus will be on IGC and XPS only. The former technique was used extensively by the author to determine the surface energy of conducting polymers before and following coating with flexible conventional polymers (see Sec. V.B), and the latter was developed to probe acid-base properties of flat polymer surfaces using adsorbed solutes (see Sec. V.C).

A. Use of the Molecular Probe Technique in XPS

XPS has been extensively used in adhesion science and technology to study the locus of failure of adhesive joints and to determine molecular orientation following failure [69,70], monitor the uptake of specific ions [71,72] at the interface, and identify molecular species segregating at a polymer–metal oxide interface [73]. The success of XPS lies in its surface specificity (analysis depth of about 5 nm), low degree of degradation of tested materials, quantitative aspect, and the detection of all elements (except hydrogen) and their chemical shifts. The so-called chemical shift is the cornerstone of XPS because it enables the surface scientist not only to recognize the chemical environments of elements but also to derive material properties such as refractive indices of thin optical layers [74] and the nondispersive component of the surface energy of polymers [75]. Of relevance to the present review, Martin and Shirley [76] showed that the chemical shifts were related to the acid-base properties of alcohols and amines. Burger and Fluck [77] established, for quickly frozen solutions of $SbCl_5$–Lewis base complexes in 1,2-dichloroethane, a linear relationship between the $Sb3d^{5/2}$ binding energy (BE) and DN, the donor number of the complexing Lewis bases. On this basis, Chehimi [78] showed that the $Sb3d^{5/2}$ BE was linearly correlated with the ΔH^{AB} of (base-$SbCl_5$) adduct formation calculated using Drago's equation and concluded that XPS was a potential tool for estimating Drago's parameters for polymer surfaces.

In the case of metal oxides, Delamar [79] used an XPS data bank and a compilation of IEPS values to establish a linear relationship between chemical shifts of the oxygen (ΔO) and those of the metal cations (ΔM) and the IEPS of the corresponding metal oxides [ΔO was defined as BE (O1s)-530 eV, where BE is the binding energy and 530 eV an arbitrary reference BE value, and ΔM was defined as BE($M2p$)$-$BE(M^02p), where M is the metal in the oxidized state, M^0 is the metal in the reference metallic state, and $2p$ is the core shell]. The IEPS versus ($\Delta O + \Delta M$) correlation was confirmed by Cattania et al. [80] in their electrophoretic and XPS characterizations of a series of metal oxides having the same history. Alternatively, ion exchange experiments were proposed to evaluate IEPS for hydrated metal oxides by XPS [81–83], but some caution must be exercized as this approach has severe shortcomings [84].

The molecular probe technique in combination with XPS has seldom been used since the early 1970s and has been applied mainly to zeolites [85–87]. These studies were aimed at identifying and quantifying Lewis acidic and basic sites at catalyst surfaces by monitoring the BE shifts of N1s from adsorbed pyridine [85,86] and pyrrole [87], respectively.

As far as polymers are concerned, Chehimi and coworkers [61–64,78,88] established protocols for quantitatively estimating the acid-base properties of polymer surfaces using (ad)sorbed molecular probes. In practice, a polymer is exposed to liquid vapors (solutes) of known acid-base properties for a few minutes. The polymer is then allowed to outgas the excess of solute and is transferred into the XPS equipment for surface characterization. If the polymer-solute interaction is strong enough (e.g., via acid-base interactions), then a residual amount of solute is detected and quantified.

Figure 4 depicts a survey scan of a basic aromatic moisture-cured urethane resin (ArMCU) before and following exposure to the vapors of hexafluoroisopropanol [HFIP, $CF_3CH(OH)CF_3$], a reference Lewis acid. The F1s from HFIP

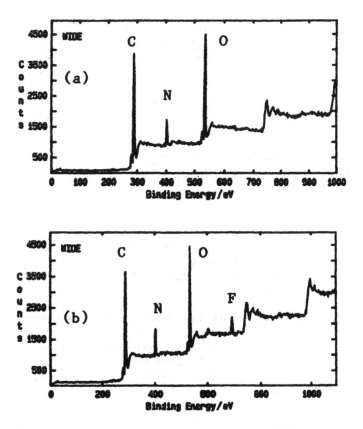

Figure 4 X-ray photoelectron survey spectra of ArMCU (a) before and (b) after exposure to HFIP. Uptake of the Lewis acid HFIP is indicated by the presence of F1s feature.

is easily detected, indicating that it was retained by ArMCU. This retention, despite the high vacuum, is believed to be governed by acid-base interactions between the OH group from HFIP and the carbamate (HN—C═O) group from the resin. The molar ratio of solute per polymer repeat unit (%S, where S stands for solute) was evaluated and used as a measure of the uptake (or retention) of solute by the host polymer. Figure 5 shows plots of %S versus AN (Gutmann's acceptor number) of the solute for the host polymers PMMA and ArMCU. The plots are S shaped, showing an increasing uptake of solute with AN, which denotes the basic character of both polymers. This XPS result agrees with FTIR studies of the Lewis basicity of PMMA [14] and ArMCU [63]. XPS inspection of a polyethylene (PE) surface following exposure to $CHCl_3$ vapor was unsuccessful because PE interacts with $CHCl_3$ via London dispersive forces only.

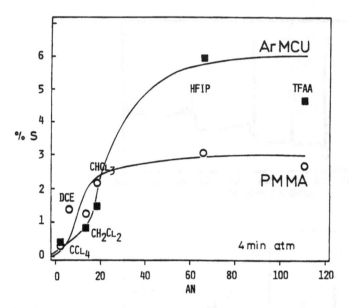

Figure 5 Uptake (%S) of Lewis acids by PMMA and ArMCU versus the acidic character of the solutes. %S is the solute per repeat unit molar ratio in the case of PMMA and
the solute per nitrogen atom in the case of ArMCU. The solutes were characterized by
AN, Gutmann's acceptor numbers: CCl_4 (2.3); 1,2-dichloroethane, DCE (6.4); dichloro-
methane, DCM (13.5); chloroform or trichloromethane, TCM (18.7); hexafluoroisopropanol, HFIP (66.3); and trifluoroacetic acid, TFAA (111). (From Ref. 61.)

Several molecular probes can be used to characterize the acid-base properties of solid surfaces by XPS. For example, chlorinated and fluorinated acidic
species probe Lewis basicity, whereas pyridine [61,64,86] and DMSO [66] are
suitable for characterizing surface acidity (Table 6).

The binding energies (BEs) of core electrons from the solutes' elemental
markers were also investigated. In the case of chloroform, Chehimi et al. [62]
established a linear $Cl2p^{3/2}$ BE-ΔH^{AB} correlation, where ΔH^{AB} is the heat of acid-
base adduct formation for the polymer-chloroform pairs in the liquid state (Fig.
6). The relationship is of the form

$$-\Delta H^{AB} \text{ (kcal mol}^{-1}) = 228.8 - 1.14BE \text{ (eV)} \tag{18}$$

and was used to determine ΔH^{AB} for $PPO:CHCl_3$ adduct formation, approximately 1 kcal/mol [45]. This result derived from XPS chemical shifts was combined with the IR data reported by Kwei et al. [55] to deduce $E_B \approx 0$ and $C_B =$
9.5 (kcal mol^{-1})$^{1/2}$ for PPO [16,45].

Table 6 Molecular Probes Used for XPS
Determination of Materials' Acid-Base Properties

Probes	Core line	Material property investigated
CHCl$_3$	Cl2p	Basicity
CH$_2$Cl$_2$	Cl2p	Basicity
HFIP	F1s	Basicity
CF$_3$COOH	F1s	Basicity
I$_2$	I3$d^{5/2}$	Basicity
Pyrrole	N1s	Basicity
Pyridine	N1s	Acidity
DMSO	S2p	Acidity

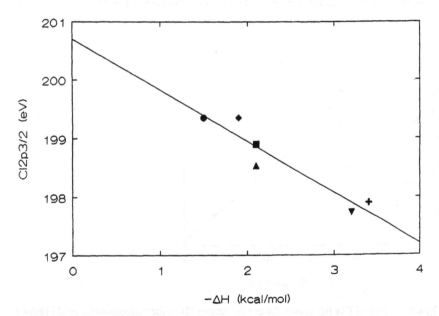

Figure 6 Cl2$p^{3/2}$ BE versus ΔH^{AB} for CHCl$_3$ sorbed in (+) ArMCU, (●) polyvinyl acetate (PVAc), (▲) PMMA, (▼) polyethylene oxide (PEO), (♦) polybutylmethacrylate (PBMA), and (■) polycyclohexylmethacrylate (PCHMA). (From Ref. 62.)

Equation (18) will be used further in the study of acid-base and adhesion properties of ammonia plasma–treated polypropylene (see Sec. V.C.2).

Another Lewis acid used to probe surface acid-base properties of polymers was CF_3COOH (TFAA), for which a relationship between F1s BE and DN (donor numbers) values of the host polymers (Fig. 7) was obtained [67,68]. The DN values were computed using Gutmann's equation and the thermochemical data for binary polymer blends of a poly(styrene-co-vinylphenyl hexafluoro dimethyl carbinol) [55]:

$$DN \text{ (kcal mol}^{-1}) = 3593 - 5.21BE \text{ (eV)} \tag{19}$$

The copolymer contained 95% styrene repeat units and its OH stretching frequency shifts were similar to those of HFIP [55]. It was thus reasonable to assign Gutmann's AN number of HFIP to this copolymer [67]. Equation (19) can thus potentially be valuable in determining DN values for polymer surfaces of a Lewis basic nature.

Finally, Fig. 8 depicts the N1sBE for pure pyridine and that sorbed in PMMA and polyvinyl butyral (PVB) as a function of the acceptor number (AN) of the pure solute and the two polymers. The positive chemical shift was interpreted in terms of donation of electron density from pyridine to the carbonyl carbon of PMMA (acidic site) [61]. The N1s chemical shift is even larger when

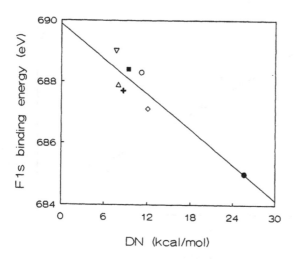

Figure 7   Plot of F1s BE versus the donor number (DN) for trifluoroacetic acid (TFAA) sorbed in polymers characterized by their DNs (see text for the estimation of polymers' DN numbers). (■) PMMA; (○) PCHMA; (+) PVAC; (△) PBAC; (●) PEO; (▽) PPO; (◇) PVME.

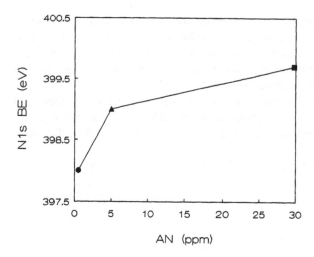

Figure 8 Plot of N1s BE versus AN for pure pyridine and pyridine sorbed in PMMA and PVB. AN = 0.5 [7], 5.3, and 29.8 for (●) pyridine, (▲) PMMA, and (■) PVB, respectively. PMMA was assigned the AN value of ethyl acetate [7], and PVB the average value of AN reported for 2-propanol [7] and the value estimated using Eq. (4), ΔH^{AB} for the ethyl acetate–PVB interaction [27], and the DN (17.1) reported for ethyl acetate [32].

pyridine is sorbed in PVB, since it is predominantly acidic due to its OH pendent groups. The N1s BE positions for pyridine-polymer complexes are higher than those of the pure pyridine because pyridine-pyridine interaction occurs via dispersive forces only, for pyridine is a monofunctional species [7,54].

To summarize, acidic (CHCl$_3$ and TFAA) and basic (pyridine) probes undergo negative and positive chemical shifts (lower and higher BEs), respectively, when they interact with host surfaces via acid-base interactions. Indeed, a Lewis acid is an electron acceptor and, upon interaction with a base via acid-base forces, electron density is transferred to the former, thus yielding a lower binding energy of its electrophilic site [61,62,45]. The opposite reasoning holds for basic interactions [61,64].

B. Inverse Gas Chromatography

Inverse gas chromatography (IGC) is now very well accepted by the adhesion community for it provides thermodynamic and morphological information on a variety of materials such as fillers, pigments, colloids, fibers, powder, wood, and polymers [18,60,89–95]. The term ''inverse'' means that the stationary phase is of interest, whereas in conventional GC the mobile phase is of interest. Its success lies in the fact that it is simple, versatile, usable over a very wide range of temperature, and very low in cost. IGC has a well-established background for the assessment of γ_s^d and acid-base parameters for polymers and fillers. Such thermodynamic parameters can further be used to estimate the reversible work of adhesion at polymer-fiber and polymer-filler interfaces [96,97].

IGC is based on the interfacial interactions between molecular probes and the stationary phase. Probes are injected at infinite dilution so that lateral probe-probe interactions are negligible and the retention is governed by solid-probe interactions only.

The net retention volume, V_N, is defined as the volume of inert carrier gas (corrected for the dead volume) required to sweep out a probe injected in the chromatographic column. At infinite dilution (zero coverage), ΔG_a, the free energy of adsorption, is related to V_N by

$$-\Delta G_a = RT \ln(V_N) + C \tag{20}$$

where R is the gas constant, T is the column temperature, and C is a constant that accounts for the weight and specific surface area of the packing material and the standard states of the probes in the mobile and the adsorbed states [98]. Treatment of ΔG_a or simply $RT \ln(V_N)$ values is a key to the assessment of dispersive and acid-base properties of, e.g., polymers, fibers, and fillers.

1. Dispersive Properties

The method of Dorris and Gray [99] is the most accepted one for determining γ_s^d values. It is based on the determination of ΔG^{CH_2}, the free energy of adsorption per methylene group, from the retention data of the n-alkane series (probes capable of dispersive interactions only). Figure 9 depicts plots of ΔG_a or $RT \ln(V_N)$ values versus the number of carbon atoms in the n-alkanes (n_c) for PPyCl, PMMA, and a PMMA-coated PPyCl (PMMA/PPy) sample prepared by adsorption of PMMA onto polypyrrole from chloroform. Each plot generates an excellent linear correlation the slope of which equals ΔG^{CH_2}. For a solid-CH_2 interaction, Eq. (9) can be rewritten as

$$W = 2(\gamma_s^d \gamma_{CH_2})^{1/2} \tag{21}$$

where γ_{CH_2} is the surface free energy of the methylene group (taken as γ values for polyethylene, since this polymer contains only methylene groups). Given that W is a free energy change per unit area, it follows that

$$W = -\Delta G_a^{CH_2}/Na_{CH_2} \tag{22}$$

where N is the Avogadro number, a_{CH_2} is the cross-sectional area of an adsorbed methylene (CH_2) group (6 Å^2). Combining Eq. (21) and (22), one can determine γ_s^d using

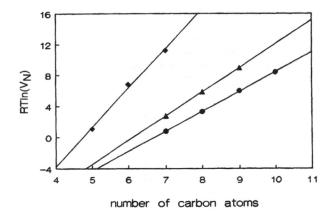

Figure 9 $RT \ln(V_N)$ versus the number of carbon atoms for *n*-alkanes adsorbed (at 48°C) onto (◆) PPyCl, (●) PMMA, and (▲) PMMA-coated PPyCl prepared in chloroform (PMMA/PPy).

$$\gamma_S^d = \left(\frac{1}{4\,\gamma_{CH_2}}\right)(\Delta G_a^{CH_2}/Na_{CH_2})^2 \qquad (23)$$

where γ_{CH_2} is temperature dependent [γ_{CH_2} (mJ/m^2) = 36.8 −0.058T (°C)], thus permitting the assessment of γ_S^d at various temperatures [99]. The validity of this approach has been established on the basis that IGC and wettability measurements led to approximately the same γ_S^d value (~40 mJ/m^2) for poly(ethylene terephthalate) [100]. In Fig. 9, the slopes yield γ_S^d values of 145, 36.6, and 55 mJ/m^2 for PPyCl, PMMA, and PMMA/PPy, respectively. The intermediate value obtained for PMMA/PPy is an indication of a patchy adsorbed layer of PMMA on the polypyrrole surface.

A similar approach for γ_S^d evaluation by IGC based on the cross-sectional area of molecular probes interacting via dispersive forces only (e.g., linear, branched, or cyclic alkanes) was proposed by Schultz et al. [96]. The method leads to values comparable to those obtained by Gray's approach at low temperature (error less than 4%) but significantly deviates at higher temperature (100°C) [18]. Hamieh and Schultz [101] proposed a temperature dependence of the cross-sectional area of the probe molecules to improve this approach. However, the refinements brought by this method to calculate γ_S^d values for a series of metal oxides were shadowed by the reproducibility of the measurements [102]. Therefore, this method is at best as valid as that of Dorris and Gray.

Table 7 provides γ_S^d values for some polymers, fillers, and fibers over a wide range of temperature, which constitutes an advantage over contact angle measurements.

Table 7 γ_s^d Values for Some Polymers, Fillers, Fibers, and Pigments

Materials[a]	γ_s^d (mJ/m^2)	Temperature (°C)	Reference
Conventional and conducting polymers			
PMMA	38.8	25	103
PVC	35.8	25	16
PET	37.9	26.5	100
LLDPE	28.8	30	104
PEEK	40	50	97
PPyCl	145	48	103
PPyTS	88.5	25	41
PANI	87.3	68	105
PPyCl-SiO$_2$	225	60	106
Fillers, fibers, and pigments			
C fiber	46.5	29	96
Carbon black	42.8	30	107
Graphite	129	44.5	108
CaCO$_3$	44.6	30	104
MgO	95.6	25	93
Al$_2$O$_3$	42–100	110	109
SiO$_2$ sol	60	60	106
Rutile	23.2–25.6	60	95
Monastral green	43.0	60	95

[a] LLDPE, low-density poly(ethylene); PEEK, poly(ether ether ketone); PPyCl, chloride-doped poly(pyrrole); PPyTS, tosylate-doped poly(pyrrole); PANI, poly(aniline). Monastral green is a phthalocyanine type of organic pigment.

Three important points must be borne in mind concerning the IGC determination of γ_s^d values:

> For heterogeneous high-energy surfaces characterized by IGC at infinite dilution, solutes preferentially probe the high-energy sites and the technique thus leads to γ_s^d values higher than those obtained by contact angle measurements [18,110]. For example, γ_s^d values determined for conducting polymers by IGC were always higher than those estimated by wettability [111].

> The γ_s^d values reported for, e.g., Al$_2$O$_3$ strongly depend on the nature and the concentration of metal oxide impurities such as silica, which are likely to segregate to the surface and thus affect the surface energy [109].

Papirer et al. [90] have also shown that the impurities significantly modify the acid-base properties of fillers such as alumina and thus considerably affect their adsorptive capacities toward polymers.

IGC yields very high values of γ_S^d for microporous and lamellar materials by comparison with the reference amorphous ones [91]. Table 7 shows an extremely high γ_S^d value of 225 mJ/m^2 for polypyrrole-silica nanocomposite at 60°C, which is much higher than those of the reference silica sol and bulk polypyrrole powder (60 and 145 mJ/m^2, respectively). The apparent surface energy of the nanocomposite [106] was interpreted in terms of microporosity of these raspberry-structured colloidal materials [112].

2. Acid-Base Interactions

For polar probes interacting via acid-base interactions with the stationary phase, ΔG_a is expected to have a contribution from such specific interactions. Assuming that dispersive and acid-base interactions are additive, ΔG_a^{AB}, the acid-base contribution to the free energy of adsorption, is substracted from ΔG_a:

$$-\Delta G_a^{AB} = -(\Delta G_a - \Delta G_a^d) = RT \ln(V_N/V_{N,\mathrm{ref}}) \tag{24}$$

where V_N and $V_{N,\mathrm{ref}}$ are the net retention volumes of the polar probe and a hypothetical reference n-alkane having the same property (e.g., boiling point, logarithm of the vapor pressure), respectively. There are various approaches for the assessment of ΔG_a^{AB} in which the probes can be characterized by the boiling point [113]; the logarithm of the vapor pressure [114]; $a(\gamma_L^d)^{1/2}$ [96] (where a and γ_L^d are the cross-sectional area of the probe and the dispersive contribution to the surface energy, respectively); the deformation polarizability [108]; and $\Delta H_{\mathrm{vap}}^d$, the dispersive contribution to the heat of vaporization [115]. All these methods have advantages and shortcomings that have been discussed elsewhere [115].

In practice, ΔG_a^{AB} is determined as shown in Fig. 10, where the variation of $RT \ln(V_N)$ versus $\Delta H_{\mathrm{vap}}^d$ is plotted for p-toluenesulfonate-doped polypyrrole (PPyTS), a conducting polymer, at 35°C.

The n-alkane data lead to a linear relation that defines the dispersive interactions for the PPyTS-probe pairs. For "polar" probes interacting via acid-base interactions, the corresponding markers lie above the reference line with a vertical distance that accounts for ΔG_a^{AB}. In Fig. 10, the markers corresponding to the polar probes lie significantly above the reference line defined by the n-alkanes, indicating that PPyTS behaves amphoterically. However, ΔG_a^{AB} values are significantly much higher for the Lewis bases (EtAc, THF, and diethyl ether) than for the acidic species $CHCl_3$ and CH_2Cl_2, an indication that PPyTS has a predominantly acidic character.

The ΔH_a^{AB} is usually determined from the temperature dependence of

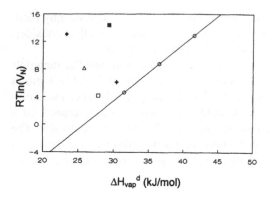

Figure 10 $RT \ln(V_N)$ versus ΔH_{vap}^d for alkanes and polar probes adsorbed onto tosylate-doped polypyrrole (PPyTS) at 35°C. (O) C6–C8; (□) CH$_2$Cl$_2$; (+) CHCl$_3$; (△) diethylether; (■) EtAc; (◆) THF. (Adapted from the data in Ref. 41.)

ΔG_a^{AB}, but alternative approaches have also been proposed [40,42,116]. Combining ΔH_a^{AB} values with Drago's constants or Gutmann's DN and AN values permits determining acid-base constants for the material under test. For example, Saint Flour and Papirer [114] suggested an extension of Gutmann's equation to determine the acidic and basic constants of the solid under test:

$$-\Delta H_a^{AB} = DN\, K_A + AN^*\, K_D \tag{25}$$

Practically, this equation is rewritten as

$$-\Delta H_a^{AB}/AN^* = (DN/AN^*)K_A + K_D \tag{26}$$

and for each probe $-\Delta H_a^{AB}/AN^*$ is plotted versus (DN/AN^*). This usually results in a linear relation with slope K_A and intercept K_D (see Fig. 11).

Kuczynski and Papirer [93] as well as Chehimi et al. [103] found that it was also simple to derive K_A and K_D values from the following equation:

$$-\Delta G_a^{AB} = DN\, K_A + AN^*\, K_D \tag{27}$$

Although fundamentally this approach is not correct because it relates ΔG^{AB} values to Gutmann's numbers derived from ΔH^{AB} terms, it has proved to be a fast and effective semiquantitative approach to monitoring changes in the surface properties of fillers and polymers [92,103,114,117,118]. It does not necessarily mean that the entropic term is ignored so that $\Delta G^{AB} = \Delta H^{AB}$. It simply produces different scales of K_A and K_D constants, but at one given temperature. In our laboratory, we found it very suitable for materials, such as conducting polypyrrole, that may degrade quite rapidly in the course of IGC characterization [105].

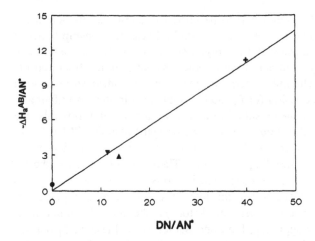

Figure 11 Plot of $-\Delta H_a^{AB}/AN^*$ versus DN/AN^* for tosylate-doped polypyrrole (PPyTS). The slope and the intercept permit the determination of K_A and K_D, respectively, for PPyTS. (●) Chloroform; (▼) EtAc; (+) THF; (▲) diethyl ether. (From Ref. 41.)

An alternative method for the determination of acid-base characteristics of solids was proposed by Lara and Schreiber [95], who defined

$$K_A = -\Delta G_a^{AB}(\text{THF}) \tag{28a}$$

and

$$K_D = -\Delta G_a^{AB}(\text{CHCl}_3) \tag{28b}$$

This is another empirical simple approach to assessing acid-base properties of polymers and fillers. Of course, one can use other reference acids and bases if they are more suitable for the solid under test.

Using K_A and K_D constants for polymer matrices (m) and fillers or fibers (f), one may define pair-specific interaction parameters (I_{sp}):

$$I_{sp} = K_A^f K_D^m + K_A^m K_D^f \tag{29a}$$

$$I_{sp} = (K_A^f K_D^m)^{1/2} + (K_A^m K_D^f)^{1/2} \text{ in kJ/mol} \tag{29b}$$

Equation (29a) is based on Eq. (25) and was proposed by Schultz et al. [96], whereas Eq. (29b) derives from Eq. (28a) and (28b) [95].

In their study of carbon fiber-epoxy composites, Schultz et al. [96] determined K_D and K_A parameters [derived from Eq. (25)] for untreated and treated carbon fibers and an epoxy matrix (diglycidyl ether of bisphenol A with 35 parts by weight of diaminodiphenyl sulfone, DGEBA-DDS) and used the specific interaction parameter defined by Eq. (29a) to describe the acid-base interaction between the fiber and the matrix. The authors found a linear relationship between the interfacial shear resistence τ and I_{sp} (Fig. 12) and concluded that interfacial adhesion resulted mainly from acid-base interactions between the fiber and the matrix. By contrast, an attempt to correlate the interfacial shear strength with either the total work of adhesion (or I_{sp}, author's attempt) in the case of single-carbon fiber–PEEK composites resulted in a very scattered graph, possibly due to the existence of a crystalline boundary layer between the fiber and PEEK [119]. The thickness of such a layer can depend on both the physicochemical fiber-matrix interactions and the molding conditions. Therefore, in this case, the interfacial shear strength cannot be interpreted in terms of acid-base interactions alone, since the physical properties of the interphase layer are as important as the characteristics of PEEK poly(ether ether ketone) at the fiber-PEEK interface.

Following a similar approach, Lara and Schreiber [95] related I_{sp} defined by Eq. (29b) to the adsorbed amount of polyester dispersion and binder resins onto rutile and organic pigments. Figure 13 depicts the relationship between adsorption and I_{sp} for an amine-modified polyester dispersion resin adsorbed onto mineral and organic pigments. It provides strong evidence that acid-base interac-

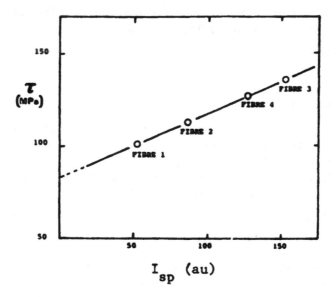

Figure 12 Interfacial shear resistance τ versus I_{sp}, specific interaction parameter for carbon fibers and an epoxy matrix (DGEBA-DDS). The carbon fibers were untreated (1), oxidized (2), sized (3), and commercial sized (4) PAN-based fibers. (From Ref. 96.)

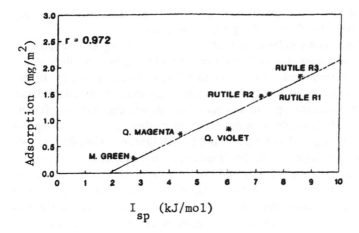

Figure 13 Amount of a polyester dispersion resin adsorbed onto organic and mineral pigments versus the pair-specific interaction parameter defined in Eq. (29b). (From Ref. 95.)

tions are dominant in determining the adsorption behavior of polymer-pigment combinations.

V. SOME PRACTICAL APPLICATIONS OF ACID-BASE INTERACTIONS

The knowledge of the acid-base properties of polymers, fibers, and fillers is an important step toward the choice of the components used to design high-performance materials (e.g., composites, blends). However, it is also very important to interrogate the effectiveness of acid-base interactions in adsorption, wetting, and adhesion these phenomena being of relevance to the performance of the end products. It is the aim of this section to address such questions. The role of acid-base interactions is emphasized in the adsorption (from the liquid phase) of a conventional untreated homopolymer of predominant basicity onto an untreated rigid inherently conducting polymer of predominant acidity. This choice was driven by the fact that adsorption is one of the most important phenomena in adhesion [1] and that one of the most important properties of polymers is their ability to become readily adsorbed onto solid surfaces [120]. For this reason polymers are widely used in coatings and adhesion applications. Indeed, these materials have a very large number of repeat units, which enable them to adopt various conformations, have several interacting sites, and therefore have a very high surface activity that favors adsorption at interfaces.

Although adhesion can be good without adhesive or any pretreatment, in several practical situations it was shown conclusively that the reliability might be poor and that pretreatments of fibers or metal oxides were necessary. In this regard, it is well documented that grafting appropriate surface acidic and/or basic functional groups leads to composites or coatings with good and durable polymer-substrate adhesion. Conversely, it is necessary to modify the surface of apolar polymers to achieve, e.g., adequate and durable metallization. The effect of surface pretreatments of polymers will be illustrated in the case of the wetting and adhesion characteristics of plasma-treated polypropylene. However, prior to the manifestations of acid-base interactions at the macroscopic scale (polymer wetting, adsorption, and metallization), the importance of such specific interactions will be discussed in terms of the determination of the adhesion forces at the molecular scale using interfacial force microscopy.

A. Measuring Adhesion Forces at the Molecular Scale

Thomas et al. [20] have reported the possibility of measuring at a molecular scale adhesion forces between organic films over separations ranging from 10 nm to repulsive contact using interfacial force microscopy (IFM). This microscope uses a self-balancing force-feedback system to avoid the mechanical instability encountered in atomic force or scanning tunneling microscopy (AFM and STM, respectively) due to the use of deflection-based force sensors.

Quantitative measurement of the adhesion forces between organic films was achieved by chemically modifying a gold substrate and a gold tip with self-assembling organomercaptan molecules, SAMs (Fig. 14), having either the same or different end groups (—CH_3, —NH_2, and —COOH). Figure 15 shows representative force profiles for some terminal group combinations. The arbitrary zero displacement represents the point where the interaction force goes through zero while the probe is in contact with the sample. The force axis is normalized to the probe radius and the same scale is used for direct comparison of the different chemical interactions. The peak value of the attractive force from the unloading curve (pull-off force) was used to evaluate the work of adhesion, W, for the various combinations using

$$W = F/2\pi R \tag{30}$$

where R is the tip radius. The W values were found to be 60 ± 32, 100 ± 24, 228 ± 54, and 680 ± 62 mJ/m^2 for the CH_3—CH_3, NH_2—NH_2, COOH—COOH, and NH_2—COOH combinations, respectively. These values qualitatively scale with those expected for van der Waals, hydrogen bonding (for NH_2—NH_2 and COOH—COOH pairs), and acid-base interactions. The high

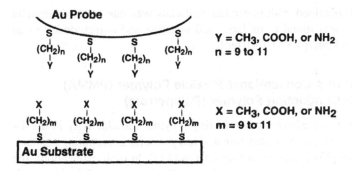

Figure 14 Schematic illustration of the experimental configuration. A gold probe and a gold substrate are covered by self-assembled monolayers of n-alkanethiol molecules with different combinations of end groups. The force of interaction between these films is recorded as a function of relative separation up to repulsive contact. (From Ref. 20.)

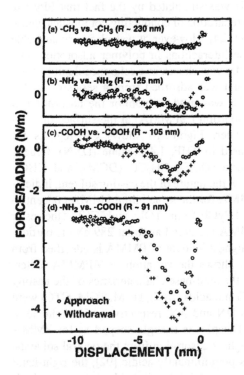

Figure 15 Force-displacement curves taken between (a) two methyl-terminated self-assembled monolayers (SAMs), (b) two amine-terminated SAMs, (c) two carboxylic acid–terminated SAMs, and (d) an amine- and a carboxylic acid–terminated SAM. The force is normalized to R, the tip radius. (From Ref. 20.)

work of adhesion obtained with dissimilar materials was due to the interfacial energy. The latter was found to be large and negative and corresponded to an NH_2—COOH bond energy of 67 kJ/mol.

B.   Adsorption of a Conventional Flexible Polymer (PMMA) onto a Stiff Conducting Polymer (Polypyrrole)

Polypyrrole (PPy) is an inherently conducting polymer (conductivity in the 1–300 S/cm range) that has a rigid structure and has potential in preparing conducting polymer blends [121] and conducting textiles [122]. In such applications, the interfacial properties may have a major effect on the properties of the end products. It was established via IGC measurements [41,92] that PPy is a high-energy organic polymer (see Table 7) and has a dominant Lewis acidic character [103,123]. As an extension of the study of molecular probe–conducting polymer interactions, the study of the adsorption of a flexible conventional polymer onto conducting PPy was undertaken. This was stimulated by the fact that PPy can be used as a conducting filler for the preparation of conducting composites whose threshold of conductivity is low (7%) [124]. However, it was also very desirable to check Fowkes' theory of acid-base interactions in polymer adsorption onto conducting PPy. For this reason, PMMA was chosen as a reference flexible polymer of predominant Lewis basicity and was adsorbed from a series of acidic, basic, and neutral solvents [125]. XPS was used to monitor the adsorption of PMMA and determine its relative proportion (%PMMA) at the surface of chloride-doped polypyrrole (PPyCl) powder. Figure 16 shows Cls regions for PMMA-coated PPyCl powders prepared in THF, 1,4-dioxane (DXN), toluene (TOL), CCl_4 (CTC) 1,2-dichloroethane (DCE), CH_2Cl_2 (DCM), and $CHCl_3$ (TCM). The Cls signal shape varies with the nature of the casting solvent, indicating different amounts of adsorbed PMMA for the same initial PMMA concentration. When PMMA adsorbs onto PPyCl from TOL and CCl_4, the ester —COOCH$_3$ Cls component from PMMA (centered at about 289 eV) is particularly prominent, suggesting that the highest amount of PMMA is adsorbed from these two neutral solvents. Figure 17 shows the variation of %PMMA (determined using Cls and Nls peaks) with the acid-base characteristics of the casting solvents. Lewis bases (THF and DXN) and acids (DCE, DCM, TCM, CTC) were characterized by Gutmann's numbers DN and AN, respectively. For toluene, it was assumed that DN = AN = 0. The plot is triangle shaped as for PMMA adsorption onto silica [11] with the highest adsorption from the neutral solvents. As PMMA is basic [27] and PPyCl is predominantly acidic [92], the right-hand side of Fig. 17 shows competition between solubility in acidic solvents and adsorption of PMMA onto PPyCl, and the left-hand side indicates competitive adsorption of basic solvents and PMMA.

The average thickness of PMMA coatings was estimated using the attenua-

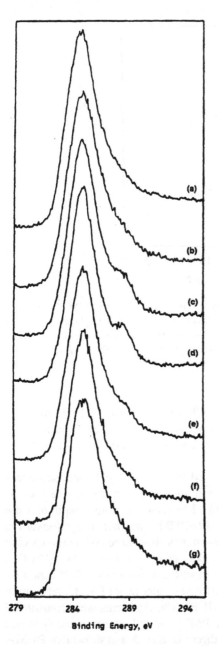

Figure 16 C1s region for PMMA-coated PPyCl powders prepared in (a) THF, (b) 1,4-dioxane, (c) toluene, (d) CCl$_4$, (e) 1,2-dichloroethane, (f) CH$_2$Cl$_2$, (g) CHCl$_3$. (From Ref. 125.)

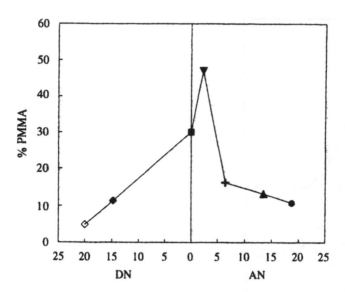

Figure 17 XPS evaluation of the relative proportion of PMMA (%PMMA) at the surface of PPyCl versus the acid-base characteristics of the solvents. The plot has the shape of the Fowkes triangle. DNs are in kcal/mol and ANs (corrected for van der Waals interactions [7]) in ppm. (◇) THF; (◆) DXN; (■) TOL; (▼) CTC; (+) DCE; (▲) DCM; (●) TCM. (From Ref. 125.)

tion of C12p and Cl$_{LMM}$ peaks from PPyCl and was found to be in the 0.2–0.9 nm range [125]. Depending on the acid-base properties of the solvents, a submonolayer or one or two monolayer(s) can be adsorbed from THF, DXN, and CCl$_4$, respectively.

The XPS findings were complemented by time-of-flight static secondary ion mass spectrometry (ToFSSIMS), a technique that is more surface specific than the former because it provides chemical information from a depth of 1 nm only. Moreover, whereas the assessment of %PMMA was not straightforward in XPS, ToFSSIMS detects fragments of known mass-to-charge (m/z) ratio specific to the material investigated. The mass peaks centered at $m/z = 31^-$ (CH$_3$O$^-$) and $m/z = 26^-$ (CN$^-$) were chosen as diagnostic ions for PMMA [126] and PPy [127], respectively, and were used to determine isotherms for PMMA adsorption onto the inherently conducting polymer (ICP). The isotherms were constructed by plotting $K_{31} = I_{31}/I_{31} + I_{26}$ versus the PMMA initial concentration (I stands for the peak area). Figure 18 displays the direct isotherms and shows that PMMA adsorption is more favorable from CCl$_4$ than from CHCl$_3$ [128], in agreement with XPS results [125]. In addition, the isotherms have distinct shapes; the isotherm in $CHCl_3$ has a rounded shoulder, and that obtained in CCl_4 is a high-affinity isotherm consisting of an initial sharply rising part followed by a plateau, an indication that PMMA strongly binds to hydrogen sulfate–doped PPy ($PPyHSO_4$).

The Langmuir equation was adapted for ToFSSIMS analysis and applied to the data in Fig. 18 in the form

$$C/\Gamma = (1/\Gamma_m) \times (1/K) + C/\Gamma_m \tag{31}$$

where C is the polymer initial concentration, Γ is the PMMA adsorption as measured via ToFSSIMS (K_{31}), Γ_m is the total number of adsorption sites for the solute, and K is a constant related to the adsorption equilibrium constant. Figure 19 shows plots of the C/K_{31} ratio versus C for PMMA adsorption onto polypyrrole from CCl_4 and $CHCl_3$. Excellent linear plots are obtained, which is an indication that the isotherms are of the Langmuir type. The constants K determined using

$$K = \left(\frac{\Gamma}{\Gamma_m - \Gamma}\right) \times \frac{1}{C} \tag{32}$$

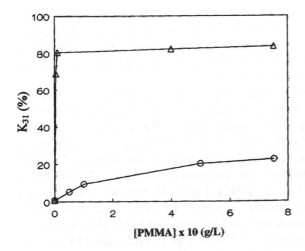

Figure 18 ToFSSIMS adsorption isotherms of PMMA onto HSO_4-doped PPy ($PPyHSO_4$) from $CHCl_3$ (O) and CCl_4 ($\triangle$). The isotherms were constructed by plotting the K_{31} ratio against [PMMA], the initial PMMA concentration. $K_{31} = I_{31}/(I_{31} + I_{26})$ ratio is in % where I_{31} and I_{26} are the intensities of mass peaks 31 and 26 corresponding to $^-OCH_3$ from PMMA and CN^- from polypyrrole, respectively. (From Ref. 128.)

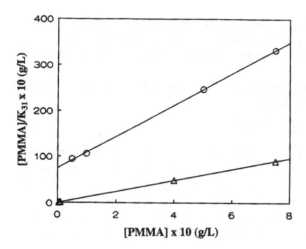

Figure 19 [PMMA]/K_{31} ratio versus [PMMA], for PMMA adsorption onto PPyHSO$_4$, where [PMMA] is the initial PMMA concentration. The linear plots indicate that the adsorption isotherms from both CHCl$_3$ (O) and CCl$_4$ ($\triangle$) are of the Langmuir type. (From Ref. 128.)

were 46 and 5300 (1/mol) for the adsorption from CHCl$_3$ and CCl$_4$, respectively. This is quantitative evidence that the adsorption from a neutral and noncompeting poor solvent is much stronger than that from a good acidic solvent. The low K value obtained in the acidic solvent is due to a strong CHCl$_3$-PMMA acid-base interaction that hinders polymer adsorption.

The XPS and ToFSSIMS studies were complemented by the use of IGC as a tool for determining the effect of the solvents' acid-base properties on the morphology of the resulting PMMA coatings. Indeed, IGC is a very surface sensitive technique based on adsorption measurements [103] and is thus sensitive to the chemical species present at the outermost surface of the substrate. Values of γ_s^d, K_D (basicity), and K_A (acidity) were determined using Eqs. (23) and (27) for PMMA, PPyCl, and PMMA-coated PPyCl. The K_D/K_A ratio was used as a measure of the overall acid-base character. Figure 20 shows the variation of the dispersive and acid-base properties described by γ_s^d (a) and K_D/K_A (b), respectively, versus the surface relative proportion of PMMA for PMMA-coated PPyCl powders. For the coating and the conducting substrate, γ_s^d and K_D/K_A values are (38.8 mJ/m^2, 5.5) and (140 mJ/m^2, 1.3), respectively. In Fig. 20a DXN and DCM lead to γ_s^d values that do not fit the linear plot. The former and the latter are below and above the expected value, respectively. In Fig. 20b, a linear plot is obtained if the casting solvents DCM and TCM (good solvents) are excluded. For approximately the same adsorbed amount of PMMA, DXN leads to a γ_s^d value matching

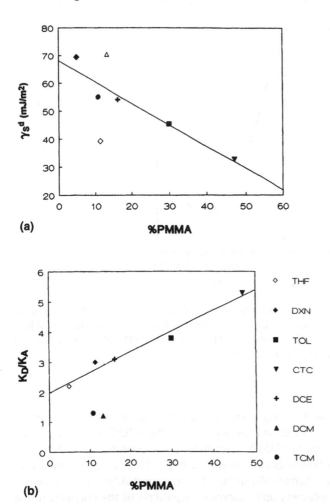

Figure 20   Plots of γ_s^d (a) and K_D/K_A (b) against %PMMA for PMMA-coated PPyCl. In (a): (◆) THF; (◇) DXN; (△) DCM. In (b): (◇) THF; (◆) DXN; (▲) DCM. In both (a) and (b): (■) TOL; (▼) CTC; (+) DCE. (From Ref. 103.)

that of PMMA and an intermediate K_D/K_A ratio, whereas DCM and TCM lead to a γ_s^d value significantly higher than that of PMMA and K_D/K_A ratios matching that of PPyCl.

In Fig. 21, K_D/K_A is plotted versus DN and AN of the casting solvents for the various PMMA-coated PPyCls. It is noteworthy that the materials' acid-base properties drawn from IGC data result in a plot whose shape is the "Fowkes

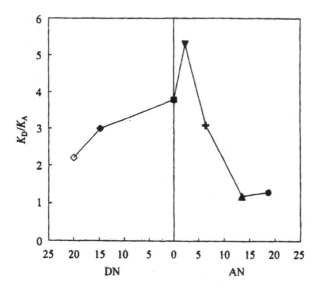

Figure 21 Fowkes' triangle obtained by plotting K_D/K_A derived from IGC versus the acid-base characteristics of the casting solvent for PMMA-coated polypyrrole powders. DNs are in kcal/mol and ANs in ppm. ($\diamond$) THF; ($\blacklozenge$) DXN; ($\blacksquare$) TOL; ($\blacktriangledown$) CTC; (+) DCE; ($\blacktriangle$) DCM; ($\bullet$) TCM. (From Ref. 103.)

triangle'' [11]. This is another way to emphasize the effect of acid-base properties of the solvents, the adsorbent, and the adsorbate. These IGC results suggested that for a given minimum of PMMA adsorption, poor solvents (e.g., DXN) lead to uniform coatings whereas good solvents such as $CHCl_3$ lead to nonuniform coatings (as illustrated schematically in Fig. 22), most probably caused by a competition between the solubility and adsorption equilibria of the polymer. In the case of the casting solvent DXN, since the dispersive and acid-base properties of PMMA-coated PPyCl are very comparable to those of bulk PMMA [103] and the adsorbate thickness determined via XPS matches that of one PMMA monolayer [11], it follows that DXN leads to the formation of approximately one uniform monolayer of PMMA on the surface of PPyCl.

The effect of acid-base properties of the casting solvents was also found to have an impact on the long-term stability of the PMMA coatings: a dewetting of PMMA was observed in the case of the coatings prepared in $CHCl_3$ [129]. Therefore, not only adsorption from this solvent but also the long-term adhesion of PMMA to PPy powder was poor. This is probably due to unfavorable orientation effects of functional groups such as the ester from PMMA at the PMMA-polypyrrole interface.

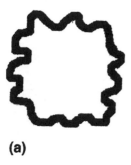   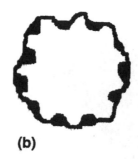

(a)   **(b)**

Figure 22 Schematic representation of uniform (a) and nonuniform (b) PMMA coatings on PPyCl particle surface. Poor, but not good, solvents lead to uniform coatings that assume the surface roughness.

In order to check the hypothesis drawn from IGC results on the coating morphology, we solvent-casted thin PMMA films (~16 nm) onto PPyTS flat surfaces from THF and $CHCl_3$. AFM was used in the contact mode to image and to measure the average surface roughness (R_a) of the bare and PMMA-coated PPyTS surfaces prepared in THF and $CHCl_3$ (PPyTS/PMMA/THF and PPyTS/PMMA/$CHCl_3$, respectively) [130]. Qualitatively, Fig. 23 shows that PPyTS and PPyTS/PMMA/THF surfaces are much rougher than that of PPyTS/PMMA/$CHCl_3$. Quantitatively, R_a was 3.19, 1.74, and 0.34 nm for PPyTS, PPyTS/PMMA/THF, and PPyTS/PMMA/$CHCl_3$, respectively. Because the same mass of PMMA was coated onto the two PPyTS specimens, the resulting differing roughness is likely to be due solely to the nature of the solvent.

The solvent effect on R_a can be understood as follows (Fig. 24): THF is a poor solvent and is expected, from IGC, to yield uniform coatings in contrast to $CHCl_3$ (good acidic solvent). Therefore, PMMA film cast from THF, but not from $CHCl_3$, assumes the surface roughness of PPyTS. In the case of $CHCl_3$, strong polymer-solvent interactions may compete with film formation and contribute to removing PMMA in the solvent bulk until total evaporation of the latter and cure of the film. Therefore, it is likely that the adsorption-solubility competition results in an enhanced vortex circulation that yields local defects in the surface morphology of the film. These defects compensate for the substrate roughness of PMMA-coated PPyTS and the average surface roughness is thus relatively lower when PMMA is cast from $CHCl_3$.

The AFM results were further interpreted in thermodynamic terms by the assessment of K values from ToFSSIMS adsorption measurements. These measurements yielded a K value [see Eqs. (31) and (32)] 18 times higher when PMMA was adsorbed (onto $PPySO_4$ powders) from THF than from $CHCl_3$. For thinner PMMA coatings (~2 nm) the effect of the acid-base properties of the

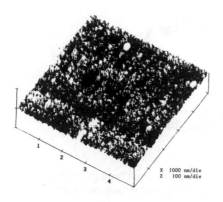

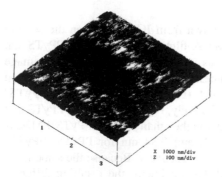

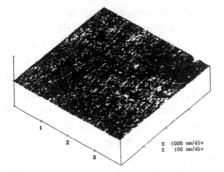

Figure 23 Three-dimensional AFM images and section analysis: (a) untreated PPyTS, field of view (FOV) of 5 μm; (b) PPyTS/PMMA/THF, FOV of 4 μm; (c) PPyTS/PMMA/CHCl₃, FOV of 4 μm. (From Ref. 130.)

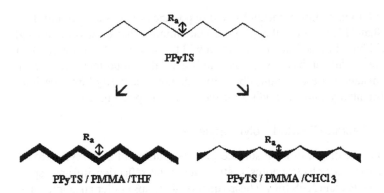

Figure 24 Schematic modification of tosylate-doped polypyrrole (PPyTS) surface average roughness (R_a), indicated by double-headed arrows, following PMMA film casting. PMMA coating assumes the surface morphology when it is cast from the poor solvent THF.

casting solvents was much more pronounced, since R_a for PPyTS/PMMA/THF matched that of PPyTS (8 and 6.5 nm, respectively) but was almost three times larger than that of PPyTS/PMMA/CHCl$_3$ (2.8 nm) [131].

C. Wettability and Adhesion Properties of Plasma-Treated Polypropylene

This section is devoted to the effect of oxygen and ammonia plasma modifications on the acid-base, wetting, and adhesion properties of polypropylene (PP) surfaces. PP is a polyolefin type of polymer, entirely built of hydrocarbon units that confer apolar and hydrophobic characteristics. PP has excellent mechanical and dielectric properties, good chemical resistance, and thermal stability. It finds applications mainly in packaging and the automobile industry. However, due to its chemical inertness, PP has a very poor adhesion property. This can be improved by modifying the surface of PP using chemical or physical methods [132]. Chemical methods include the use of strong oxidants such as sulfuric acid, nitric acid, and a sulfo-chromic acid mixture, and physical methods include flame and plasma treatments, the former being preferably used for thick objects. In the case of thin sheets of isotactic PP, surface modification was achieved by the plasma technique [133], a dry process that can tailor polymer surfaces by grafting specific surface functional groups (acidic, basic, polar) [132–135]. For example it is well established that oxygen plasma oxidizes the surface of PP and renders it more acidic [66,134], whereas nitrogen or ammonia plasmas introduce basic nitrogen-containing groups [65]. These ''polar'' groups modify the acid-base properties and adhesion of PP to materials of technological importance such as aluminium. Finlayson and Shah [136] showed that polymer-aluminium adhesion was dominated by Lewis acid-base interactions. However, it will be shown in the following that control of the acidity or basicity of the surface is also important because, in certain circumstances, Lewis acidity (basicity) and not necessarily Lewis basicity (acidity) substantially improves adhesion (see, for example, Fig. 2).

1. Oxygen Plasma–Treated Polypropylene

Oxygen plasma treatment of a PP surface permits grafting of carboxylic groups as shown by capillary electrophoresis, attenuated total reflection infrared spectroscopy (ATR-IR), and XPS [66]. These functional groups confer an acidic character to the PP surface, which was monitored by contact angle titration versus the plasma treatment time [134]. Figure 25 shows plots of W_{SL} against the aqueous solution pH for various PP-solution systems. Although the pH has no particular effect on the wettability of untreated PP, the plasma treatment yields a substantial decrease of contact angles when PP is treated. For a given test solution, W_{SL} increases with treatment time. Moreover, for each contact angle titration W_{SL} increases for the high-pH solution, an indication that the treated surface is predominantly acidic.

The contact angle titration method was confirmed by a combination of the molecular probe technique and XPS. In this regard, DMSO was used as a refer-

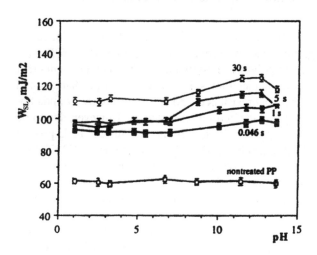

Figure 25 Variation of the total work of adhesion (W_{SL}) as a function of pH for oxygen plasma–treated polypropylene surfaces. W_{SL} increases with treatment time. (From Ref. 134.)

ence Lewis base to probe the surface acidity of oxygen plasma–treated PP. S$2p$ was used as the diagnostic core line for adsorbed DMSO. Figure 26 shows S$2p$ signals from DMSO adsorbed onto PP for different treatment times. With the untreated PP, no S$2p$ from DMSO was detected, whereas the intensity of this feature readily increased with treatment time longer than 3 s. Because oxygen plasma–treated PP bears acidic sites, it is likely that acid-base interactions are effective in retaining DMSO adsorbed onto the polymer surface in the high vacuum. This is supported by the S$2p$ BE, which was found to be in the 166.5–168 eV range. The S$2p$ BE is higher than the value of 165.8 eV reported for $(C_6H_4)_2S{=}O$ and $CH_3(C_6H_4)S{=}O$ [137], indicating a positive chemical shift for the S$2p$ core line. This positive chemical shift is due to electron density transfer from the basic DMSO to the acidic surface sites of the treated PP. Therefore, XPS provides direct evidence of acid-base adduct formation for the polymer-probe pair investigated. This, in turn, was checked by the determination of the work of adhesion of DMSO to treated PP by contact angle measurements of DMSO drops. Figures 27 and 28 depict the acid-base contribution to the DMSO-PP work of adhesion and the uptake of DMSO molecules (S/O atomic ratio) by PP, respectively: both W and %S increase with the oxygen plasma treatment time. However, the asymptotic value of W is reached very rapidly (less than 1 s), whereas that of the uptake of DMSO vapors is reached in 15–20 s. This delay in reaching the plateau value of the uptake of DMSO was attributed to the fact that the surface becomes enriched in fumaric acid species only after 15 s of plasma treatment [66].

2. Ammonia-Plasma Treated Polypropylene

Metallization of PP by an adherent thin aluminum coating ($\sim$20 nm) can be achieved if the PP surface is treated with an ammonia or nitrogen plasma prior to metallization [133]. For example, the ammonia plasma treatment results in grafting groups containing nitrogen and/or oxygen (from the residual reactor atmosphere) as judged by the increase in the N/C and O/C atomic ratios [134]. Shahidzadeh-Ahmadi et al. [134] showed via wettability measurements that ammonia plasma–treated polypropylene (APTPP) was predominantly basic, most probably due to the grafted nitrogen-containing functional groups (e.g., amine and amide). This characteristic was further assessed by XPS using chloroform as an acidic molecular probe to interrogate the acid-base properties of the treated polymer surfaces [65]. Figure 29 shows that the APTPP surface is labeled by the Cl$2p$ core line from chloroform (Fig. 29b) but not the untreated PP (Fig. 29a). The retention of chloroform by the surface of APTPP was interpreted in terms of a strong polymer-probe acid-base interaction: $-NH_2 \ldots HCCl_3$. This was supported by the low Cl$2p^{3/2}$ BE (196.9–198 eV) due to a large electron density transferred to the probe. The Cl$2p^{3/2}$ BE—ΔH^{AB} correlation [Eq. (18), Fig. 6]

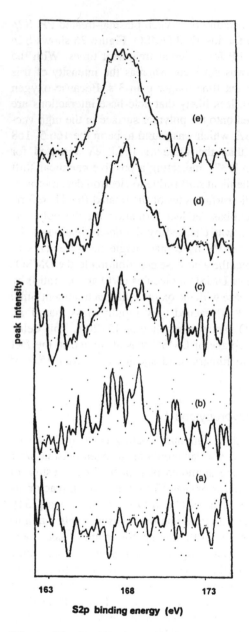

Figure 26   S2*p* signals from DMSO adsorbed onto oxygen plasma–treated polypropyl-ene versus the treatment. (a) Untreated PP. Treatment times (sec.): (b) 3; (c) 8; (d) 15; (e) 30. (From Ref. 66.)

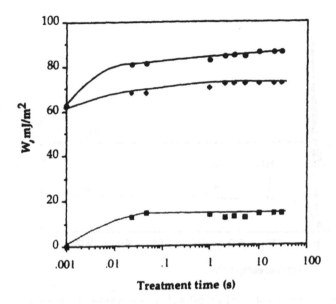

Figure 27 Variation of the total (T) work of adhesion and its dispersive (d) and acid-base (ab) contributions for the DMSO-polypropylene system versus oxygen plasma treatment time of polypropylene. (●) $W^T_{DMSO-pp}$; (◆) $W^d_{DMSO-pp}$; (■) $W^{ab}_{DMSO-pp}$. (From Ref. 66.)

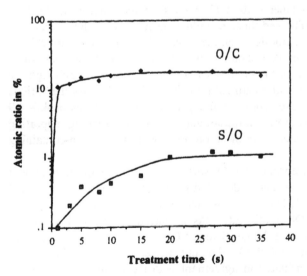

Figure 28 O/C and S/O atomic ratios (in %) versus the oxygen plasma treatment time of polypropylene. (From Ref. 66.)

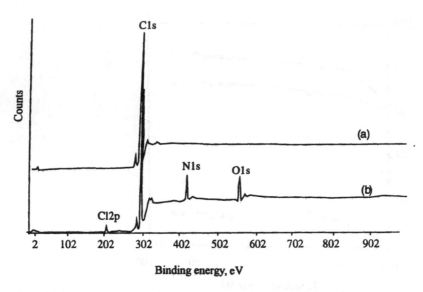

Figure 29 XPS survey scans of untreated PP (a) and 0.7 s ammonia plasma–treated polypropylene (APTPP) (b) following 10 min exposure to $CHCl_3$ vapors. (From Ref. 65.)

was used to assess ΔH^{AB} for the chloroform-APTPP pairs. The limiting values are shown in Table 8 together with ΔH^{AB} values for the chloroform-ArMCU system determined via FTIR and those calculated using Drago's four-parameter equation (3) for a series of chloroform-amine adducts. The ΔH^{AB} value for chloroform-APTPP remains roughly constant around 4.1 kcal/mol with treatment time, indicating that the strength of the bond does not vary significantly although the N/C atomic ratio increases with treatment time. The ΔH^{AB} values for the various chloroform-APTPP complexes compare fairly well with the values of 4.3–4.9 kcal/mol obtained for chloroform-amine adducts. This provides strong supporting evidence for the specific interaction of chloroform with the nitrogen-containing basic sites.

The uptake of chloroform by the PP surface as a function of the treatment time was of particular interest and is depicted in Fig. 30, which was constructed by plotting the molar $CHCl_3$/N ratio versus treatment time. For a treatment period shorter than 1 s, $CHCl_3$/N atomic ratio has a plateau value equal to 5–6%, whereas for a plasma treatment longer than 1 s $CHCl_3$/N atomic ratio decreases sharply to reach a value of 0.7% at 30 s. The unexpected variation of the $CHCl_3$/N ratio with time is nevertheless in agreement with the results obtained from contact angle measurements [134] and especially with the study of the adhesion

Table 8 ΔH^{18} Values for $CHCl_3$:Amine and $CHCl_3$:APTPP Complexes

	C_A	E_A	C_B	E_B	$-\Delta H^{AB}$ (kcal/mol)[a]
Chloroform	0.159	3.02			
NH_3			3.46	1.36	4.7
CH_3NH_2			5.88	1.3	4.9
$(CH_3)_2NH$			8.73	1.09	4.7
$(CH_3)N$			11.54	0.81	4.3
$(CH_3CH_2)_3N$			11.09	0.99	4.8
APTPP, 0.7 s					4.3
APTPP, 25 s					3.1
ArMCU					3.4

[a] ΔH^{AB} values for chloroform-amine complexes were calculated using Drago's E and C parameters for chloroform and amines in Eq. (3) [31]; ΔH^{AB} for (chloroform-ArMCU) adduct was determined by FTIR [63].

of thin aluminum coatings to APTPP [133]. APTPP was metallized by thermal evaporation of aluminum [133]. The metallized polymer film was subsequently subject to a U-peel test and the amount of peeled metal assessed by an image processing system [133]. Figure 31 shows a plot of the percentage of aluminum peeled off against the treatment time. For the untreated PP, 95% of the aluminum was peeled off, a result that signifies poor adhesion of aluminum to the untreated polymer. For plasma treatment as short as 0.023 s, the percentage of peeled aluminum sharply decreases to 4–5% and remains constant with increasing plasma treatment time, an indication of a strong metal-polymer adhesion. However, the amount of peeled metal increases again to reach the value of 20% for treatment times exceeding 1 s. The similarity between the XPS study and the adhesion measurements shown in Fig. 30 and 31, respectively, is striking and the peel test measurements were interpreted in terms of acid-base interactions. Indeed, Arefi-Khonsari et al. [138] showed that aluminum was present in the alumina form at the aluminum-PP interface, and this metal oxide is amphoteric with predominant acidity as judged from contact angle titrations. The amphoteric behavior of alumina is in line with the 4.6–9 value range of its IEPS [80,139].

The XPS, wettability, and adhesion studies are thus in agreement and indicate the important effect of acid-base forces in the interaction of APTPP with either an acidic molecule such as chloroform or an amphoteric metal oxide such as alumina. Attempts to metallize oxygen plasma–treated PP with aluminum resulted in poor metal-polymer adhesion, probably due to a less favorable acid-acid interaction [140].

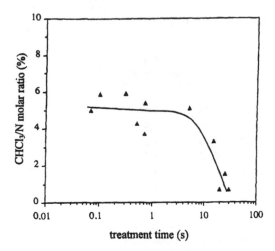

Figure 30 Chloroform-to-nitrogen molar ratio (in %) versus the ammonia plasma treatment time of polypropylene. The plot shows that the optimal treatment time is less than 1 s. (From Ref. 65.)

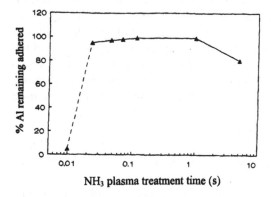

Figure 31 Adhesion of aluminum evaporated onto polypropylene versus ammonia plasma treatment time of polypropylene. Adhesion was measured as percentage of aluminum remaining adhered to polypropylene following U-peel test. (Adapted from Ref. 133.)

Following our XPS strategy, Bodino et al. [141] have characterized the acid-base properties of N_2 and ammonia plasma–treated PP surfaces (N_2-PP and NH_3-PP, respectively) using chloroform and iodine (I_2) as reference Lewis acidic molecular probes. Their XPS-derived thermochemical data are summarized in Table 9. It should be noted that although the ΔH^{AB} values for chloroform were obtained using Eq. (18), there was no mention of the establishment of a binding energy–ΔH^{AB} plot for iodine.

The chloroform data agree fairly well with what we reported for APTPP (see Table 8) [65]. They also reflect a stronger interaction of chloroform with ammonia plasma–treated PP (a basic surface) than with N_2-PP (an amphoteric surface). Using Drago's equation (3) and E and C constants for chloroform and iodine (see Table 2), the author estimated E_B and C_B constants (not reported in Ref. 141!) for the treated polymer surfaces: $E_B = 1.46$, $C_B = 0.33$, and $C_B/E_B = 0.226$ for N_2-PP and $E_B = 1.65$, $C_B = 0.91$, and $C_B/E_B = 0.552$ for NH_3-PP (E and C in kcal$^{1/2}$/mol$^{1/2}$). Whereas the E_B constants are in line with those reported for basic species in general (see Table 2), the C_B values and consequently the C_B/E_B ratios are too low for nitrogen-containing Lewis bases. Nevertheless, with this in hand and the E_A and C_A constants for silica (see Table 3), ΔH^{AB} values for N_2-PP/silica and NH_3-PP/silica pairs were predicted to be 6.8 (6.2 in Ref. 141) and 8.3 kcal/mol, respectively. These values correlate well with the larger sticking coefficients of evaporated $SiO_{1.7}$ (acidic ceramic) on the NH_3-PP surface as compared with N_2-PP, O_2-PP (acidic surface, see above), and the untreated PP. Additional depth profiling of the $SiO_{1.7}$ deposits indicated a very sharp ceramic-PP interface for the untreated PP, whereas an interphase with an increasing degree of diffuseness was observed for O_2-PP, N_2-PP, and NH_3-PP.

The work reported by Bodino et al. [141] confirmed that our XPS strategy has potential in assessing surface acid-base properties of polymers [45,61,62,65, 66,78,138] and provided additional evidence for the importance of acid-base interactions in controlling the metal-PP adhesion.

Table 9 XPS Derived ΔH^{AB} Values (kcal/mol) for Chloroform and Iodine Adsorbed onto N_2- and NH_3 Plasma–Treated PP Surfaces

	Polymer surfaces	
Molecular probes	N_2-PP	NH_3-PP
$CHCl_3$	4.5	5.1
I_2	1.8	2.6

Source: Ref. 141.

VI. CONCLUSION

The background of acid-base interactions has been reviewed with emphasis on both simple organic and inorganic molecules and more complex materials such as polymers, fillers, fibers, and pigments. When these specific exothermic interactions operate at interfaces they have a significant impact on adsorption, wettability, and adhesion phenomena as predicted by the late Professor Fowkes in the early 1970s and subsequently confirmed by several other research laboratories over the past two decades.

Such findings resulted in the establishment of an impressive array of methods that enable an adhesionist to determine acid-base scales for materials. Various experimental techniques are now fairly well established and routinely used to interrogate the acid-base characteristics of polymers and other materials. Two of them have been described: (1) inverse gas chromatography, which is very popular in the adhesion community for its versatility and applicability to finely divided materials over a wide temperature range, and (2) XPS studies of acidic and basic probes adsorbed onto polymers under test. The latter method is shown to have great promise in the assessment of acid-base characteristics of material surfaces.

Some practical aspects of acid-base interactions, selected from the author's work and the current literature, have been described. In the case of untreated materials, it is shown that acid-base interactions have an impact on polymer adsorption and, more important, on polymer coating morphology. It is also emphasized, in the case of weakly interacting or incompatible materials, that surface modification enabling the grafting of functional groups with specific acid-base characteristics is an important step toward strong and durable adhesion bonds.

Beyond these selected examples, it is hoped that the acid-base properties of materials will routinely be taken into account when designing high-performance multicomponent systems (e.g., miscible polymer blends, organic or ceramic coatings, composites).

ACKNOWLEDGMENTS

I am greatly indebted to Dr. F. Arefi-Khonsari and Prof. J. Amouroux (Université Pierre et Marie Curie) for their very efficient and friendly collaboration in extending the XPS "molecular probe technique" to plasma-modified polymer surfaces. I would like to thank Dr. J.F. Watts (University of Surrey, UK) for his help in the determination of polymer adsorption isotherms by ToFSSIMS. I also appreciate the fruitful discussions with Prof. M. Delamar (Université Paris 7—Denis Diderot) and his critical reading of the manuscript.

APPENDIX

Acid-Base Interaction Studies and Related Phenomena for Several Polymers Studied by Various Groups[a]

Polymers	Phenomena	References	Polymers	Phenomena	References
P(E-co-AA)	Wettability	8	PPyCl	GS	40,103
PMMA	Adsorption	11,103,125,12,7		Interaction with PMMA	103,125
	Mixing	14,46,27	PPyTS	GS	41
	Wettability	10,43,38	PVB	GS	This work
	Coating morphology	130	PP	Plasma treatment	133–135,138,140,141
PBAC, PVME	GS	67		Wettability	66,134,138
PVAc, PBMA, PCHMA,	GS	62		GS	65,66,138,140,141
PMDA-ODA PAA	Adhesion	52		Adhesion	65,133,138,140,141
Polyimide	Adhesion	51		Metallization	133,138,141
ArMCU	Mixing	63,62	PE	Wettability	10
	GS	61,62	PAI	Wettability	58
PVdF	Mixing	47	Epoxy resin	Adhesion to CF	29
PPO	GS	45		Mixing	14
PEO	Mixing	46	Phenoxy resin	Mixing	46,
	Wettability	15	Dextran	Wettability	15
	GS	62	PS	Wettability	10
CPVC	Adsorption	11,38	PEEK	Adhesion to CF	29,119
	Mixing	27,38		GS	119
PVC	Wettability	10	Polyester resin	GS and adsorption	95
DGEBA-DDS	GS and adhesion to CF	96			

[a] GS, gas-solid interaction; CF: carbon fiber; P(E-co-AA), poly(ethylene-co-acrylic acid); CPVC, chlorinated PVC; DGEBA-DDS, diglycidyl ether of bisphenol A–diamino diphenyl sulfone; PBAC, poly(bis-phenol A carbonate); PCHMA, poly(cyclohexyl methacrylate); PVAc, poly(vinyl acetate); PBMA, poly(n-butyl methacrylate); PVB, poly(vinyl butyral).

REFERENCES

1. AJ Kinloch. J Mater Sci 15:2141, 1980.
2. MF Vallat, M Nardin. J Adhesion, 57:115, 1996.
3. AN Gent, J Schultz. J Adhesion, 3:281, 1972.
4. FM Fowkes. Ind Eng Chem 56(12):40, 1964.
5. GC Pimentel, AL McClellan. The Hydrogen Bond. San Francisco: W H Freeman, 1960.
6. FM Fowkes. J Adhesion Sci Technol 1:7, 1987.
7. FL Riddle Jr, FM Fowkes. J Am Chem Soc 112:3259, 1990.
8. FM Fowkes In: L-H Lee, ed. Adhesion and Adsorption of Polymers. New York: Plenum, 1980, Part A, p 43.
9. FM Fowkes. In: Vol 2. KL Mittal, ed. Physicochemical Aspects of Polymer Surfaces. New York: Plenum Press, 1983, p 583.
10. RJ Good, MK Chaudhury, CJ van Oss. In L-H Lee, ed. Fundamentals of Adhesion. New York: Plenum Press, 1991, chap 4.
11. FM Fowkes, MA Mostafa. Ind Eng Chem Prod Res Dev 17:3, 1978.
12. J Adhesion Sci Technol 4(4,5,8), (1990); 5(1), 1991.
13. KL Mittal, HR Anderson, Jr, eds. Acid-Base Interactions: Relevance to Adhesion Science and Technology. Utrecht: VSP, 1991.
14. FM Fowkes. J Adhesion Sci Technol 4:669, 1990.
15. CJ van Oss, RJ Good. Langmuir 8:2877, 1992.
16. MM Chehimi. In: NP Cheremisinoff, PN Cheremisinoff, eds. Handbook of Advanced Materials Testing. New York: Marcel Dekker, 1995, chap 33.
17. JC Berg. In: JC Berg, ed. Wettability. New York: Marcel Dekker, 1993, chap 2.
18. P Mukhopadhyay, HP Schreiber. Colloids Surfaces A 100:47, 1995.
19. DM Dwight. In: WJ van Ooij, HR Anderson, Jr., eds. Adhesion Science and Technology (Festschrift in Honor of KL Mittal). Utrecht: VSP, 1998.
20. RC Thomas, JE Houston, RM Crooks, T Kim, TA Michalske. J Am Chem Soc 117:3830, 1995.
21. L -H Lee. J Adhesion Sci Technol 7:583, 1993.
22. CJ van Oss, RJ Good, MK Chaudhury. Langmuir 4:884, 1988.
23. JA Brydson. In: AD Jenkins, ed. Polymer Science. Vol 1. Amsterdam: North Holland, 1972, chap 3.
24. DW van Krevelen. Properties of Polymers. Amsterdam: Elsevier, 1990.
25. DR Paul, JW Barlow, H Keskkula. In Encyclopedia of Polymer Science and Engineering. Vol 12. New York: Wiley, 1988, p 399.
26. DJ Walsh. In: C Booth, C Price, eds. Comprehensive Polymer Science Vol 2. Oxford: Pergamon Press, 1989, p 135.
27. FM Fowkes, DO Tischler, JA Wolfe, LA Lannigan, CM Ademu-John, MJ Halliwell. J Polym Sci Chem Ed 22:547, 1984.
28. W Gutowski. In: L-H Lee, ed. Fundamentals of Adhesion. New York: Plenum Press, 1991, p 87.
29. M Nardin, J Schultz. Langmuir 12:4238, 1996.
30. RG Pearson. J Am Chem Soc 85:3533, 1963.
31. RS Drago. Struct Bonding, 15:73, 1973.

32. V Gutmann. The Donor-Acceptor Approach to Molecular Interactions. New York: Plenum Press, 1978.
33. JC Bolger, AS Michaels. In: P Weiss, GD Cheever, eds. Interface Conversion for Polymer Coatings. New York: Elsevier, 1968, p 3.
34. RS Drago. J Chem Ed 51:300, 1974.
35. LH Lee. In Ref 13 p 25 and in Fundamentals of Adhesion. L-H Lee, ed. New York: Plenum Press, 1991, p 1.
36. FL Riddle, Jr, FM Fowkes. ACS Polym Prepr 29:188, 1988.
37. FM Fowkes, DW Dwight, DA Cole, TC Huang. J Noncryst Solids, 120:47, 1990.
38. TB Lloyd. Colloids Surfaces A 93:25, 1994.
39. F Chen. Macromolecules 21:1640, 1988.
40. E Pigois-Landureau, MM Chehimi. J Appl Polym Sci 49:183 1993.
41. MM Chehimi, SF Lascelles, SP Armes. Chromatographia, 41:671, 1995.
42. AC Tiburcio, JA Manson. J Appl Polym Sci 42:427, 1991.
43. FM Fowkes, MB Kaczinski, DM Dwight. Langmuir 7:2464, 1991.
44. MB Kaczinski, DW Dwight. J Adhesion Sci Technol 7:165, 1993.
45. JF Watts, MM Chehimi. Int J Adhesion Adhesives 15:91, 1995.
46. DA Valia. PhD thesis, Lehigh University, 1987.
47. Wan-Jae Myeong, Son-Ki Ihm. J Appl Polym Sci 45:1777, 1992.
48. ST Joslin, FM Fowkes. Ind Eng Chem Prod Res Dev 24:369, 1985.
49. JA Manson. Pure Appl Chem 57:1667, 1985.
50. YY Lim, RS Drago. Inorg Chem 11:202, 1972.
51. TS Oh, LP Buchwalter, J Kim. In Ref. 13, p 287.
52. YH Kim, GF Walker, J Kim, J Park. J Adhesion Sci Technol, 1:331, 1987.
53. DK Owens, RC Wendt. J Appl Polym Sci 13:1741, 1969.
54. FM Fowkes, FL Riddle Jr, WE Pastore, AA Weber. Colloid Surfaces 43:367, 1990.
55. TK Kwei, EM Pearce, F Ren, JP Chen. J Appl Polym Sci Polym Phys 24:1597, 1986.
56. MD Vrbanac, JC Berg. J Adhesion Sci Technol 4:255, 1990.
57. PM Costanzo, RF Giese, CJ van Oss. In Ref. 13, p 135.
58. CJ van Oss, RF Giese Jr, RJ Good. Langmuir, 6:1711, 1990.
59. KL Mittal, ed. Contact Angle, Wettability and Adhesion. Zeist: VSP, 1993.
60. DR Lloyd, TC Ward, HP Schreiber, eds. Inverse Gas Chromatography Characterization of Polymers and Other Materials. ACS Symposium Series No 391. Washington, DC: American Chemical Society, 1989.
61. MM Chehimi, JF Watts, SN Jenkins, JE Castle. J Mater Chem 2:209, 1992.
62. MM Chehimi, JF Watts, WK Eldred, K Fraoua, M Simon. J Mater Chem 4:305, 1994.
63. MM Chehimi, E Pigois-Landureau, M Delamar, JF Watts, SN Jenkins, EM Gibson. Bull Soc Chim Fr 129:137, 1992.
64. JF Watts, MM Chehimi, EM Gibson. J Adhesion, 39:145, 1992.
65. N Shahidzadeh, MM Chehimi, F Arefi-Khonsari, J Amouroux, M Delamar. Plasmas Polym 1:27, 1996.
66. N Shahidzadeh, MM Chehimi, F Arefi-Khonsari, N Foulon-Belkacemi, J Amouroux, M Delamar. Colloids Surfaces A 105:277, 1995.
67. MM Chehimi, M Delamar, N Shahidzadeh-Ahmadi, F. Arefi-Khonsari, J

Amouroux, JF Watts. In Organic Coatings. P-C Lacaze, ed. Woodbury, NY: AIP Press, 1996, p 25.
68. MM Chehimi, M Delamar. Analysis 23:291, 1995.
69. JF Watts. Surf Interface Anal 12:497, 1988.
70. JF Watts. Vacuum 45:653, 1994.
71. RW Paynter, JE Castle, DK Gilding. Surf Interface Anal 7:63, 1985.
72. CA Baillie, JF Watts, JE Castle. J Mater Chem 2:939, 1992.
73. JF Watts, AM Taylor. J Adhesion 55:99, 1995.
74. RH West, JE Castle. Surf Interface Anal 4:68, 1982.
75. AS Kinloch, GKA Kodokian JF Watts. J Mater Sci Lett. 10:815, 1991; Roy Soc Philos Trans Lond A338:83, 1992.
76. RL Martin, DA Shirley. J Am Chem Soc 96:5299, 1974.
77. K Burger, E Fluck. Inorg Nucl Chem Lett 10:171, 1974.
78. MM Chehimi. J Mater Sci Lett 10:908, 1991.
79. M Delamar. J Electron Spectrosc Relat Phenom 53:c11, 1990.
80. MG Cattania, S Ardizzone, CL Bianchi, S Carella. Colloids Surfaces A 76:233, 1993.
81. GW Simmons, BC Beard. J Phys Chem 91:1143, 1967.
82. JF Watts, EM Gibson. Int J Adhesion Adhesives 11:105, 1991.
83. G Kurbatov, E Darque-Ceretti, M Aucouturier. Surf Interface Anal 18:811, 1992.
84. M Delamar. J Electron Spectrosc Relat Phenom 67:R1, 1994.
85. C Defossé, P Canesson, PG Rouxhet, B Delmon. J Catal 51:269, 1978.
86. R Borade, A Sayari, A Adnot, S Kaliaguine. J Phys Chem 94:5989, 1990.
87. M Huang, A Adnot, S Kaliaguine. J Catal 137:322, 1992.
88. MM Chehimi, JF Watts. J Adhesion 41:81, 1993.
89. Z Al-Saigh. Polym News, 19:269, 1994.
90. E Papirer, J-M Perrin, B Siffert, G Philipponneau. J Colloid Interface Sci 144:263, 1991.
91. E Papirer, H Balard. Prog Org Coat 22:1 1993.
92. MM Chehimi, M-L Abel, E Pigois-Landureau, M Delamar. Synth Met 60:183, 1993.
93. J Kuczynski, E Papirer. Eur Polym J 27:653, 1991.
94. DP Kamden, SK Bose, P Luner. Langmuir 9:3039, 1993.
95. J Lara, HP Schreiber. J Coat Technol 63:81, October, 1991.
96. J Schultz, L Lavielle, C Martin. J Adhesion 23:45, 1987.
97. M Nardin, EM Asloun, J Schultz. Polym Adv Technol 2:109, 1991.
98. EF Meyer. J Chem Ed 57(2):120, 1980.
99. GM Dorris, DG Gray. J Colloid Interface Sci 77:353, 1980.
100. J Anhang, DG Gray. In: KL Mittal, ed. Physicochemical Aspects of Polymer Surfaces, Vol 2. New York: Plenum Press, 1983, p 659.
101. T Hamieh, J Schultz. J Chim Phys 93:1292, 1996.
102. T Hamieh, M Rageul-Lescouet, M Nardin, J Schultz. J Chim Phys 93:1332, 1996.
103. MM Chehimi, M-L Abel, Z Sahraoui. J Adhesion Sci Technol 10:287, 1996.
104. HP Schreiber, J-M Viau, A Fetoui, Zhuo Deng. Polym Eng Sci 30:263, 1990.
105. MM Chehimi, M-L Abel, Z Sahraoui, K Fraoua, SF Lascelles, SP Armes. Int J Adhesion Adhesives 17:1, 1997.

106. C Perruchot, MM Chehimi, M Delamar, SF Lascelles, SP Armes. J Colloid Interface Sci, 193:190, 1997.
107. HP Schreiber, F St Germain. J Adhesion Sci Technol 4:319, 1990.
108. JB Donnet, SJ Park, H Balard. Chromatographia, 31:434, 1991.
109. E Papirer, JM Perrin, B Siffert, G Philipponneau. Prog Colloid Polym Sci 84:257, 1991.
110. M Fafard, M El-Kindi, HP Schreiber, G DiPaola-Baranyi, Ah-Mee Hor. Polym Mater Sci Eng 70:470, 1994.
111. MJ Liu, K Tzou, RV Gregory. Synth Met 63:67, 1993.
112. S Maeda, SP Armes. Synth Met 73:151, 1995.
113. DT Sawyer, DJ Brookman. Anal Chem 40:1847, 1968.
114. C Saint Flour, E Papirer. J Colloid Interface Sci 91:69, 1983.
115. MM Chehimi, E Pigois-Landureau. J Mater Chem 4:741, 1994.
116. P Koning, TC Ward, RD Allen, JE McGrath. ACS Polym Prep, 26:189, 1985.
117. E Papirer, H Balard, H Vidal. Eur Polym J 24:783, 1988.
118. AM Taylor, JF Watts, ML Abel, MM Chehimi. Int J Adhesion Adhesives 15:91, 1995.
119. M Nardin, EM Asloun, J Schultz. Surf Interface Anal 17:485, 1991.
120. R Roe. In: L-H Lee, ed. Adhesion and Adsorption of Polymers, Part B. New York: Plenum Press, 1980, p 629.
121. H-L Wang, JE Fernandez. Macromolecules 25:6179, 1992; RA Zoppi, MI Felisberti, M-A De Paoli. J Appl Polym Sci 32:1001, 1994.
122. CL Heisey, JP Wightman, EH Pittman, HH Kuhn. Textile Res J 63:247, 1993; C Forder, SP Armes, AW Simpson, C Maggiore, M Hawley. J Mater Chem 3: 563, 1993.
123. MM Chehimi, M-L Abel, E, Pigois-Landureau, M Delamar. Le Vide Les Couches Minces 268 (suppl):105, 1993.
124. H-L Wang, L Toppare, JE Fernandez. Macromolecules 23:1053, 1990.
125. M-L Abel, MM Chehimi. Synth Met 66:225, 1994.
126. NM Reed, JC Vickerman. In: L Sabbatini, PG Zambonin, eds. Surface Characterisation of Advanced Polymers. Weinheim: VCH, 1993, chap 3.
127. M-L Abel, SR Leadley, AM Brown, J Petitjean, MM Chehimi, JF Watts. Synth Met 66:85, 1994.
128. M-L Abel MM, Chehimi, AM Brown, SR Leadley, JF Watts. J Mater Chem 5: 845, 1995.
129. M-L Abel. Personal communication.
130. M-L Abel, J-L Camalet, MM Chehimi, JF Watts, PA Zhdan. Synth Met 81:23, 1996.
131. MM Chehimi, M-L Abel, M Delamar, JF Watts, PA Zhdan. In: P-C Lacaze, ed. Organic Coatings. Woodbury, NY: AIP Press, 1996, p 351.
132. F Garbassi, M Morra, E Ochiello. Polymer Surfaces from Physics to Technology. Chichester: Wiley, 1994.
133. F Arefi, M Tatoulian, V André, G Lorang, J Amouroux. In: KL Mittal, ed. Metallized Plastics 3: Fundamental and Applied Aspects. New York: Plenum, 1992, p 243.
134. N Shahidzadeh-Ahmadi, F Arefi-Khonsari, J Amouroux. J Mater Chem 5:2, 1995.

135. F Arefi, V Andre, P Montazer-Rahmati, J Amouroux. Pure Appl Chem 64:715, 1992.
136. MF Finlayson, BA Shah. J Adhesion Sci Technol 4:431, 1990.
137. JF Moulder, WF Stickle, PE Sobol, KD Bomben. In: J Chastain, ed. Handbook of X-Ray Photoelectron Spectroscopy. Eden Prairie, MN: Perkin-Elmer Corp., 1992.
138. F Arefi-Khonsari, M Tatoulian, N Shahidzadeh, MM Chehimi, J Amouroux D Leonard, P Bertrand. In: WJ van Ooij, HR Anderson Jr, eds. Adhesion Science and Technology (Festschrift in Honor of KL Mittal). Utrecht: VSP, 1998.
139. GA Parks. Chem Rev 65:177, 1965.
140. N Shahidzadeh. PhD thesis, Université Pierre et Marie Curie, Paris, 1996.
141. F Bodino, N Compiegne, L Köhler, JJ Pireaux, R Caudano. Adhesion Society Preprints, 1997, p 41.

3
Molecular Mechanics-Dynamics Modeling and Adhesion

A. Pizzi
ENSTIB, University of Nancy I, Epinal, France

I. INTRODUCTION

Molecular mechanics in the broader sense of the term is a computational technique that is, among others, particularly suited for determining at the molecular level the interactions at the interface of well-defined polymers. It has already been used in many fields—for instance, to calculate the most stable conformation (i.e., the conformation of minimum energy) of biological materials such as proteins; for the effects of oxygen, carbon monoxide, and carbon dioxide on the functioning of the heme of respiratory proteins; for the design and activity forecasting of pharmacological drugs or other biologically active materials to fit the active sites of enzymes; to determine the structure of a variety of high-technology materials, and to determine the structure and properties of a variety of synthetic and natural polymers; and even to model homogeneous and heterogeneous catalytic processes. The variety and number of applications of this technique in the past few years have indeed been great and it has positively influenced many fields of science.

What exactly is molecular mechanics? It is the study of the interactions of noncovalently bonded atoms in one or more molecules that determine the spatial conformation of such a structure or its change of conformation induced by a neighboring molecule. In short, it is the modeling of the structures of molecules, their structural interactions and modifications, and hence their macroscopic and microscopic properties derived from the molecular level according to first principles in physics and physical chemistry. Its mundane appearance is that of a computational technique, and today extensive computation is always included, but it is indeed much more than this: it is a technique par excellence for explaining our physical world from first, molecular, and atomic principles.

Although it has now been used for more than 30 years in many other fields, the application of this technique in the field of adhesion and adhesives, namely to theoretical and applied problems of adhesion and to the optimization of adhesion, is still relatively in its infancy. A few notable applications of this technique to adhesion and adhesives do, however, exist, and this chapter is aimed at describing them and their relevant consequences without pretending to be either exhaustive or limiting regarding any further, future applications. As molecular mechanics and molecular dynamics are really "going back to basics" techniques aimed at explaining at the molecular level the behavior of materials, there is no doubt that their use is also bound to grow in the field of adhesion as rapidly and effectively as it has occurred in other scientific and technological fields, once the potential of such a technique is understood.

In the field of adhesion, in its broadest sense, several different pioneering trends are already on record:

1. Studies of the adhesion of generalized particles to generalized surfaces or of generalized particles to generalized particles.
2. Studies of the adhesion of polymers well defined at the molecular level to surfaces equally well defined at the molecular level.
3. Studies of the dynamic, differential, competitive adsorption, hence adhesion, of molecularly well-defined oligomers to an equally molecularly well-defined surfaces in the presence of solvents, such as in the modeling of chromatography.

This chapter addresses these three sectors of activity.

II. ALGORITHMS USED IN MOLECULAR MECHANICS

Different molecular mechanics systems and programs exist. There are programs that allow simultaneous variation of bonds and bond angles as well as bond rotation, and there are programs in which instead all the covalent bond lengths and bond angles between covalently bonded atoms are fixed to specific values without allowance for their adjustment or modification during computation. It cannot be said that one system is better than the other, as either of the two systems can be more apt at solving a particular problem; it might then be necessary to choose according to the problem at hand.

The former, by an unconstrained force field approach, is more comprehensive but suffers from the limitation of the size of molecules that can be investigated due to the extent of computations needed. It is thus very apt for the study of smaller molecules or systems of molecules up to 40–60 atoms, but this limitation is also fictitious because it really depends on the capacity and calculation rate of the computer used. Such unconstrained force field programs also tend to suffer from the problem that the automatic search for the minimum of energy might lead the program to minimize on a local rather than total minimum, and if particular attention is not exercised completely false results can be obtained (the "black box" syndrome).

The latter, or constrained force field, approach is generally taken to render computation more rapid. It is particularly useful when big molecules such as polymers are involved. All these programs are based on the finding that conformational studies in the field of biological macromolecules have shown that the conformational energy of a molecule can be represented with accuracy even when bond lengths and angles between covalently bonded atoms are prefixed [1] and represented by a sum of four types of contributions, namely

$$\mathbf{E}_{tot} = \mathbf{E}_{vdW} + \mathbf{E}_{H\text{-bond}} + \mathbf{E}_{elec} + \mathbf{E}_{tor} \tag{1}$$

where $\mathbf{E}_{tot}$ represents the total conformational energy of the molecule as a function of all the internal angles of rotation and $\mathbf{E}_{vdW}$ represents the contribution to the total energy due to van der Waals interactions between all the couples of unlinked atoms whose relative position depends on one or more internal bond rotational angles ($\phi°$, $\psi°$) (in degrees). This contribution can be expressed by Buckingham-type functions

$$\mathbf{E}_{vdW} = \sum_{ij} [a_{ij} \exp(-b_{ij}r_{ij}) - c_{ij}r_{ij}^{-6}] \tag{2}$$

where the coefficients a, b, and c depend on the couple i, j of atoms, or by functions of Lennard-Jones type

$$\mathbf{E}_{vdW} = \sum_{ij} [(d_{ij}/r_{ij}^{12}) - (c_{ij}/r_{ij}^{6})] \tag{3}$$

Both types of functions are commonly used. Several sets of a, b, c, and d coefficients are available [1–3]. Equally good results can be obtained using Lennard-Jones-type functions alone or Buckingham-type functions alone or mixtures of Lennard-Jones and Buckingham functions [4]. The attraction coefficients c_{ij}'s are generally, but not always, calculated with the formula of Slater and Kirkwood [5]:

$$\frac{[^3\!/_2 \exp(h/m^{1/2}) \, \alpha_i \alpha_j]}{[(\alpha_i/N_i)^{1/2} + (\alpha_j/N_j)^{1/2}]} \tag{4}$$

where α_i and α_j are the values of the polarizability of the atoms i and j, respectively and N_i and N_j are the numbers of effective electrons.

In Eq. (2), b_{ij} is fixed to a constant value [1,6–10] and a_{ij} is determined by imposing the minimum at a distance that is the sum of the van der Waals radii of the atoms or groups considered [1,6–10]. It must be pointed out that the van der Waals interaction of any chemical grouping is calculated as the sum of the single interactions between individual couples of unlinked atoms.

The term E_{elec} describes the electrostatic contribution to the total energy. Dipolar momenta are expressed, in the so-called monopolar approximation, by means of partial charges whose values are fixed in such a manner as to reproduce the dipolar momenta of both bonds as well as the total momentum. Using partial charges, the dipolar interactions can be calculated with Coulomb-type law of the form $E_{elec} = \Sigma_{ij}(q_i q_j)/(\varepsilon r_{ij})$, where q_i and q_j are the charges of the two atoms i and j, r_{ij} is the distance between them, and ε is the dielectric constant. The term E_{tor} describes the contribution to the total energy due to hindered rotation around skeletal bonds. The formulas generally used for the torsional potentials are those of Brant and Flory [2,6–10] in which the torsional barriers used can be of different values [2,6–10]. It is necessary to point out, however, the limits imposed on the rotational degrees of freedom of single bonds conjugated with bonds that have very high torsional barriers (C$=$C, C$=$O), as in the case of polypeptides [11].

The term $E_{H\text{-bond}}$ represents the hydrogen bond (H-bond) contribution between couples of noncovalently bonded atoms. Several functions, even very simplified and empirical ones, have been used, often with good success. The H-bond is, however, at best a difficult interaction to describe through a function that is capable of both giving good results and taking into account the physicochemical reality of its interaction. For this reason, there is a multitude of empirical simplified functions for its calculation. Where the hydrogen bond is of little or no importance, the molecular mechanics calculations are often done on the basis of only the van der Waals interactions (and not with bad results, as in the earlier days of protein structure refinement). The writer has always preferred, when examining molecular systems in which the H-bond contribution is important or determining to the results, to use a more complicated but more comprehensive function proposed by Stockmayer, which has already been found to give very representative results for polypeptide sequences [11] and cellulose systems [12]:

$$E_{H\text{-bond}} = 4\varepsilon[(\sigma/r^{12}) - (\sigma/r^6)] \qquad (5)$$
$$- (\mu_a \mu_b/r^3)[2 \cos \Theta_a \cos \Theta_b - \sin \Theta_a \sin \Theta_b \cos(\gamma_a^\circ - \gamma_b^\circ)]$$

and which takes into consideration the angular dependence of the hydrogen bond. The first term in Eq. (5) describes the interaction between the hydrogen atom and the oxygen atom participating in the H-bond, and it is just a Lennard-Jones potential with the expression in simplified form. The second term describes the H-bond as an electrostatic interaction between two point-like dipoles of magnitudes μ_a and μ_b centered on the oxygen and hydrogen atoms. The directional

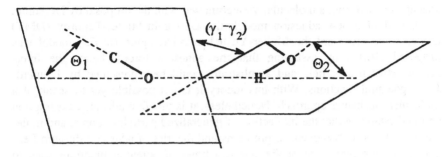

Figure 1 Dihedral angle of importance in the calculation of H-bond energy and showing the importance of directionality in this type of interaction.

character of the H-bond is ensured by the angular dependence of this function, and Θ_a and Θ_b are the angles that the bonds C—O and O—H form with the C—O $\cdots$ H—O segment linking the hydrogen and oxygen atoms (Fig. 1). The term $(\gamma_a^o - \gamma_b^o)$ (in degrees) is the angle between the planes containing the H-bond and the O—H and C—O bonds (Fig. 1). The ε, σ, and μ are obtained by minimizing the first term of Eq. (5) at the van der Waals distance between the hydrogen and oxygen atoms and the whole function at an H-bond distance of 2.85 Å with aligned C—O and O—H bonds.

In many of the unconstrained force field type programs today, similar expressions for the H-bond based on a Lennard-Jones first term as above or a Buckingham first term [13] followed by a term describing the dipolar and angular dependence of the hydrogen bond are used. However, there are also programs in which the H-bond is described by just a Buckingham function, without any consideration of the directionality of the H-bond, or even by simpler expressions. There is nothing very wrong with these simplified approaches in cases in which the hydrogen bond is not of fundamental importance to the study, but they cannot be used reliably in cases in which the H-bond is of determining importance.

III. GENERALIZED PARTICLE-SURFACE AND PARTICLE-PARTICLE MODELS

The mechanics of particle adhesion and the deformations resulting from the stresses generated by adhesion forces have been studied both experimentally and theoretically for a long time. Most of the approaches taken on the subject stem from a thermodynamic rather than a molecular viewpoint, i.e., from the so-called Johnson-Kendall-Roberts (JKR) model [14]. The first time such a type of problem was approached from a molecular viewpoint was with the proposal by Derjaguin et al. [15] of a new adhesion model, the Derjaguin-Muller-Toporov (DMT) model. Soon afterward the Muller-Yushchenko-Derjaguin (MYD) model was proposed [16,17] by assuming that the adhesion forces, hence the interaction between a particle and a substrate, could be represented by Lennard-Jones potential functions. With this theory it is not possible yet to speak of a molecular mechanics approach. Nonetheless, it is this first educated assumption and understanding that the interaction of generalized particles, atoms, and molecules can be described even for problems and theories of adhesion through Lennard-Jones functions, one of the classical types of potential functions used to describe interactions in molecular mechanics, that perhaps opened the way to the subsequent use of molecular mechanics in the field of adhesion. The JKR and DMT models have since been shown to be particular subsets of the more general MYD theory. Equally, the dynamics of particle adhesion and particle-surface adhesion has also been extensively studied [18–21] and models have been presented.

Although all these theories have helped our understanding of particle adhesion, all of them suffer from the considerable drawback of treating the mechanical response of materials as something totally independent of the molecular level parameters influencing adhesion. As the intermolecular potential of a material determines, or at least strongly influences, both its mechanical properties and its surface energy, the possibility exists that a more holistic adhesion model could be conceived by going back to the drawing board and starting from first principles. This, added to the fact that the theories briefly referred to above do not always predict the correct value of the power law dependence of the contact area on particle radius, prompted some attempts in this direction. Notable in this respect are two investigations, and only here for the first time in this chapter can one really talk of a molecular mechanics approach to some form of adhesion. The first of them, by Pucciariello et al. [22], was based on molecular mechanics calculations of the interaction of acrylic-type monomers with an idealized model surface composed of a rectangular parallelepiped of generalized, idealized atoms treated as spheres arranged at regular nodes of a square grid network. It constitutes the first example ever of this type of approach. The second one [23–25], a molecular dynamics study along very similar lines four years later, defined the interaction between the two surfaces of such parallelepipeds of generalized, idealized atoms treated as spheres and went further. This latter study will be briefly presented and discussed, with all its advantages and limitations, because it is a more clear-cut case of a generalized model of particle adhesion based on molecular mechanics. The earlier model by Pucciariello et al. [22] will not be discussed further as it really constitutes a hybrid case between the type of approach presented in the following section of this chapter dealing with examples of even earlier but conceptually more correct nongeneralized models and the type of approach based on particle adhesion proper.

Before getting more involved in the finer points of particle adhesion by molecular mechanics, it must be pointed out that such an approach suffers from considerable drawbacks. Molecular mechanics and dynamics by definition involve interactions between clearly defined types of atoms, with clearly defined atomic characteristics, placed in clearly defined molecular structures; thus, a generalized, fictitious surface of only, let us say nitrogens, or even worse of generalized spheres, appears by definition somewhat out of context in a molecular mechanics and dynamics investigation. Such a drawback needs to be pointed out to put in perspective and understand the limitations inherent in a model pretending to describe atomic and molecular interactions in real systems by a molecular mechanics approach oversimplified at the physical level. Nonetheless, valuable information has been gathered by this type of approach. It is furthermore a very good approach for the description of particle-particle interactions when the particles themselves are composed of well-defined atom types interacting with each other both at the particle-particle interface and within the body of the particle itself.

The more advanced work today on particle adhesion [23–25] then builds on the assumption that particles interact through a Lennard-Jones-type potential function, namely [25]

$$\mathbf{E} = -4\varepsilon[(\sigma/r)^6 - (\sigma/r)^{12}] \tag{6}$$

where ε is the binding energy between one atom and its nearest neighbor and σ is the distance between the two atoms when the value of the potential energy represented by this function is neither attractive nor repulsive (at the crossover, intersection point of the function with the axis). The authors of the theory recognized that the choice of this potential was purely empirical [25].

At first a flat surface of atoms was generated in a stepwise manner allowing the energetics associated with the creation of a surface to be determined; then two of these surfaces were brought together and allowed to form a bond. The pairs of mated surfaces were then separated in a constrained tension test to form two fracture surfaces. The potential energy of the system and the axial stresses used to produce the displacements observed were monitored. The molecular mechanics computational modeling part consisted of assembling a parallelepiped of generalized atoms represented as spheres, arranged in a predetermined regular array. The study was mainly a molecular dynamics one and considered the velocity components of each atom obtained by a seeded random number generator, transformed to give a Maxwell-Boltzmann distribution of initial velocities. The atoms were then released and allowed to interact through the Lennard-Jones potential function. Thus, when the atoms are initially released the system is not at equilibrium but the energy is all in the form of kinetic energy, which rapidly equipartitions into kinetic and potential energies. The computational model used in tension, compression, and shear modes allows examination of the stresses produced when free surfaces approach one another. The parallelepiped of atoms forming each surface was constituted of 768 atoms aligned parallel to the X, Y, and Z axes of the reference system. The parallelepiped was then constituted of 24, 8, and 8 layers of atoms in the X, Y, and Z directions, respectively. By the time the first 100 iterations were terminated, the system temperature had halved as a consequence of the equipartitioning of the energy into kinetic and potential contributions. The lateral dimensions of the surface that was about to be created were then fixed by putting the atoms back in their previous position after each computation, inducing, as a consequence, a gradual increase in the gaps between atoms without actual movement of the atoms. This effectively suppressed the usual atomic motions, allowing the authors to follow computationally the variation of the apparent potential energy of the system as the surfaces were separated. Once the size of the interatomic gaps had increased to the size of the cutoff radius, there was no further increase in potential energy with increasing gap size. Once the gap was established, the atomic motions and temperature dependence were reactivated to allow the system to relax into its new state of equilibrium.

When the free surfaces in equilibrium were computationally generated, they were brought closer to one another by a very slow approach rate of 10,000 iterations to reduce the gap between surfaces of one cutoff radius. The slow approach was necessary to minimize or eliminate complications arising from impact energies. All this allowed changes in potential energy and the resulting surface tractions that developed at constant temperature and constant dimensions to be followed. During the approach the mutual attraction increased monotonically until a certain critical stress level. At such a critical stress the two surfaces leaped into mutual contact as the strain energy increased because the rate of energy storage due to elastic deformation equalled the rate at which energy was provided by the attraction between the surfaces. Thus the model was able to reproduce, at least qualitatively, the leap-to-contact effect between a particle and a planar surface observed experimentally using atomic force techniques [26–28]. In the cases in which the surfaces were pulled, rather than leaped into contact, the stresses were not uniform and traveling waves were generated. Additional traveling waves were also generated when the surfaces struck one another. These waves, which interacted with each other, correspond to atomic level kinetic energy and can be interpreted as an increase in temperature. As a consequence, the authors came to the interesting conclusion that in a real system this implies that energy loss occurs even if only elastic deformations resulting from the forces of adhesion are considered; hence, not all the energy is recoverable on surface separation as not all the energy of the system is stored elastically.

Study of the model during subsequent separation of the two surfaces, corresponding to a test in tension, showed clearly the existence of hysteresis effects. This hysteresis might account for the effect of Young's modulus on particle adhesion, which is not considered in the JKR model. In the separation process it was not until the interatomic spacing of some regions exceeded the critical value, to an extent that further separation suddenly reduced the energy, that fracture of the model finally occurred. This appeared to be a rather sudden and marked process that occurred over a very small number of iterations. Even in the case of surface separation, waves were generated that decayed as a function of time, thus generating thermal energy, which was then lost by the modeled system. Thus, even during elastic deformation consequent to surface separation, energy loss mechanisms exist. The maximal stress experienced in the leap to contact decreased with increasing temperature, whereas the average stress, shortly after leap-to-contact, was much less sensitive to the temperature. Finally, there was a distinct offset between the initial and final potential energies due to microstructural changes in the interfacial regions as the two layers of atoms near each of the two surfaces contained numerous atom site vacancies.

IV. MODELS OF ADHESION BETWEEN WELL-DEFINED POLYMERS AND WELL-DEFINED SURFACES

Contrary to the generalized approach already presented, models describing the adhesion between a polymer well defined at the molecular level and another, equally molecularly well-defined substrate also exist. These are models in which molecular mechanics and dynamics are applied in their more accepted role as described in the Introduction. It must be realized that such models derive from a need different from that which prompted the development of the generalized models already described. They stem from the need to solve some applied problem of adhesion or to upgrade the applied performance of some adhesive systems in situations in which the use of an experimental method would take too long or is not able to give any clear-cut results. For this reason, such models are bound to use the most precise and well-defined information available on the molecules involved as well as using sets of potential functions that describe in the most accurate conceptual manner the molecular behavior of the chemical species involved. All the research work that uses this approach then is applied to "real" cases, not to idealized models, and is of considerable sophistication. Furthermore, this type of research work is most commonly supported by direct or indirect experimental and even industrial results which prove that the molecular mechanics forecasts predicted well the improvements that needed to be implemented to upgrade adhesion or an adhesive system. Notwithstanding its applied ancestry, the sophistication of this type of approach has yielded very interesting results on the fundamental side of the science of adhesion and of the interface, and it offers considerable promise and opportunities for more progress in the future.

The first of such studies [8] appeared in 1987 and thus preceded by a couple of years the first of the generalized approach studies [22]. It concerned the adhesion of phenol-formaldehyde polycondensates and resins to cellulose, hence to wood. It was followed by other studies of the adhesion to crystalline and amorphous cellulose of urea-formaldehyde resins [7], of more complex phenol-formaldehyde oligomers [10], of water [29,30], and of chromates [31]. It was finally followed by the more complex case of ternary systems in which two interfaces exist, namely in the situation of adhesion to cellulose and wood of an acrylic undercoat composed of a photopolymerizable primer on which is superimposed a schematic alkyd-polyester varnish [32,33]. The algorithms used for all these studies are the general molecular mechanics ones presented at the beginning of this chapter, with the difference that in the last of the studies mentioned hydrogen bonding was disregarded in the calculation because of its negligible importance in that particular system. Before the first of these studies, an understanding of the phenomenon of adhesion between a well-defined pair of adhesive and adherend had never been attempted by means of calculation of all the values of secondary interactions between the noncovalently bonded atoms of the two molecules involved. This approach was rendered possible by the codification, again by molecular mechanics (or conformational analysis, as this technique was known in earlier days) from the data of earlier X-ray diffraction studies, of the spatial conformation of native crystalline cellulose (or cellulose I) [12], of the several mixed conformations possible for amorphous cellulose [12], and also of phenol-formaldehyde (PF) oligomers [34].

This initial molecular mechanics calculation was limited to the interaction with crystalline cellulose I of all the three possible PF dimers in which a methylene bridge links two phenol nuclei: *ortho-ortho, para-para*, and *ortho-para*. As not much was known of how the system would react, the investigation was instead very extensive as regards the routes of approach of the PF oligomers to an elementary cellulose I crystallite. As cellulose constitutes as much as 50% of wood, where its percentage crystallinity is as high as 70%, this study also implied applicability to a wood substrate. As even dried wood always contains a certain amount of water, the influence of the latter was taken into account by introducing the effect of a parameter related to the dielectric constant of water in the calculations.

The results obtained clearly indicated that adhesion of PF resins to cellulose was easily explained as a surface adsorption mechanism, a fact that, while well accepted today, was not evident in the wood gluing field and literature up to that time. The results also indicated that the interaction of the PF dimers with cellulose on all possible sites was more attractive than the average attraction by the cellulose molecule for the sorption of water molecules. In only a few cases, the interaction of water molecules with the few strongest sorption sites of cellulose is more attractive than that of PF dimers. This implied that, in general, even for the more difficult-to-wet crystalline cellulose in a solvated state, the PF dimers, and by inference also higher PF oligomers, were likely to displace water to adhere to the cellulose surface. This is an important finding as it did show for the first time by numerical values that in wood bonding the adhesion of the polymer resin to the wood must be considerably better than the adhesion of water molecules to the wood. It is of importance first for "grip" by the adhesive of the substrate surface and second, in the cured adhesive state, in partly determining the level of resistance to water attack of the interfacial bond between adhesive and adherend. This result added a new dimension to the well-known water and weather resistance of PF-bonded lignocellulosic materials: it is due not only to the imperviousness to water of the cured PF resin itself, as believed up to then, but also to the imperviousness to water of the adhesive-adherend interfacial bond, a bond, furthermore, exclusively formed through secondary forces. A further deduction, with some applied inference, from the results was that a PF resin used to impregnate wood was likely to depress the wood, and cellulose, water sorption isotherm according to the number of substrate sorption sites that, on curing, have been denied to water by sorption of the resin, a deduction later confirmed experimentally.

The most important result, however, was that there are significant differences in the values of minimal total energy in the interaction of the three PF dimers with cellulose. It is possibly the more important conclusion, because it also had a more immediate industrial application. The *ortho-ortho* and *ortho-para* dimers had much greater average relative E_{tot} of interaction with the cellulose surface than the *para-para* dimer. In general, the distribution of the methylene linkage in standard commercial PF resins at that time indicated a higher proportion of the *ortho-para* and *para-para* linkages over the *ortho-ortho* linkages. Experimental results confirmed this finding [35]. To maximize the proportion of *ortho-ortho* and *ortho-para* linkages and to decrease the relative proportion of the *para-para* coupling is easily done in PF resins manufacturing by the addition of *ortho*-orientating additives, through now well-known mechanisms [36–40]. As a consequence, the adhesion and performance improvement caused by a shift in methylene bridge coupling renders possible a reduction by about 10%, at parity of performance, or to improve the performance at parity of quantity, of the quantity of PF adhesive resins used in a product such as wood particleboard—not a bad result if one considers that approximately 2 million tons of PF resins are used for wood bonding each year! It is not claimed here that the molecular mechanics result converted the PF resins industry to maximizing *ortho* coupling, but the theoretical justification it offered contributed to greatly accelerate the already existing empirical trend in such a direction. Maximization of *ortho*-coupling in commercial PF resins is now a much more common practice. Further confirmation of this, and this time more scientifically exact and acceptable, was later obtained by studies of dynamic, differential, and competive adsorption [9], which are discussed in the next and latter sections of this chapter.

The findings also contributed to the visualization of the conformation of minimal energy of a resin on a substrate: the equivalent of a static, schematic photograph of the conformation of the two molecules at an interface. An example of this is shown in Fig. 2. It also contributed to the understanding, although this came from later work [7], that not only the energy at an interface but also the conformation of minimal energy of a molecule on a substrate is quite different

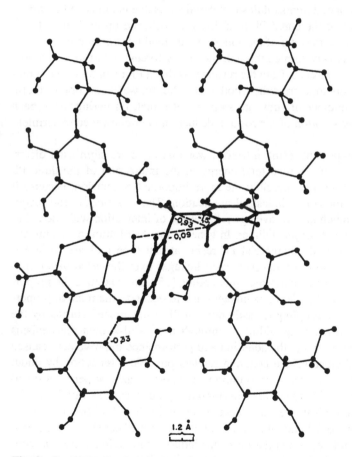

Figure 2 Example of a planar projection of one of the configurations of an *ortho-para* PF dimer on the surface of a schematic cellulose crystallite showing phenolic dimer conformation of minimal energy and main dimer-cellulose hydrogen bonding.

from the conformation of minimal energy of the same molecule when alone or when on a different substrate. This was confirmed later by X-ray studies of the level and/or lack of crystallinity of hardened urea-formaldehyde resins in the presence and absence of cellulose [41]. It is also for this reason that idealized models are limited to never being able to compete with "real" models to solve adhesion problems.

Further molecular mechanics investigations in the same direction but for urea-formaldehyde (UF) resins also followed, with equally interesting results. In these the efficiency of resin adhesion to both amorphous and crystalline cellulose was computed by combining the adhesion of the resin oligomers formed at each step of the synthesis of the resin. Thus, it was obtained by calculating by molecular mechanics the adhesive-adherend interactions with the two types of cellulose of each isomeride produced through the reaction of urea with formaldehyde up to the level of trimer (thus, not only of urea, monomethylene diureas, and dimethylene triureas but also of their monomethylolated, dimethylolated, and trimethylolated species) [6,7]. All the results found correspondence in already existing experimental results [42–44]. Among them, of interest to compare with what was already found for PF resins, was that the lack of water and weather resistance of lignocellulosic materials bonded with UF resins does not appear to be due, to any large extent, to failure of their adhesion to cellulose, although this might in some cases be a contributory factor. It confirmed that the weakness to water attack of these resins resides in the hydrolysis of their amidomethylenic bond, a fact since confirmed experimentally. More important is the finding, later confirmed by X-ray diffraction [41], that when UF resins interact with cellulose some of the conformations that would be forbidden when the UF resin is cured alone become possible and are allowed. The same experimental study also confirmed that the secondary forces binding together linear chains, not cross-linked, of UF oligomers with cellulose are stronger than the intermolecular forces between the UF oligomers themselves. The consequence of this study was the development of a method for evaluating comparatively the performance of UF resins prepared according to different procedures starting from the relative abundance of the various UF oligomers in each resin and their molecular mechanics calculated energy of interfacial interaction [6,7,45]. Similar results were also obtained with a more comprehensive investigation of the oligomers of PF resins [10].

It is, however, with the more complex and comprehensive investigation of ternary systems—i.e., systems composed of three molecular species, namely the cellulose substrate, a photopolymerizable primer resin, and a top coat alkyd-polyester varnish, presenting two interfaces—that this molecular mechanics approach started to yield results of greater interest on the fundamental principles of adhesion [32] (Fig. 3). This work was started mainly to address the concept of flexibility of a surface finish system on lignocellulosic materials, but it also led to some

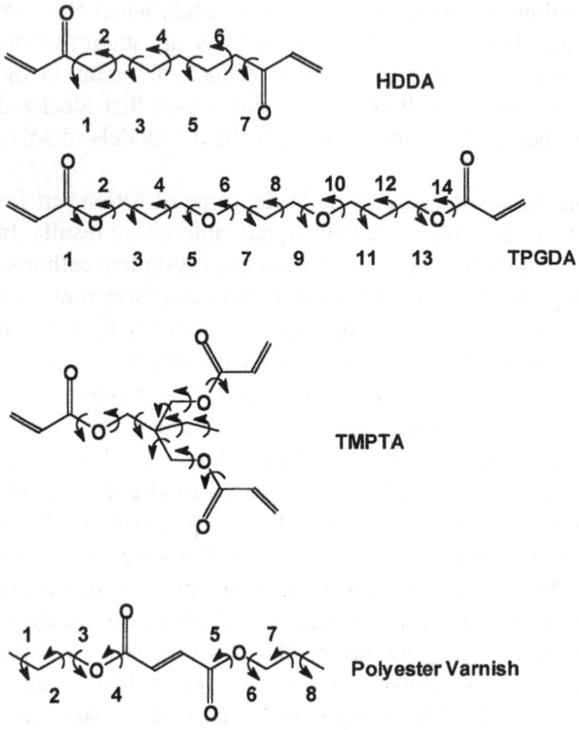

Figure 3 Acrylic monomers and schematic polyester showing the number of degrees of rotational freedom *m*.

unexpected and rewarding results on adhesion. Examples of the visualization of the conformations of minimal energy of ternary systems are shown in Figs. 4 and 5.

Three photopolymerizable primer monomers, namely the linear hexanediol diacrylate (HDDA), the branched trimethylol propane triacrylate (TMPTA), and the linear tripropyleneglycol diacrylate (TPGDA), and a model of a linear unsaturated polyester-alkyd varnish repeat unit were used for the study (Fig. 3). A model of the two top chains of an elementary cellulose I crystallite, the two parallel chains considered composed of four glucose residues each, the refined conformation of which had already been reported [12], was used as the substrate.

The number of degrees of freedom for such calculations is considerable and the following technique, already used in previous work [7–10], was employed to facilitate the computation. The central carbon of each of the primer monomers in their extended flat configuration with the monomer parallel to the longitudinal

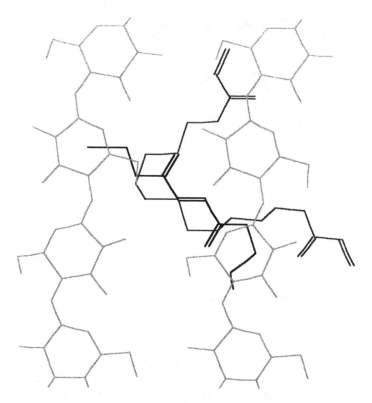

Figure 4 Example of planar projection of the minimal energy configuration of a ternary system composed of cellulose crystallite surface, photopolymerizable acrylic primer (TPGDA), and polyester finish.

axis of the cellulose crystallite surface was aligned with the central point of the assembly of two cellulose chains. The latter is defined as the intersection of the axis of the crystal surface (parallel to the longitudinal axis of the crystal) and the line connecting the two oxygen atoms of each central β-glucosidic bridge of the two cellulose chains. Each primer monomer was aligned at a distance of 12 Å and then allowed to approach the cellulose crystallite surface by increments of 0.5 Å. At each distance all the bonds that could rotate in the primer molecule were allowed to rotate, plus the monomer was rotated across 360° in the plane parallel to the cellulose crystallite surface as well as being rotated 360° in the plane normal to it. All rotations were done simultaneously in increments of 120° to find the optimal distance of the primer molecule from the cellulose surface. At the optimized distance the 360° rotations were performed simultaneously for

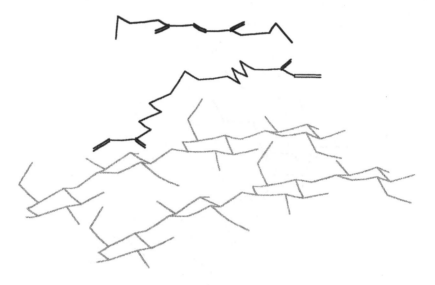

Figure 5 Example of perspective view of the ternary system in Fig. 4.

all the degrees of internal rotational freedom (bonds of the molecule rotated) and for all the degrees of external rotational freedom (rotation of the whole molecule in the two planes defined above), at first in steps of 120°, then in steps of 60°, and finally in steps of 30°. Further refinements were not possible because of the already very considerable volume of calculations. The calculations yielded the energy of the system composed of the primer and the cellulose surface, E_1 (with the cellulose internal energy taken as 0, as the cellulose and its groups were not allowed to move, because of the energy and configuration stability of the crystallite). However, this is not representative of the energy of interaction at the interface, as the gain or loss of energy due to movement of the primer from its configurations of minimal internal energies have not been taken into consideration. As a consequence, the internal energies of the primer or varnish molecule, both in its configuration of minimal internal energy when standing alone (E_2) and in its configuration of minimal internal energy on the surface of the cellulose (E_3), were also calculated. The overall energies of interaction (E) of primers with the surface of cellulose were then obtained from $E = E_1 + (E_3 - E_2)$. The procedure was repeated for one of the primers on the most common short-chain configuration of four glucose residues of amorphous cellulose [12].

    The same procedure was repeated for the calculation of the energies of interaction at the second interface, i.e., between varnish and primer. For this purpose the energy- and conformation-minimized primer-cellulose assembly was used as the substrate on which the repeat unit of the unsaturated polyester-alkyd varnish model was allowed to move and be energy minimized. The same expression as above for the calculation of the interfacial energy of interaction was used.

Finally, the total energy-minimized assembly of varnish, primer, and cellulose was allowed to adjust and minimize the energy of its configuration.

The main computational program used for the calculations of the secondary force interactions between primers and substrate and between varnish and the primer-cellulose assembly was a constrained force field model [12] that had been checked many times against automatic unconstrained force field models for its accuracy for proteins [11], cellulose [12,46], and cellulose interfaces with other polymers or simpler molecules [6,10,29,30] and whose results were also checked by other authors to obtain good interfacial forecasts of experimental phenomena [47–50].

The applied part of this study relied on a standard peel test in which closely set vertical and horizontal cut lines were incised on the specimen surface and on a dynamic thermomechanical analysis of finish flexibility at constant temperature. Two of the primers were tested dynamically by thermomechanical analysis (TMA). Samples of beech wood alone and of beech wood treated with photopolymerized layers of the two primers 350 μm thick were tested isothermally at 25°C in three-point bending by exercising a force cycle of 0.1 N/0.5 N on the specimens with each force cycle of 12 s (6 s/6 s). The classical mechanics relation between force and deflection $Y = [L^3/(4bh^3)][\Delta F/(\Delta f_{wood} - \Delta f_{finish})]$, where L, b, and h are respectively length span, width, and thickness of the specimen and ΔF is the force cycle applied, allows the calculation of the Young modulus Y for each of the cases tested. As the deflections Δf obtained were proved to be constant and reproducible and are proportional to the flexibility of the assembly, the relative flexibility as expressed by the Young modulus of the two primers was calculated for the two primer-wood assemblies through the relationship $Y_1/Y_2 = \Delta f_2/\Delta f_1$ in relation to wood alone [32,33,51].

A. Adhesion, Networks, Work of Adhesion, Viscoelastic Energy Dissipation, and Flexibility

The behavior of a complete varnish-primer-cellulose surface finish should be mostly reflected by the behavior of the weakest boundary layer of the system. Thus, the several layers of a complete surface finish can be schematically described as follows: the varnish, the varnish-primer interface, the primer, and the primer-cellulose (or wood) interface. In reality, the last one is often not a clear-cut interface because there is often penetration of the primer into the wood; the interface is then a primer-wood composite that gradually becomes again the primer alone on the wood surface. The situation is different using free-standing cellulose as the model, because a clear primer-cellulose interface can be identified. Thus, whereas at a macroscopic level flexibility or rigidity of a single material is a well-documented concept [52], at the interfaces the situation is less clear. The elasticity and flexibility of a molecule containing in its backbone chain a number of methylene (—CH$_2$—) and/or oxygen (—O—) groups should depend on a variety of factors that can be schematically represented as:

The molecule can be considered similar to a spring in which covalent bond vibrational stretching (symmetrical and asymmetrical), covalent bond angle stretching (symmetrical and asymmetrical), and covalent bond rotations can all contribute to the elasticity and flexibility of the spring. Furthermore, the interactions of secondary forces within the molecule and between the molecule and the surrounding molecules, particularly the substrate, must also be considered. Thus, when the predominant contribution due to viscoelastic energy dissipation is not considered or eliminated, a concept of notional flexibility could be defined as the sum of different energies internal to the molecule:

$$\text{Notional flexibility} = E_{\text{bond stretching}} + E_{\text{angle stretching}} \qquad (7)$$
$$+ E_{\text{bond rotation}} + E_{\text{secondary forces}}$$

It is also clear that to translate this to a macroscopic level it is necessary to minimize the very important contribution of crack tip propagation during breaking, at the interface or not, as it could be so preponderant as to "drown" most of the contributions in the preceding equation. The standard test chosen [53], based on physically cutting on the surface to be examined a grid of small squares, minimizes the effects of crack tip propagation by limiting it to the extent of each small square; this should bring to the fore the contribution of the four terms in the preceding equation and hence of the molecular level contributions to adhesion.

Unfortunately, a total concept of flexibility in a complete, complex finish was beyond the experimental data available. Such data derive exclusively from the results of the standard test [53] already described, of thermomechanical analysis (TMA), and of infrared (IR) spectroscopy. Crack tip propagation has been minimized by the type of tests used and the cohesion strength of the polymerized materials used had already been shown to be greater than interfacial adhesion; the interference of these two important factors was thus minimized in one case and eliminated in the other.

From the results of IR spectra [54], the contribution of bond and angle stretching to the flexibility of a primer in contact with the wood surface or of a varnish in contact with the primer is first of all very small and second not easily calculable [54]. The contribution of the rotational forces of the molecule is not inconsiderable, but more important is the contribution of intramolecular and particularly intermolecular secondary force interactions. The rotational barriers and their influence are taken into account by the molecular mechanics approach and translated directly in the number of degrees of freedom allowed in the primer molecules for the primer-cellulose interface and in the varnish for the varnish-primer interface. The influence of the secondary forces, particularly the intermolecular ones, on flexibility needs to be examined in more depth. A prepolymerized primer on wood would form a layer held by secondary forces to the wood and the layers of which are also held together by secondary forces along the length of each monomer, except at the covalent cross-linking sites of the polymer network repeat unit. Thus, any stress-induced dimensional variation of the timber substrate should be counterbalanced.

1. By the primer by both stretching and contracting through the monomer rotational degrees of freedom, and
2. By the ability of the linear non-cross-linked segments of the polymer network held together by secondary forces to partly break such secondary forces and partly disentangle in the regions between cross-link nodes only, as well as by their ability to re-form such intermolecular secondary forces and be partly re-established once the stress induced in the timber, and its consequent deformation, is released.

With regard to calculating flexibility, flexibility was defined as the inverse of the calculated interfacial energy of interaction as a function of the number of degrees of freedom of the segments of the molecule between two covalently cross-linked nodes of the polymer network. Why the inverse? Because the stronger the secondary energy of interaction between molecules, both at the interfaces and within the same polymer, the lower the flexibility or, better, the lower the capability of the material for an infinitesimally small separation between the chains segments to occur. Furthermore, the higher the number of rotational degrees of freedom of the molecular segments in between cross-linking nodes, the greater the flexibility of the system at the interface. The number of degrees of freedom is equal to the number of bonds of the main chain that can be rotated without overcoming excessive energy barriers (hence —CH_2—CH_2— and —CH_2—O—) if the monomer is linear. Thus, a single formula for defining this notional flexibility can be expressed as

$$\text{Notional flexibility} = -(m/n)/(E/n) = -m/E \qquad (8)$$

where m is the number of degrees of freedom, n is the number of atoms of the monomer studied, and E is the interaction energy of the monomer molecule with the substrate. This equation is also valid if the calculation is preferred on the basis of the molecular weight of the primer. Thus, for a molecular weight M the equation would be

$$\text{Notional flexibility} = -(m/M)/(E/M) = -m/E \tag{9}$$

It is possible to see then that the concept of molecular mass is incorporated in the equation even if the mass itself does not appear in it.

The preceding equation clearly indicates that to define flexibility at a molecular level the two important factors are the total energy of interaction of the monomer and the number of bond rotational degrees of freedom. This formula was found to satisfy all the results obtained for linear primers and finishes but did not satisfy the comparison of branched primers with linear ones obtained experimentally, which showed quite obvious discrepancies

It was found finally that to take into account the differences in flexibility modes presented by branched and linear primers the important parameter was the coefficient of networking α, similar to but not the same as Flory's coefficient of branching for polycondensates [55]. This means that in their polymerized state, and both TMA and adhesion tests for these primers have been carried out on their polymerized state, the number of degrees of freedom of the primer (or of a finish) should be curtailed as the system is not as free as when the primer is not polymerized. A linear primer doubly linked at each of its two ends would present for the chain segments attached at the interface to the substrate by secondary forces a very different extent of freedom than a branched primer with three double reaction sites, one at each of its three ends. Thus, if a coefficient α of reactive sites is introduced, similar to Flory's coefficient of branching for polycondensates [55], then $\alpha = 1/(f - 1)$ where f is the functionality of the monomer in terms of number of reactive carbons in a $C{=}C$ bond (two carbons for each unsaturation), and the relationship of flexibility as a function of the energy of interaction at the interface and of the number of degrees of freedom of the segments in between cross-links changes to

$$\text{Notional flexibility} = -(m/\alpha)/E = -m/(\alpha E) \tag{10}$$

where notional flexibility is the relative flexibility of the primer and primer interface.

It remained to ascertain whether α was really the equivalent of Flory's coefficient of branching or was another, related, parameter. It was found that if α was taken as the coefficient of branching by reactive sites, hence by reactive carbons in each monomer, not only was the trend of flexibility of all the cases respected but also the flexibility calculated by molecular mechanics according to the last equation proposed indicated a difference of 32% in flexibility between the linear and branched primers used in the experiment. This compares well with the difference of 29% obtained experimentally by thermomechanical analysis. The small difference could be due to additional entanglement effects. It is important to point out that the flexibility expressed by the preceding equation is only the interfacial flexibility and not the cohesive flexibility of the system. For the latter the monomer-monomer interaction needs to be calculated instead.

As the preceding equation correlates flexibility and interaction energy, it can be used to determine the energy of interfacial interaction starting from a measure of TMA flexibility, or vice versa. It is then a useful experimental tool. For this purpose, it is necessary to introduce a multiplicative constant k to be able to compare numerically the values of relative flexibility expressed in the experimental form of relative TMA deflection with the actual values of the energy of interaction in kcal/mol. As a consequence,

$$\mathbf{E} = -km/(\alpha f) \quad \text{and conversely} \quad f = -km/(\alpha \mathbf{E}) \tag{11}$$

where f is the relative deflection obtained for the system by thermomechanical analysis and is related to the calculated notional flexibility F through the expression $F = f/k$. From the TMA and calculated results the numerical value of k was always found to be approximately 16 when $\mathbf{E}$ was expressed in kcal/mol. This value of 16 is clearly valid only for the dimensions of the samples used in the particular experiments carried out and the conditions of use of the particular TMA equipment; the value might well be different for different adhesive-substrate systems or different experimental conditions.

As the experiments that led to the definition of the preceding formula were done at a fixed amount of adhesive and with fixed dimensions of the bonded joint due to the geometry of the sample needed to fit the TMA equipment available, it is necessary to point out that a clear dependence on the quantity of the material applied and on the dimensions of the sample, namely bonded area and thickness, also exists. As the original experiments were conducted at equal quantities of synthetic polymer applied to equal areas of substrate samples, hence at equal spread per unit area, this was not at first realized. As a consequence, if Eq. (11) needs to be generalized, a more appropriate form of it is

$$f = -k[m/(\alpha \mathbf{E})] \ \phi(\text{area, quantity, joint thickness}) \tag{11a}$$

where ϕ(area, quantity, joint thickness) is a function of the inverse of the three parameters between parentheses.

The results and effectiveness of Eq. (11) were also checked for other, quite different polymers, namely the polycondensates of resorcinol-formaldehyde, of melamine-urea-formaldehyde (MUF), of phenol-formaldehyde (PF), and of quebracho and pine polyflavonoid tannins hardened with formaldehyde. The energies of interaction obtained with measures of TMA flexibility and use of this formula compared well with the results obtained for their energy of adhesion with crystalline cellulose in previous work [6–10]. It appears then that the formula also works for other polymer systems on the surface of cellulose. More interesting is a result that was obtained as a consequence of this work in the field of formaldehyde-hardened polycondensation networks [56]. In this case the following three regression equations relating only two of the parameters above, hence capable of yielding the more difficult to obtain experimental parameters in an easier manner, were obtained [56].

$$m = 0.919\alpha f - 0.579 \tag{12}$$

$$E = km/(0.1081m + 0.683) \tag{13}$$

$$E = k/(0.919 - 0.579/\alpha f) \tag{14}$$

with coefficient of correlation $r = 0.996$. These regression equations may have better practical use than the generalized formula because they express any of the two more difficult or time-consuming parameters to obtain, namely m and E, as a function of only one of the other parameters. The results and correlations presented lead to the conclusion that it is possible to follow through the variation of the value of m, in a quantitative manner, the progress of networking after the gel point of a polycondensation.

Furthermore, since $m = 2DP_n$ and $DP_n = 1/(1 - p)$ then $m = 2/(1 - p)$, hence there is a relationship between the degree of conversion p and m. That this is valid from $p = 0$ and $m = 2$ (no polymerization) to $p = 1$ (complete conversion) is just nothing else than expressing the Carothers [57] equation in m rather than in DP_n; but as at $p = 1$, $m = \infty$, thus a molecule infinitely long, it means that after the gel point p can become greater than 1 in order to be able to describe the process of hardening. This is a logical conclusion as, for instance, if for a polycondensation between a trifunctional and a difunctional monomer the gel point intervenes at complete conversion, hence at $p = 1$ then also $p_{gel} = 1$: this means that after the gel point the reaction continues to form a tridimensional network and after the gel point $1 < p < 2$ and $\infty > m > 2$. The preceding formula can then be written just as $m = 2/(1 - p)$ with m yielding a negative value only after the gel point to indicate that it is in the domain between gel and complete hardening, or otherwise $m = $ absolute value of $2/(1 - p)$. As the gel point rarely intervenes at $p = 1$ and as the relative proportions of the two reactive groups taking part in the polycondensation are important to define p, p_{gel}, and m, the expression can be modified to take into account the relative proportions of reactive groups. As according to Carothers $DP_n = 2/(2 - fp)$ so $m/2 = 2/(2 - fp)$, and

$$m = 4/(2 - fp) \tag{15}$$

and vice versa, $p = (2m - 4)/fm$, where f is the Carothers average functionality of the whole polycondensation system. It must be pointed out that such an equation functions exclusively for the Carothers-defined value of the average degree of functionality f and thus for the Carothers-defined $p_{gel} = 2/f$ value. If the p_{gel} is calculated by Flory's method [58], which takes into account the functionality in the calculation of p_{gel}, then Eq. (15) is not valid, and a more general equation (16) is valid for both theories

$$m = 2/(1 - p) \tag{16}$$

and vice versa $p = (m - 2)/m$.

Equation (16) indicates that in theory at the exact instant of the gel m is infinite, thus a linear polymer infinitely long with no cross-links. As the system proceeds over the gel point, cross-linking has started and m starts to describe the average number of degrees of freedom of the average length segment between cross-linking nodes from the initial to the final tridimensional network. Thus in Eq. (15) before and up to the gel point $0 < p < p_{gel}$ and $2 < m < \infty$, and after the gel point to complete hardening $p_{gel} < p < 2$ and $\infty > m > -2$, m presenting a negative value to indicate that it is in the region after the gel point. These equations, as well as the $DP_n = 1/(1 - p)$ equation, are valid both before and after the gel point of the system, i.e., from the polycondensation beginning through to its gel point to formation of the complete, final hardened network. From these expressions the values of p, DP_n, gel temperature (t^0_{gel}), and p_{gel} can be obtained with ease by a simple TMA measure of the deflection and the calculation of m according to regression equation $m = 0.919\alpha f - 0.579$.

It was also interesting to relate what had already been discussed to the existing models relating adhesion strength and adhesion energy. In the rheological model [52,59–63] the peel adhesion strength **G** is simply equal to the product of the adhesion energy **E** and a loss function Φ that corresponds to the energy irreversibly dissipated in viscoelastic or plastic deformations in the bulk materials and at the crack tip and which depends on both peel rate v and temperature T. Thus

$$\mathbf{G} = \mathbf{E}\Phi(v, T) \tag{17}$$

The value of Φ is usually far higher than that of **E**, and the energy dissipated can then be considered as the major contribution to the adhesion strength **G**. It is more convenient in this equation to use the intrinsic fracture energy $\mathbf{G_0}$ of the interface in place of **E** to have $\mathbf{G} = \mathbf{G_0}\Phi(v, T)$. When viscoelastic losses are negligible, i.e., where tests minimizing their influence have been chosen, Φ tends to 1 and **G** must tend toward **E**. However, the resulting threshold value $\mathbf{G_0}$ is generally a few orders of magnitude higher than **E**. Carré and Schultz [64] have concluded that the value of $\mathbf{G_0}$ can be related to **E** for cross-linked elastomer-substrate assemblies through the expression

$$\mathbf{G_0} = \mathbf{E}g(M_c) \tag{18}$$

where g is a function of molecular weight M_c between cross-linked nodes and corresponds to a molecular dissipation.

This leads to a few interesting considerations regarding the results obtained by molecular mechanics for the primer-cellulose interfaces. From the equation obtained to relate the flexibility at the interface to the interaction energy it is evident that

$$\mathbf{E} \sim m/(\alpha f) \tag{19}$$

(the negative sign of $\mathbf{E}$ obtained by molecular mechanics is a convention to indicate attraction rather than repulsion), and as the concept of M_c is intrinsic in the $(m/n)/\alpha$ ratio relating the number of degrees of freedom m per number of atoms n of the segments between cross-linking nodes as determined by α, it means that $g(M_c)$ can be represented by $m/(\alpha f)$ and hence

$$\mathbf{G}_0 \sim \mathbf{E} m/(\alpha f), \text{ thus } \mathbf{G}_0 \sim \mathbf{E}^2 \tag{20}$$

This is an important aspect and would, at least partly, explain in a manner somewhat different from the more accepted ones why $\mathbf{G}_0$ is generally 100 to 1000 times higher than the thermodynamic work of adhesion, W [52]. It indicates then that $\mathbf{G} \sim \mathbf{E}\Phi(v, T)m/(\alpha f)$ or, differently expressed, $\mathbf{G} \sim \mathbf{E}^2 \Phi(v, T)$. Apart from this, the interesting consideration that the flexibility at the interface is inversely proportional to both the intrinsic fracture energy and the peel adhesion strength still holds, at least where the preponderant effect of Φ is minimized.

This, however, is not the whole story and presents an interesting but slanted view of the true situation. Thus, from the equation obtained to relate the flexibility at the interface to the interaction energy, considering that a contribution from the cohesive energy of the two materials forming the interface is also present, it is then evident that also

$$\mathbf{E} \sim E = (W + E_{coh}) \sim m/(\alpha f) \sim m/\alpha \tag{21}$$

and as the concept of M_c is intrinsic in the $(m/n)/\alpha$ ratio relating the number of degrees of freedom m per number of atoms n of the segments between cross-linking nodes as determined by α, it means that $g(M_c)$ can be represented by $m/(\alpha f)$ and hence $M_c \sim m/\alpha$, thus

$$\mathbf{G}_0 = Wg(M_c) \sim Wg(m/\alpha) \tag{22}$$

and as a consequence

$$\mathbf{G}_0 \sim Wg(\mathbf{E}) = Wg(W + E_{coh}) \tag{23}$$

and

$$\mathbf{G}_0 \sim W^2 + WE_{coh} \tag{24}$$

and in tests in which viscoelastic energy dissipation is eliminated ($E_{coh} = 0$) or at least strongly minimized (E_{coh} tends to 0), as in the case at hand

$$\mathbf{G}_0 \sim W^2 \tag{25}$$

which is indeed the case as E calculated by molecular mechanics is by definition equal to W.

This is an important aspect and would at least partly explain in a manner somewhat different from the more accepted ones why G_0 is generally 100 to 1000 times higher than the thermodynamic work of adhesion W [52]. It indicates then that $G \sim W\Phi(v, T)m/(\alpha f)$, or differently expressed $G \sim W^2 \Phi(v, T)$. Apart from this, the interesting consideration that the flexibility at the interface is inversely proportional to both the intrinsic fracture energy and to the peel adhesion strength still holds, at least where the preponderant effect of Φ is minimized.

Because from cross-linked and non-cross-linked rubbers the rubber plateau modulus G_0 using the theory of rubber elasticity [65,66] is

$$G_0 = (\rho/M_c)RT \tag{26}$$

where ρ is the density, M_c is as above, and R and T are, respectively, the gas constant and the absolute temperature, in cases in which viscoelastic energy dissipation has not been eliminated or minimized

$$G_0 \sim W^2 + WE_{coh} \sim (\rho/M_c)RT \sim [\rho/(m/\alpha)]RT \sim [(\rho\alpha)/m]RT \tag{27}$$

and as from previous work [56] $M_c = M_0(m/2)$, where M_0 is nothing else than the molecular weight of the repeat unit of the polymer, or equally

$$G_0 = [(2\rho)/(mM_0)]RT \tag{28}$$

All this indicates that some of the values derived from the material characteristics also belong in G_0 and not only in the viscoelastic energy dissipation function $\Phi(v, T)$. This means that the supporters of $G_0 \sim W^2$ [67] and of $G_0 \sim W$ [68] in adhesion theory are both correct and incorrect, but to different extents, because if the component of G_0 based on viscoelastic properties of the material is maintained in G_0, then really $G_0 \sim \geq W^2$ [56], but if its viscoelastic component is removed and passed as it should be into the viscoelastic energy function $\Phi(v, T)$ as the total viscoelastic energy function $\Phi(v, E_{coh}, T)$, then $G_0 = W^2$ [56]. Conversely, in the theoretical case in which viscoelastic energy dissipation has been minimized or eliminated, $G_0 \sim W^2 \sim [(\rho\alpha)/m]RT$ and, consequently, the viscoelastic component characteristic of the material (not of crack tip propagation within the material) still being present, then $W \leq G_0 \ll W^2$, and thus most likely $G \sim W$ considering the importance of any viscoelastic energy component [56]. Although this applies strictly only to cellulosic substrates, it also indicates clearly that the contested findings by certain authors on other substrates and with other adhesive systems that $G_0 \sim W$, and by other authors instead that $G_0 \sim W^2$, are really both correct findings and nothing else than two distinct, particular aspects of the same theory [56]. It also indicates that the results reported above might well be generalizable (but not overgeneralizable) and applicable to most adhesive-substrate systems.

The foregoing discussion also brings the interesting consideration that depending on the particular situations considered

$$\mathbf{G}_0 \sim k \left(\sum_{ij} [a_{ij} \exp(-b_{ij}r_{ij}) - c_{ij}r_{ij}^{-6}] \right)^2$$

$$\text{or equally} \quad \mathbf{G}_0 \sim k \left(\sum_{ij} [(d_{ij}/r_{ij}^{12}) - (c_{ij}/r_{ij}^{6})] \right)^2 \quad (29)$$

and also

$$\mathbf{G}_0 \sim k \sum_{ij} [a_{ij} \exp(-b_{ij}r_{ij}) - c_{ij}r_{ij}^{-6}]$$

$$\text{or equally} \quad \mathbf{G}_0 \sim k \sum_{ij} [(d_{ij}/r_{ij}^{12}) - (c_{ij}/r_{ij}^{6})] \quad (29a)$$

Thus it depends on the Lennard-Jones or Buckingham expressions used for the calculation of the van der Waals nonbonded atom interactions, or in other words it depends indirectly on the mass of the molecule, as the greater the mass the greater is the number of interactions. However, a greater mass does not mean a greater attractive energy of interaction. Then $\mathbf{G}_0$ depends on the coefficients a, b, c, and d, which, notwithstanding the care with which they have been developed to keep the results of Lennard-Jones or Buckingham expressions as close as possible to the experimental reality [1–3], are empirically derived coefficients; hence molecular mechanics equations are only a model, although one very well experimented with. As a consequence, a Lennard-Jones equation or a Buckingham equation or a mixture of the two types [4] can give equally acceptable results, with the Buckingham expressions accepted as being more flexible than the Lennard-Jones one. The mass of the molecule then features as influencing $\mathbf{G}_0$ because it is the more accessible parameter, but in reality the number and type of atoms constituting the molecule and the other parameters on which their interactions depend are what really counts.

The preceding expressions take into account only the van der Waals interactions, but from Eq. (1) the other energies must also be considered. Although in our case $\mathbf{E}_{\text{H-bond}}$ was inconsequential due to the particular molecular species used, in others it may be a very important contribution, e.g., as it has been found and taken into account in the cases of the interaction of polycondensation polymers with a polar substrate [6–10]. Thus,

$$k\mathbf{E}_{\text{tot}} \leq \mathbf{G}_0 \leq k\mathbf{E}_{\text{tot}}^2 \quad \text{where} \quad \mathbf{E}_{\text{tot}} = k[\mathbf{E}_{\text{vdW}} + \mathbf{E}_{\text{H-bond}} + \mathbf{E}_{\text{elec}} + \mathbf{E}_{\text{tor}}] \quad (30)$$

where the molecular mass is represented by the molecular degrees of freedom, the type of atoms involved, the coefficient of molecular branching or cross-linking, the atom polarizability, the angle and direction of the interactions, the electrostatic charges, the number of effective electrons participating, the dipolar momenta, and even the molecular weight of the monomer. The mass is then used, incorrectly, only as a simplified blanket parameter covering all this. Furthermore, to be conceptually correct, even symmetrical and asymmetrical bond or angles stretching movements and molecular translational movements, even if their contribution is quite small, should be considered.

From an experimental point of view the flexibility of the primer-cellulose interface can also be defined by its glass transition temperature (T_g). The higher the value of T_g, the greater the rigidity of a material will be at a given temperature. It has been determined, for instance, that the effectiveness and durability of a primer or a finish such as wood surface finishes are connected to its glass transition temperature, and mathematical relationships for it have been developed [69]. A definition of the glass transition temperature [70] is that T_g is the temperature at which translational movements (also induced by covalent bond rotations) of main chain segments start or cease. This means that in the case of surface finishes the value of T_g is linked to interchain secondary force interactions between the primer (or finish) chains as well as primer (or finish) and substrate (at the molecular interface); hence T_g is linked as well to the degrees of rotational freedom of chain backbone covalent bonds and to their energy barriers. The consequence is that the flexibility of a primer or a finish applied on a substrate has to be different from the flexibility of the same finish when taken in isolation. Furthermore, more far reaching is the conclusion that the value of T_g of the same finish or primer is different if it is alone or applied on a substrate, as in the latter case the start of ceasing of the chain translational movements is influenced by the interaction forces across the interface, hence it varies as a function of the energy of adhesion at the interface. It is also interesting to note that subsequent work indicated that both the DiBenedetto [71] and Couchman [72] relationships could be expressed as a function of the number of degrees of freedom m for the "tightening" of an already formed covalently linked network [56]. This was shown to be valid for polycondensation networks [56] but is also valid for other covalently linked networks [56]. Thus,

$$\frac{[(m_{end}/m_{start})p]}{[1 - (1 - (m_{end}/m_{start}))p]} = 1 - e^{-kt} \tag{31}$$

From this equation it is possible to calculate the rate of hardening of a polycondensation network at any predetermined temperature [56]. This also means that T_g can be measured through the value of m obtained by molecular mechanics calculations and, vice versa, that m can be calculated from the value of T_g obtained experimentally, for instance, from differential scanning calorimetry, through the expression

$$\frac{[(m_{end}/m_{start})p]}{[1 - (1 - (m_{end}/m_{start}))p]} = \frac{(T_g - T_{g0})}{(T_{g,infin} - T_{g0})} \tag{31a}$$

or equally

$$\frac{[(m_{end}/m_{start})\,(m-2)/m]}{[1 - (1 - (m_{end}/m_{start}))\,(m-2)/m]} = \frac{(T_g - T_{g0})}{(T_{g,infin} - T_{g0})} \tag{31b}$$

where p can be expressed as a function of m by substituting in the former of the two equations [Eq. (31a)] the relationship $p = (m-2)/m$ [56], to yield a direct relationship between the number of degrees of freedom and the glass transition temperature. Here T_{g0} is the glass transition temperature of the reactive blend at time 0, thus at a zero degree of conversion; $T_{g,infin}$ is the glass transition temperature for a completely cured polymer, hence extrapolated to maximal theoretical degree of conversion; and T_g is the glass transition temperature corresponding to the degree of conversion p.

A further calculation of interest that was carried out in the same investigation [32] was that of the energy of interaction between primer monomer and substrate when the monomer reactive sites were maintained fixed in the position of minimal energy and were thus not allowed to move with the rest of the molecule (the monomer is thus constrained, as in a tridimensional network, to reproduce the situation in which the photopolymerizable primer network has already formed). The trends in relative flexibility and energies of interfacial interaction obtained between the two primers tried were the same but the relative values of flexibility changed and differed from those obtained previously. First of all, the interfacial energies of adhesion by secondary forces are lower, as would be expected from a covalently cross-linked network as the freedom of the system to reach a conformation of minimal energy is now severely constrained. Second, the conformation of minimal energy obtained is different, as would also be expected. Third, the number of degrees of rotational freedom is maintained unaltered and thus does not correspond linearly to the decrease in energy of interaction; consequently, the numerically absurd result of obtaining greater flexibility in the constrained system is obtained. This is obviously incorrect, and its cause can be traced to the fact that if one talks about degrees of rotational freedom, then what is also meant is the value of the energy barriers to rotation around each single rotatable bond. Calculating back the degrees of freedom left in the constrained case from the energies of interaction and the relative values of flexibility, it became clear that cross-linking the system depressed in both cases the number of degrees of freedom to just about 40% of the number of degrees of freedom of the unconstrained system. This is an important conclusion as it indicates the extent to which the primer system loses flexibility on networking. It is a confirmation that such photopolymerizable primers improve the flexibility of the total finish by presenting a gradient of degree of polymerization as a function of their depth of penetration, being heavily polymerized on the wood surface and decreasing rapidly in degree of polymerization the deeper they have penetrated into the wood.

All the calculations reported up to now were carried out by maintaining blocked the structure of the elementary cellulose I crystallite in its conformation of minimal energy derived from X-ray diffraction data [73–78], refined and minimized to the atomic coordinates and charges of Pizzi and Eaton [12,47–50]. The blocking of the cellulose crystallite surface in a fixed, predetermined conformation of minimal energy is a very acceptable assumption given the energetic stability of the crystallite itself. However, it is also of interest to investigate what influence the application of a primer or of a finish can have on the surface conformation of a cellulose crystallite as forecast by a molecular mechanics method. As the calculations involved are considerable, a simpler algorithm was used [79] for the calculations. Thus the conformational variation of a system composed of two parallel chains of the elementary crystallite of cellulose I and of the primer and of the finish superimposed on the cellulose in their configurations of minimal energy already calculated, in which all the component molecules were allowed to move, was followed. The results obtained were of two types: (1) The stabilization obtained in terms of total energy of the system indicated that the longer the segment that relied on secondary forces for adhesion to the cellulose surface, the better was the stabilization of the system by secondary forces; logically, this result should have been expected. (2) The conformation of the two chains of cellulose changed, but what was unexpected is that they changed to the conformation of the crystallite of cellulose II [46], a different, more stable crystalline morphology! This is an interesting result which implies that treatment of a lignocellulosic surface with a surface finish or other polymers might well alter irreversibly the conformation of the structure of some of the wood constituents on the wood surface. Experimental results based on X-ray diffraction [41,80] and ^{13}C nuclear magnetic resonance (NMR) of a PF resin-cellulose system [80] and of a UF resin-wood system [41] have confirmed the possibility of surface cellulose conformational changes in the presence of a synthetic polymer. It implies that to discuss adhesion in terms of only modification of the conformation of the applied polymer without taking into account the variations induced in the substrate itself by the applied polymer might give a very partial view of the process of adhesion at the molecular level. Previous, generalized but still very acceptable models [22] in which the substrate is taken as homogeneous surface of hard or soft spheres can describe very well cases in which the substrate is constrained in such a way that it cannot modify its configuration but cannot hope to explain well the cases in which the substrate molecule changes its configuration as a consequence of interfacial interactions with another polymer species.

Notwithstanding this, the situation for crystalline cellulose I is only partially one of these cases. In a full-scale crystallite the stabilization given by the minimization of energy that maintains the crystal in its configuration is such that the secondary forces at the interface exerted by another polymer species are too feeble to induce reorganization of the crystallite to another, even if more stable, configuration. However, in amorphous cellulose and in small crystallites, of which there is an abundance in elementary and subelementary cellulose fibrils [81], in which the interfacial secondary forces exerted by the applied polymer can easily be more intense at the localized level than the forces holding the crystal together, the interfacial interactions might well be more than sufficient to induce part of the substrate to overcome the energy barrier and to fall into a different configuration. To conclude, in the case of lignocellulosic substrates any crystallite at the microfibrillar or elementary fibrillar level or of greater dimensions is not likely to be affected, but any amorphous holocellulose and crystallites belonging to subelementary fibrils are likely to change in configuration as a consequence of the interfacial secondary forces of adhesion induced on such parts of the substrate by the presence of a finish or of an adhesive, a far-reaching conclusion.

It is also evident that in the case of the primer and substrate studied in which the primer is highly cross-linked and the cellulose crystallite is a highly crystalline solid, no diffusion mechanisms at the molecular interface are likely. The situation might well be different when one deals with a molecular interface where reorganization of the substrate as a consequence of the interfacial interaction forces induced by the finish is possible (see earlier). This is the case of a nonlinked primer monomer, or even a primer of a low degree of polymerization or cross-linking, on amorphous cellulose or even on a subelementary cellulose crystallite area. In this case the inverse dependence of the peel energy $\mathbf{G}$ of the system on, among others, the inverse of the molecular mass $(1/M)^{2/3}$ applies [52,82], which appears again to be confirmed by the dependence on m and α of the surface finish; these parameters clearly linked to the molecular mass of the finish monomer, and of the segments between cross-linking nodes when the latter exists. That m, n, and α are the key parameters representing the molecular mass was confirmed by previous authors from the determination of the dependence of peel strength on a 2/3 exponent of the molecular mass, clearly pointing to the constraints of rotational degrees of freedom (and rotational energy barriers) found [32] for monomers whose ends are constrained. Furthermore, the apparent proportionality of the diffusion coefficient of the movement of reptation to M^{-2} [52,83–85] appears to be confirmed by the direct proportionality of the interfacial flexibility of the system to the number of degrees of freedom m of the system that has been found. The more flexible the system, hence the greater the number of its degrees of freedom per unit mass (or per atom), the easier is interdiffusion. Again, it is not the molecular mass of the chemical species as such that will determine either the coefficient of diffusion or the relaxation time of reptation but its undoubted relation to the flexibility of the system, which depends on the parameters m, n, α, and $\mathbf{E}$, especially on the per-atom values m/n and $\mathbf{E}/n$.

V. DYNAMIC ADHESION MODELING OF MOLECULARLY WELL-DEFINED SYSTEMS

The molecular mechanics approach described above functions well, but the manner in which it has been used in the previous section is rather limited to static situations; dynamic situations can also be described well, although by a series of finite steps of "before and after" static calculations to ascertain the changes that have occurred or are occurring. Even the most modern molecular mechanics and dynamics programs still work in this manner. It might appear otherwise to a user, but any automatic molecular mechanics or dynamics computational program still works by a sum of small static situation steps, even if infinitesimally small. There is nothing wrong in such an approach, of course, as it has been shown to work rather well even in the most complex situations. However, it is interesting to examine how such an approach works for systems in which movement at the molecular rather than atomic level is inherent in the definition of the system itself.

As regards adhesion in the broader sense of the word, only two series of studies fall in this category, and the most important of them is just on the boundary between adhesion proper and physical chemistry applied to the biological sciences. These two series of studies set out to model the process of chromatography, although in widely different contexts. There is not doubt that chromatography is a clear case of differential, competitive sorption and hence a case of differential, competitive adhesion.

The first record of a study in this direction goes back to 1987 [47–49] and consisted of modeling and giving a solid theoretical base to a little understood, but very much used, chromatography process in the field of biological sciences: the chiral discrimination and separation of amino acid enantiomers by crystalline and amorphous celluloses and by many other polymeric substrates with a structure presenting a helix configuration. This is the model that will be discussed more in depth, as it presents some unusual and very interesting advanced characteristics.

The second model is more recent [9], and it is applied to explaining the achiral (hence a more general case) paper chromatographic separation of three dihydroxydiphenylmethanes (the three PF dimers discussed earlier in this chapter) on crystalline cellulose and checking whether the relative R_f values obtained by experimental chromatography corresponded to the interaction energies calculated at the interface. The algorithms used were the same as those used for the previous approach. The results obtained were excellent, showing not only a trend correspondence between experimental R_f values and calculated energy values but also a very close numerical correspondence with the actual relative values of R_f for the three compounds. One of the most interesting findings was that in the case of the interaction with a substrate of homologous series of chemical compounds the mobile phase, the chromatography solvent or mixture of solvents, could be easily modeled by just varying the dielectric constant used in the model according to the solvent or mix of solvents used. Such a result is of importance because for many ternary systems it spares having to model the third component of the system, namely the solvent or the water present either in the polymer or in the substrate (such as in wood and cellulose).

The work was continued on the modeling of paper chromatography of UF oligomers [86], and in this series of experiments the limits of the molecular mechanics system finally started to become apparent. Although good trend correspondence with experimental R_f values was again obtained within each of the two series of UF oligomers tested, correspondence was lost when one tried to compare the compounds within a series with the compounds of the other series. Thus, excellent correspondence existed within the homologous series urea, methylene-bis-urea and dimethylene-triurea, and within the second homologous series monomethylol urea, N, N'-dimethylol urea, trimethylol urea, monomethylol methylene-bis-urea, and N, N'-dimethylol methylene-bis-urea., but not between the two series. It became evident that in order to compare two nonhomologous series of compounds it would be necessary to model the water in the system as a third species. Thus, the conclusion was that wise use of the dielectric constant to spare modeling a ternary system is very valid, but only when the molecules to be compared belong to a homologous series of compounds. If they do not, one needs to model the solvent as a separate species [86]. There is no doubt that this can be achieved, either by a classical molecular mechanics method or, even better, by modeling the solvent through a molecular dynamics approach in which the solvent layer is modeled as exposed for the generalized system discussed at the beginning of this chapter while the substrate and the polymer are modeled by more classical molecular mechanics approaches. As the UF adhesives chromatography problem did not warrant such an extensive further investigation, the next investigation centered instead on the more difficult ternary systems but for a totally different set of molecules, i.e., the work on varnish-primer-cellulose systems reported earlier in this chapter.

More interesting is the chiral separation model of amino acid enantiomers on cellulose [47–49], particularly the theory that was derived for the interaction, hence for the adhesion between biological monomers and a polymer possessing a helical configuration, of whatever nature. The interaction potential of substrate and enantiomers was treated as a series of achiral and chiral terms of interactions. Thus

$$\mathbf{E}_{tot} = \mathbf{E}_{ach} + \mathbf{E}_{chi} \qquad (32)$$

The achiral terms appeared as a combination of hydrogen-bonding $\mathbf{E}_{H\text{-bond}}$, ionic $\mathbf{E}_i$, dipolar $\mathbf{E}_d$, and induction contributions $\mathbf{E}_{Pl}$ as

$$\mathbf{E}_{ach} = \mathbf{E}_{H\text{-bond}} + \mathbf{E}_i + \mathbf{E}_d + \mathbf{E}_{Pl} \qquad (33)$$

where a simplified nondirectional Lippincott-Schroeder type of potential was used to describe the hydrogen-bonding term, and sophisticated, but conceptually thorough, expressions, which are beyond the scope of this chapter, were used for the other three terms [49]. The two terms E_i and E_d are in reality achirals, although the substrate helix dipole momenta have a chiral structure. The chiral interactions were represented by two potentials, namely a Lennard-Jones term E_{L-J} and an induction term E_{P2}, as

$$E_{chi} = E_{L-J} + E_{P2} \qquad (34)$$

The Lennard-Jones interaction was expanded in a Fourier series, with the first term of the series corresponding to the continuum approximation of the helix. Because of the particular molecular system under examination, the first term of the series provided more than 90% of the total expansion and the remaining terms could be disregarded. The second chiral term corresponds to the polarization of the amino acid molecule by the dipoles of the helix. Discriminatory sorption ratios were presented and the differential contribution and variability of the chiral potential were accurately described. The study, showing excellent agreement with experimental results, concluded that in the dynamic differential interfacial discrimination between enantiomers and a helix the following three types of interaction appeared to have been identified.

1. Hydrogen bond interactions, probably due to the use by the authors of a simplified nondirectional function, are nearly insensitive to the geometry of the interacting systems and are achiral and mainly local. They obviously provide a fundamental contribution to the interaction potential and to the discriminatory ratio as they freeze the position and orientation of the amino acid molecule. They are the strongest of the interactions in the particular system investigated, and this notwithstanding the nondirectional function used (they might be even higher otherwise).

2. The electrostatic interactions, on the contrary, strongly depend on the conformation of the substrate and of the amino acid. They are not generally chiral but they freeze the amino acid orientation and reveal chirality. Their magnitude, although slightly smaller than that of the hydrogen bond, is similar to the magnitude of the third kind of interactions.

3. Interactions of the third kind (Lennard-Jones and induction terms) are chiral and depend crucially on the mutual geometry of the two interacting systems, since they characterize the spatial distribution of atomic matter.

The influence of the solvent was, in the words of the authors, crudely schematized through the formation of a zwitterion (still giving good results), but it could be taken into account by considering a third polarizable species, eventually a polar and ionic medium embedding the amino acid molecule, and thus a ternary system as already described for the previous investigation.

VI. CONCLUSIONS

The brief discussion presented here should give the reader a brief overview of what has already been achieved in the field of molecular mechanics and dynamics to improve adhesion or to explain adhesion phenomena. The approaches presented are not exhaustive of what can be achieved; they are rather a beginning instead.

In the molecular dynamics method, Newton's equations of motion are numerically integrated for all the atoms of the system. This yields not only an atomistic description of the dynamical behavior of a system but also useful thermodynamic information [87]. There are, however, a number of difficulties associated with this approach for polymer systems. One deals here with the simulation of very high molecular weight materials in which complex potentials are needed to represent properly the interactions along and between the polymer chains. As polymer relaxation times still exceed by far those that can be simulated using current computer resources, the choice of a physically reliable starting configuration will also be a determining factor. Another aspect of molecular dynamics simulations in polymers is that the dynamic step is of the order of 10^{-15} s, thus restricting the analysis to short-time-scale properties in spite of long computer runs. This means that, at this stage at least, practical problems can be solved much more rapidly and approached with more ease by using more sophisticated molecular mechanics methods rather than using simple molecular dynamics approaches. As regards molecular dynamics, it is necessary furthermore to point out that attention needs to be paid to the types of programs and algorithms used, as simple, generalized molecular dynamics methods tend to give unacceptable results.

More will surely be achieved in times to come by the application of such tools to adhesion problems both of a fundamental and of a more applied nature. Molecular mechanics and dynamics then present an incredibly powerful tool that should be given due cognizance in the field of adhesion.

REFERENCES

1. P De Santis, E Giglio, AM Liquori, A Ripamonti, Nature 206:406, 1967.
2. DA Brant, PJ Flory. J Am Chem Soc 87:2791, 1965.
3. AI Kitaygorodsky. Tetrahedron 14:230, 1961.

4. DA Rees, RJ Skerrett. Carbohydr Res 7:334, 1968.
5. JC Slater. Quantum Theory of Matter. New York: McGraw-Hill, 1951.
6. A Pizzi. Advanced Wood Adhesives Technology. New York: Marcel Dekker, 1994.
7. A Pizzi. J Adhesion Sci Technol 4:573, 589, 1990.
8. A Pizzi, NJ Eaton. J Adhesion Sci Technol 1:191, 1987.
9. A Pizzi, G de Sousa. Chem Phys 164:203, 1992.
10. A Pizzi, S Maboka. J Adhesion Sci Technol 7:81, 1993.
11. A Damiani, P De Santis, A Pizzi. Nature 226:542, 1970.
12. A Pizzi, NJ Eaton. J Macromol Sci Chem Ed A21:1443, 1984; A22:105, 1985; A22: 139, 1985.
13. NL Allinger, YH Yuh. Operating instructions for MM2 and MMP2 programs, 1978.
14. KL Johnson, K Kendall, AD Roberts. Proc R Soc Lond A324:301, 1971.
15. BV Derjaguin, VM Muller, YP Toporov. J Colloid Interface Sci 53:314, 1975.
16. VM Muller, VS Yushchenko, BV Derjaguin. J Colloid Interface Sci 77:91, 1980.
17. VM Muller, VS Yushchenko, BV Derjaguin. Colloids Surfaces 7:251, 1983.
18. LN Rogers, J Reed. J Phys D Appl Phys 17:677, 1984.
19. J Reed. In: Particles on Surfaces 2: Detection, Adhesion and Removal. (KL Mittal, ed.) New York: Plenum, 1988, pp 3–18.
20. S Wall, W John, SL Goren. In: KL Mittal, ed. Particles on Surfaces 2: Detection, Adhesion and Removal. New York: Plenum, 1988, p 19.
21. KL Johnson, HM Pollock. J Adhesion Sci Technol 8:1323, 1994.
22. R Pucciariello, N Bianchi, R. Fusco. Int. J Adhesion Adhesives, 9:205, 1989.
23. DJ Quesnel, DS Rimai, LP DeMejo. Solid State Commun 85:171 1993.
24. DJ Quesnel, DS Rimai, LP DeMejo. Phys Rev B 48:6795 1993.
25. DJ Quesnel, DS Rimai, LP DeMejo. In: DS Rimai, LP DeMejo, KL Mittal, eds. Fundamentals of Adhesion and Interfaces. Utrecht: VSP, 1995, p 281.
26. HA Mizes, K-G Loh, RJD Miller, SK Ahuja, EF Grabowski, Appl Phys Lett 59: 2901, 1991.
27. DM Schaeffer, M Carpenter, R Reinfenberger, LP DeMejo, DS Rimai. J Adhesion Sci Technol 8:197, 1994.
28. DM Schaeffer, M Carpenter, B Gady, R Reinfenberger, LP DeMejo, DS Rimai. J Adhesion Sci Technol 9:1049, 1995.
29. A Pizzi, NJ Eaton, M Bariska. Wood Sci Technol 21:235, 1987; 21:317, 1987.
30. A Pizzi, NJ Eaton. J Macromol Sci Chem Ed A24:1065, 1987.
31. A Pizzi. Holzforschung 44:373, 1990; 44:419, 1990.
32. A Pizzi, F Probst, X Deglise. J Adhesion Sci Technol 11:573, 1997.
33. F Probst, M -P Laborie, A Pizzi, A Merlin, X Deglise. Holzforschung 51:459, 1997.
34. R Smit, A Pizzi, C Schutte, SO Paul. J Macromol Sci Chem Ed A26:825, 1989.
35. A Pizzi, F-A Cameron. Holz Roh Werkstoff 39:463, 1987.
36. DA Fraser, RW Hall, ALJ Raum. J Appl Chem 7:676, 1957: 689, 1957.
37. DA Fraser, RW Hall, PA Jenkins, ALJ Raum. J Appl Chem 7:701, 1957.
38. A Pizzi. J Polym Sci Polym Lett 17:489, 1979; J Appl Polym Sci 24:1247, 1979; 24:1257, 1979.
39. A Pizzi, P van der Spuy. J Polym Sci Chem Ed. 18:3447, 1980.
40. A Pizzi, In: A Pizzi, ed. Wood Adhesives Chemistry and Technology, Vol 1. New York: Marcel Dekker, 1983, chap 3.

41. D Levendis, A Pizzi, EE Ferg. Holzforschung 46:260, 1992.
42. PR Steiner. Forest Prod J 23:32, 1973.
43. M Inoue, M Kawai. Res Rep Nagoya Munic Ind Res Inst 12:73, 1954.
44. JI de Jong, J de Jonge, HAK Eden. Trav Chim Pays Bas 72:88, 1953; 73:118, 1954.
45. A Pizzi. Holzforsch Holzverwert 41:31, 1989.
46. A Pizzi, NJ Eaton. J Macromol Sci Chem Ed A24:901, 1987.
47. E Alvira, L Vega, C Girardet. Chem Phys 118:233, 1987.
48. E Alvira, V Delgado, J Plata, C Girardet. Chem Phys 143:395, 1990.
49. E Alvira, J Breton, J Plata, C Girardet. Chem Phys 155:7, 1991.
50. RM Wolf, E Francotte, L Glasser, I Simon, HA Scheraga. Macromolecules 25:709, 1992.
51. Wood Handbook: Wood as an Engineering Material. Agriculture Handbook 72. Madison: USDA Forest Service, 1987.
52. J Schultz, M Nardin. In: A Pizzi, KL Mittal, eds. Handbook of Adhesive Technology. New York: Marcel Dekker, 1994.
53. French Norm NF T 30-038 on peel adhesion strength.
54. F Probst, A Pizzi. Optimisation d'un système de finition par la méchanique moléculaire. ENSTIB report, Epinal, France, 1996, pp 1–128.
55. PJ Flory. Principles of Polymer Chemistry. Ithaca, NY: Cornell University Press, 1953.
56. A Pizzi. J Appl Polym Sci 63:603, 1843, 1997.
57. WH Carothers. Collected Papers. New York: Interscience, 1940.
58. PJ Flory. J Am Chem Soc 61:3334, 1939; 62:2261, 1940; 63:3083, 1941.
59. AN Gent, J Schultz. Proc 162nd ACS Meet 31(2):113, 1971.
60. AN Gent, J Schultz. J Adhesion 3:281, 1972.
61. EH Andrews, AJ Kinloch. Proc R Soc Lond A332:385, 401, 1973.
62. AJ Kinloch. Adhesion and Adhesives. London: Chapman & Hall, 1987.
63. D Maugis. J Mater Sci 20:3041, 1985.
64. A Carré, J Schultz. J Adhesion 17:135, 1984.
65. O Kramer. In: K Dusek, SI Kuchanov, eds. Polymer Networks '91. Utrecht: VSP, 1992.
66. PJ Flory. Proc R Soc Lond Ser A: 351, 1976.
67. HR Brown. Macromolecules 24:2752, 1991.
68. AN Gent, A Ahagon. J Polym Sci Polym Phys Ed 13:1285, 1975.
69. L Podgorski, A Merlin, X Deglise. Holzforschung 50:282, 1996.
70. F Rodriguez. Principles of Polymer Systems, 3rd ed. New York: Hemisphere, 1989.
71. JP Pascault, RJ Williams. J Polym Sci Phys Ed 28:85, 1990.
72. PR Couchman. Macromolecules 20:1712, 1987.
73. KH Meyer, H Mark. Berichte 61:593, 1928.
74. KH Meyer, L Misch. Helv Chim Acta 20:232, 1937.
75. KH Gardner, J Blackwell. Biopolymers 13:1975, 1974.
76. W Claffey, J Blackwell. Biopolymers 15:1903, 1976.
77. SSC Chu, GA Jeffrey. Acta Crystallogr B24:830, 1968.
78. JT Ham, DG Williams. Acta Crystallogr B26:1373, 1970.
79. Alchemy II. Tripos Associates USA 1990.
80. S So, JW Teh, A Rudin, WJ Tchir, CA Fyfe. J Appl Polym Sci 39:531, 1990.

81. RB Hanna, WA Coté Jr. Cytobiology 10:102, 1974.
82. RM Vasenin. In: Adhesion: Fundamentals and Practice, London: MacLaren, 1969, p 29.
83. PG de Gennes. J Chem Phys 55:572, 1971.
84. M Doi, SF Edwards. J Chem Phys Faraday Trans 74:1789, 1802, 1818, 1978.
85. WW Graessley. Adv Polym Sci 47:76, 1982.
86. A Pizzi. Unpublished results, 1992.
87. MP Allen, DJ Tildesley, Computer Simulation of Liquids, Oxford: Clarendon Press, 1987.

4

Application of Atomic Force Microscopy in Fundamental Adhesion Studies

Robert A. Hayes and John Ralston
*University of South Australia, Mawson Lakes, South Australia,
Australia*

I. INTRODUCTION

Adhesion arising from the connection of two materials via a film of a third material (an adhesive) is not at issue here. Instead, in this chapter the focus is on "fundamental" adhesion between two materials, or surfaces, arising from interatomic or intermolecular forces [1,2]. These forces essentially depend on the chemistry and physics of the two surfaces, typically solid, and the intervening medium, typically air or liquid.

In some applications adhesion is desirable and therefore enhancement of the adhesion is the aim. Examples are in the adhesion of thin films and coatings to solid surfaces. In water treatment, positively charged magnetite particles are used to remove naturally occurring colloidal material that is negatively charged [3]. In the flotation separation of minerals, the selective adhesion of valuable minerals to air bubbles is required [4]. In other cases adhesion is unfavorable and particle removal is critical to process quality control. Wafer cleaning and lens cleaning are two important examples in the semiconductor [5] and optical industries, respectively.

A number of theories have been proposed to describe the nature of the operative "surface" forces [6,7] and the consequent adhesion [8,9]. It is not the purpose of this chapter to regurgitate this theoretical development, except to note that these theories were developed for the interaction of physically ideal (i.e., smooth) surfaces and can therefore not be strictly (i.e., quantitatively) applied to commonly encountered surfaces. Instead, we focus on the methodology associated with the recent application of atomic force microscopy to adhesion measurement, review the work to date, discuss the practical implications, and make some suggestions for future experimental and theoretical studies.

II. AFM METHODOLOGY

The atomic force microscope (AFM) [10] is a derivative of the scanning tunneling microscope [11], also aimed at providing high-resolution (in the ideal case atomic) topographical analysis, but applicable to both conducting and nonconducting surfaces. The basic imaging principle is very simple: a sample attached to a piezoelectric positioner is rastered beneath a sharp tip attached to a sensitive cantilever spring. Undulations in the surface lead to deflection of the spring, which is monitored optically. The images obtained are critically dependent on the pressure, or force acting or applied, between tip and sample. Furthermore, optimal imaging conditions are inevitably sample dependent. It is essentially for these reasons that quantifying tip-sample interaction forces is embedded in the probe microscopy methodology, which has resulted in its application to surface force and adhesion measurements. In the optimization of the tip-sample interaction it is customary to *disable* the in-plane (x, y) motion of the sample (Fig. 1) and focus solely on sample motion normal to the sample surface (z). An example of the raw data obtained is reproduced in Fig. 2 as the piezoscanner cycles, resulting in sequential tip-sample separation and contact. The (effectively constant) speed of the piezoscanner can typically be varied over a four-decade range. With appropriate calibration (Sec. II.A and II.B) of the spring constant and piezoscanner movement one can routinely quantify both the separation dependence of the net surface force on tip-sample approach and the adhesion (pull-off force) on removal. In this respect there are some obvious parallels to be drawn with the direct force measurement device of Israelachvili and Adams [12]. Measurements with atomically smooth mica with the surface forces apparatus (SFA) have resulted in the confirmation of interaction and adhesion theories and have also led to the elucidation of a number of additional forces of varying range, magnitude, and sign that affect the fundamental interaction between surfaces in liquid media. Bearing in mind that the aim of direct force studies is to model practical interaction and/or adhesion under chemically realistic conditions, a distinct advantage of the AFM force technique is that the particles of interest can be directly attached to the end of the cantilever (adjacent to the tip) and the interaction with the sample can be studied. This refinement was pioneered by Ducker et al. [13] and Butt [14] and is referred to as *colloid probe microscopy*. The geometry of the colloid

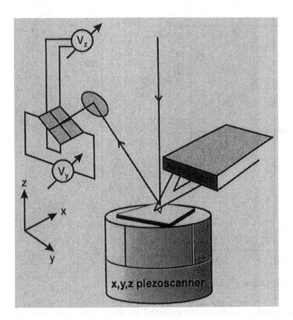

Figure 1 Schematic diagram of an AFM. The sample is placed on the piezoelectric scanner. A laser is reflected off the upper side of the cantilever and into a split photodiode via a mirror. In this way vertical (z) and horizontal (y) deflection signals can be measured. For colloid probe microscopy a well-defined particle is glued to the tip of the cantilever.

(typically spherical) is significantly better defined than a typical AFM tip* [15]. Inevitably, however, the surface of the colloid probe is not as physically ideal as a mica surface, which means that quantitative agreement between interaction theories and experiment should not be expected. The use of standard fluid cells allows the measurement of surface forces and adhesion in controlled native environments, whether liquid or gas.

A. Calibration of Spring Constants for Normal (z) Deflection

A typical AFM cantilever has a Hooke's law response for typical deflections encountered in AFM imaging or force studies. In the latter case, where quantita-

---

* AFM tips do not reduce to a single well-defined atom asperity. Electron micrographs show that silicon nitride tips have a typical radius of curvature exceeding 100 nm. While there is continuing improvement in tip fabrication technologies, these deficiencies have implications for both the spatial resolution while imaging (except in the case of atomically flat samples) and the quantification of tip-sample adhesion forces (see later).

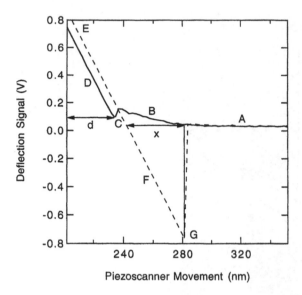

Figure 2 A typical deflection signal/z-piezotrace measured in an aqueous electrolyte. When the sphere and flat are well separated (>100 nm) there is no measurable interaction (A). As the flat approaches the sphere we observe an initial (B) electrostatic repulsion (upward deflection) followed by a jump into contact (C). Thereafter the flat and sphere are coupled (D). At E the flat is retracted. Hysteresis (F) is observed due to the adhesion between the two surfaces. At G the removal force exceeds the adhesion and the surfaces separate. In adhesion studies the key parameters are the pull-off force (F_p) and the loading force, which may be readily calculated. The x (measured along the zero force-deflection line) is the piezoelectric movement required to break sphere-flat contact and d is the piezoscanner movement after contact. (From Ref. 59.)

tive determination of the force is the aim, a routine methodology for calibration is required. In principle, the normal (out of the cantilever plane) deflection can be calculated from a knowledge of the cantilever's geometrical and material properties [16]. In early work the nominal spring constants supplied with the cantilevers were used [13,17]. However, these properties can vary between cantilevers,* requiring cantilevers to be individually calibrated. Most methods involve the attachment of calibration weights (typically tungsten spheres† to the end of the

---

* AFM cantilevers, whether triangular or beam, are typically 100 μm long and attached to the ends of a substrate (2 × 4 mm) to aid handling and mounting. Individual substrates may be readily detached from a semiconductor wafer.

† These will typically adhere to the upper side of the cantilever due to capillary forces.

cantilever). Thus the early calibration attempts involved measuring the variation in cantilever deflection when the AFM head, with the cantilever substate securely mounted inside, was inverted [18]. The advent of imaging modes involving high-frequency resonance of the AFM cantilever has led to the development of a *resonant frequency* technique for calibration [19]. In this case the resonant frequencies of the unloaded and loaded cantilevers are determined and the spring constant, *k*, can be obtained using the expression

$$k = (2\,\pi)^2 \frac{m}{(1/v_1^2) - (1/v_0^2)} \tag{1}$$

where *m* is the mass added and v_1, v_0 are the resonance frequencies of the loaded and unloaded cantilever, respectively. This method, because of its convenience and nondestructive nature, can be routinely employed before and after a series of force measurements. The major uncertainty results from the placement of the calibration sphere. Ideally, the position of the calibration mass along the cantilever should coincide with that of the tip, or colloid probe. A question has also been raised regarding the preferability of using the more gentle thermal vibration of the cantilever rather than forced vibration for calibration purposes [20].

B. Calibration of Piezoscanner

The calibration of the z-piezoscanner movement is also required for the quantitative calculation of the surface and adhesion forces. Scanning of calibration samples with well-defined features is one method that has been used. Inevitably, however, samples are more suited to the calibration of in-plane (*x*, *y*) than out-of plane (*z*) motion. Nevertheless, in early studies [13,17] a compact disc stamper, which has steps of known height, was used. An additional problem relates to how frequently the piezoscanner requires recalibration. Experience has shown that regular calibration (daily) is preferable because of the sensitivity of piezocrystals to humidity and other factors. An interferometric method of calibration, proposed by Jaschke and Butt [21], overcomes both these problems. In this method the laser beam used to measure deflection is positioned so that the beam is split between the tip of the cantilever and a mirror (e.g., a polished silicon wafer) attached to the piezoscanner. The mirror is positioned at an angle ($\sim 10°$ from the horizontal) approximately parallel to the plane of the cantilever. As the mirror is then cycled near the cantilever tip, the plot of deflection signal versus z-piezoscanner movement is a standing wave corresponding to constructive and destructive optical interference. The separation between peaks will correspond to a whole number of half-wavelengths (335 nm for a diode laser). This method of calibration is sufficiently convenient to allow the daily calibration of the z-piezoscanner. If forces are being measured in an aqueous electrolyte, then there is an important and comforting additional check on the calibration procedure. The electrical double layer interaction on approach has a characteristic decay length that can be calculated from the electrolyte concentration [2]. It is our general experience that with usual experimental care these two procedures are inevitably concordant.

C. Loading Force

An additional advantage of the AFM technique is that the loading force prior to separation is readily obtained and varied during data acquisition. The loading force can be calculated (Fig. 2) simply by multiplying the distance scanned by the z-piezoscanner after tip-sample contact by the spring constant. To vary the loading force, the starting point of the scan and the scan length can be simply adjusted. This facility has been particularly useful as the adhesion has been found to depend systematically on the loading force (Sect. III). In addition, the effect on adhesion of steadily increasing loading compared with decreasing loading, *adhesion hysteresis*, can be studied.

D. Colloid Probe Interactions

Colloid probes are typically attached to the end of AFM cantilevers by micromanipulation using an optical microscope. Colloids are glued, using either a chemically inert thermosetting resin or epoxy, or using a fiber of appropriate size and material properties (etched tungsten is ideal [22]). Smooth, spherical colloids are preferable for quantitative measurements. Particles of less well-defined geometry may also be used, but a greater degree of scatter in the measured force data can be expected. In early work the lower limit to the size of the colloid probes was dictated by the height of the AFM tip (4 μm). Particles of appropriate size and composition can be either obtained commercially or prepared in suitably small quantities. In the latter case the recipes of Matijevic and coworkers [23–32] are a good starting point. Preparation usually results in better-defined probes of higher purity, thus rewarding the additional effort. The availability of "tipless" cantilevers removed the size restriction. In recent work the attachment methodology has been refined and colloids less than 1 μm in diameter have been attached and their interactions studied [22]. Given that many dispersions are composed of submicrometer particles, this development aids the general diagnostic capability of the AFM technique. The majority of AFM force studies have focused on the interaction between a sphere and a flat surface. However, the interaction between two spheres is also readily accommodated by the technique [33–35]. Spheres can be attached to a wafer/piezoscanner either individually or by scattering a large number in a thin film of glue. In this case the concentric location of the interacting spheres is critical and can be readily achieved by utilizing the

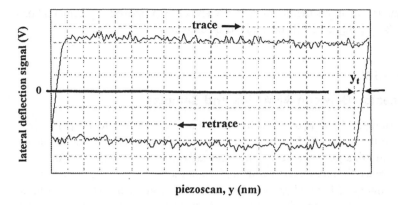

Figure 3 A typical lateral deflection signal trace measured as a function of the cyclic y-piezoscanner travel (500 nm at 5.1 μm/s) in an aqueous electrolyte. Initially there is no relative movement between the colloid probe and sample until the lateral force exceeds the static friction. The magnitude of this force can be readily calculated from the interaction geometry if the lateral spring constant, k_l, is known. (From Ref. 38.)

AFM imaging facility prior to interaction measurement. The major difficulty with the sphere-sphere interaction is in ensuring a gentle initial engagement between spheres to avoid detachment [34].

E. Frictional Forces

The advent of lateral force microscopy (LFM) has also raised the possibility of quantifying shear or frictional forces between a tip, or colloid, and a sample [36–38]. In lateral force mode the fast piezoscan direction is across the apex of the cantilever so as to maximize the cantilever twist and therefore sensitivity.* The lateral deflection of the cantilever is monitored by the horizontal photodiodes (Fig. 1). A typical trace of lateral deflection signal versus y-piezomovement is reproduced in Fig. 3. The lateral force required to initiate movement between the tip/colloid and sample may be calculated by combining the measured displacement with a lateral spring constant, k_l. The methodology for the determination of k_l is not at this stage as well established as that for the normal spring constant, k_n. Neumeister and Ducker [39] have described an elegant method for the calculation of k_l by combining an experimentally determined k_n with the more readily measured geometrical parameters for a typical triangular cantilever. How-

---

* For standard imaging the fast scan direction is parallel to the cantilever apex so as to minimize cantilever twist.

ever, comparison of this methodology with a recently proposed experimental procedure [38] highlights a still unresolved discrepancy.

III. REVIEW

A. Adhesion Between AFM Tips and Samples

In early studies, AFM tips, typically tungsten, were shown to be particularly sensitive to the nature of the sample surface [40]. High-energy surfaces such as mica and alumina gave rise to much larger adhesion forces than did poly(tetrafluoroethylene). Surfaces coated with surfactant monolayers that differed only in the functionality of the end group, CH_3 versus CF_3, yielded distinctly different adhesion forces and highlighted the sensitivity of the technique (Fig. 4). AFM tip cantilever assemblies are now prepared using standard microfabrication techniques in a range of geometries. In recent times the modification of these tips has enhanced the diagnostic ability of force techniques. Lieber and coworkers [41–43] have shown that AFM tips modified with thiols of various functionalities (e.g., COOH, CH_3, NH_2) can be used to probe a range of functionalized surfaces (Fig. 5). This approach has been termed *chemical force microscopy* (CFM). The adhesion and frictional characteristics of surfaces have been probed with these chemically modified tips. Eastman and Zhu [44] have also investigated the effect of tip modification. In this case the difference in interaction between low- and high-energy tips and mica in air can be accounted for in terms of van der Waals and capillary forces. In general, the forces measured using CFM can be highly scattered, presumably due to the variability in tip geometry on an atomic or molecular scale. The uncertainty in the geometry also makes it difficult for the forces measured to be quantitatively related to the operative surface forces. However, at the very least the technique provides a qualitative probe of surface chemistry, with high spatial sensitivity, which can only be aided by the continuing rapid development of AFM tips. Another approach to the characterization of tip-sample adhesion has been proposed by Beebe and coworkers [45] with statistical analysis of adhesion values being employed to obtain a single-bond force. The interaction between silicon nitride AFM tips and gold or mica in water has been used to demonstrate the application of this elegant methodology.

B. Adhesion Involving Colloid Probes

The variability in tip geometry can be overcome by attachment of a sphere to the end of an AFM cantilever, adjacent to the tip. The forces between the sphere, typically of colloidal dimensions, and the sample surface are then measured in the normal manner. This technique, known as *colloid probe microscopy* and pioneered by Ducker and coworkers [13,17] and Butt [14], has been widely adopted.

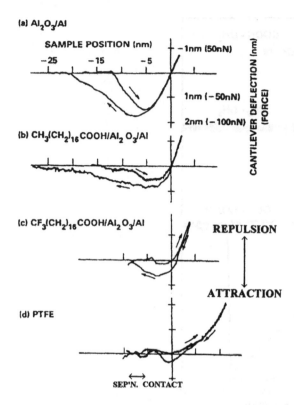

Figure 4 Representative data obtained with the AFM showing surface-force interaction in dry nitrogen between a tungsten tip and (a) the native oxide of aluminum (Al_2O_3/Al), (b) stearic acid [$CH_3(CH_2)_{16}$ COOH], (c)ω,ω,ω-trifluorostearic acid [$CF_3(CH_2)_{16}COOH$], and PTFE. The x axis is the sample position in nanometers; the y axis is cantilever deflection in nanometers relative to its rest position, or the force in nanonewtons between the cantilever and the sample. The scales for all the curves are the same. (From Ref. 40.)

It can readily be used to study the interactions between particles of applied or industrial interest in their native environment. Some of these are described in the following.

1. Interactions in Air

Most activity in this area has been driven by the interest in xerographic printing and semiconductor wafer cleaning. In the former case, it is desirable to measure the force required for detachment of toner particles. In practice, particle removal is facilitated by the application of an electric field. Toner particles, which are

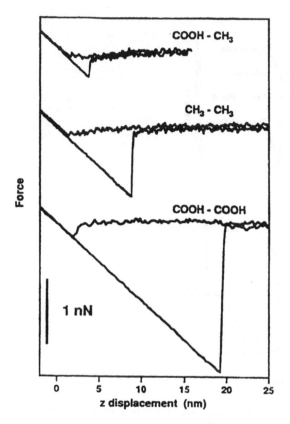

Figure 5 Representative force versus displacement curves recorded for COOH-COOH, CH_3-CH_3, and CH_3-COOH tip-sample functionalization using sharp tips ($R \sim 60$ nm). All data were obtained in EtOH solution using a fluid cell. (From Ref. 42.)

approximately 10 μm in size, have been attached to AFM cantilevers, and their interaction with a substrate has been measured in air [46–50]. The effects of particulate additives and humidity (Fig. 6), roughness (Fig. 7), and loading force (Fig. 8) on the measured adhesion were investigated. The adhesion was found to increase with loading force. Adhesion ''maps'' were also obtained with a spatial resolution of 6 nm and comparison of these maps with the topography showed that the magnitude of the adhesion correlated with local curvature. Modification of the AFM software also allowed the effect of contact time to be studied prior to separation. However, no effect of contact time on adhesion was observed for times in the range of a few milliseconds up to an hour.

The interactions of polystyrene spheres with silicon wafers in air have also

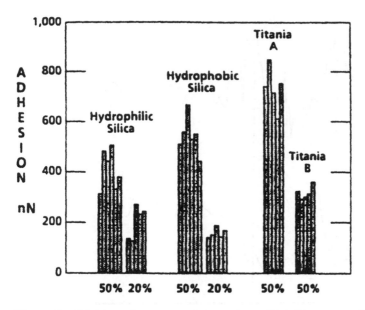

Figure 6   Adhesions of toners to a photoconductor in air. Toners coated with hydrophilic silica are measured at 50% and 20% relative humidity (RH). Toners coated with hydrophobic silica are measured at 50% and 20% RH. Toners coated with titania A (hydrophilic) and titania B (hydrophobic) are measured at 50% RH. (From Ref. 46.)

been investigated [51] (Fig. 9). The removal forces measured were about a factor of 50 times less than theoretical predictions. This difference was ascribed to the presence of microscopic asperities on the interacting surfaces.

2.   Interactions in Liquids

The vast majority of the colloid probe literature focuses on the approach data obtained between surfaces in liquids, usually an aqueous electrolyte. After conversion of the raw data into normalized force as a function of separation, a comparison is typically made with theoretical predictions [6,7]. The separation data, which yield the adhesion force or energy, have received much less attention and are usually either not reported or only described briefly [52–54]. In the latter case it was often noted that the adhesion values were highly variable [17] and that the nature of the AFM cantilever was inappropriate for the measurement of adhesion forces because it could not prevent rolling of the surfaces on separation [55]. As theory [8,9,56–58] predicts that the adhesion, or pull-off, force is single valued, there is also perhaps an expectation that experimental determinations should be single valued. A study of the interaction between iron oxide and silica

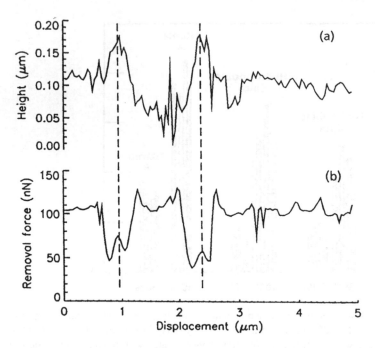

Figure 7 Cross section through a groove on a molecularly doped polycarbonate surface. The topography (a) and pull-off force (b) in air were extracted simultaneously from a series of loading curves similar to those in Fig. 3. The dashed lines are a guide to the eye and demonstrate that bumps on the surface give rise to lower adhesion. (From Ref. 47.)

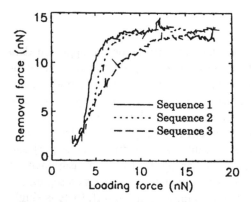

Figure 8 Variation of adhesion with loading force for contacts between a toner particle and a gold surface in air. (From Ref. 47.)

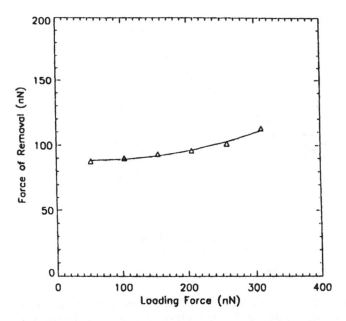

Figure 9   The force of removal for a 5-μm polystyrene sphere as a function of the loading force. By extrapolation to zero load, the force of removal for this 5-μm-diameter sphere from a silicon substrate can be estimated to be 90 nN. Each data point represents an average of three independent measurements of the removal force. (From Ref. 51.)

in dilute electrolyte where adhesion is the major focus has been reported [59]. In this case the adhesion is essentially Coulombic between the isoelectric points of the respective surfaces. The pull-off force was found to be highly variable, but a strong linear correlation with the loading force, applied prior to separation, was found (Fig. 10). This correlation was subsequently found to be general to interactions involving a wide range of iron oxide colloid–silica surface contacts. The load dependence of the pull-off force was ascribed to the presence of a deformable surface layer between the interacting surfaces. The ease with which the loading force can be varied allowed loading-unloading cycles to be routinely obtained. When the magnitude of the pull-off force was converted to an adhesion energy using the Derjaguin approximation [2] and the macroscopic radius of curvature of the iron oxide colloid (2.2 μm), the value obtained was two orders of magnitude less than the van der Waals adhesion, without consideration of the dominant electrical component. This difference was explained by the physical nonideality of the component surfaces, the topography of which was probed by AFM in imaging mode. If one calculates the adhesion energy using the radius

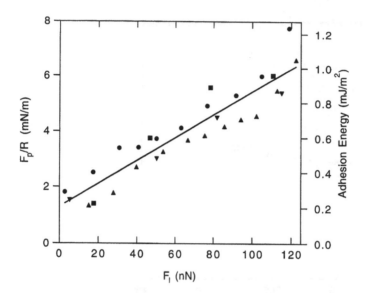

Figure 10   The reproducibility of the pull-off force (F_p/R)–loading force (F_l) relationship in water is demonstrated by the use of different iron oxide spheres and locations on the silica surface of the silicon wafer. Different symbols correspond to different sphere–contact point combinations. (From Ref. 59.)

of curvature of the microscopic asperities (approximately 100 nm), experiment and theory are essentially reconciled. The observation of attenuation of removal forces due to the presence of surface asperities is not confined to AFM studies [51]. Centrifugal measurements [60] have led to similar conclusions, with asperities leading to a dramatic reduction in contact areas.

The adhesion, or more correctly cohesion, between like spheres has also been measured with the colloid-probe technique [34]. The pull-off force obtained for ZnS spheres interacting in an aqueous electrolyte with varying pH and concentration of Cu^{2+} was compared with corresponding electrokinetic and rheological measurements. The pull-off force, measured after loading to a constant value of 100 nN, was again found to be dramatically reduced due to the roughness of component surfaces.

Interactions in nonaqueous liquids can also be studied. Mulvaney and coworkers [61] have examined the interaction between gold spheres and poly(tetrafluoroethylene) (PTFE) in a range of hydrocarbon, fluorocarbon and organohalogen liquids. The measured interactions were attributed to van der Waals forces. Smooth PTFE surfaces were prepared by compression between mica sheets at elevated temperature. The roughness of the gold-coated tungsten spheres was not reported.

Most AFM studies have focused on the interactions between two solid, nondeformable, surfaces. However the methodology has also been applied to interactions, including adhesion, involving a deformable surface, either liquid or gas. As there is no independent measure of interfacial separation with the AFM technique when deformation occurs, it is the adhesion measurements rather than the approach data that are more quantitative. Studies involving a deformable interface have focused on interactions between particles and either air bubbles [4,62,63] or oil droplets [64–66].

IV. FUTURE DIRECTIONS

The application of atomic force microscopy to surface force measurement has proceeded at a rapid pace similar to that of developments in the general probe microscopy field. However, if the number of publications is used as a guide, its use in adhesion studies has been relatively limited. In this chapter we have attempted to demonstrate the potential of the technique in fundamental adhesion studies. At the same time, because commercial AFMs are designed as imaging tools and not for surface force or adhesion measurement, there is some room for design improvement. For adhesion measurements in particular the study of single contacts of varying duration is more appropriate than the multitude of short-duration contacts currently probed. Mizes et al. [47] simply utilized software modification to allow this facility. Hlady et al. [67] have instead used a programmable signal generator to drive the z-piezoscanner with similar effect. Adhesion forces are typically found to be significantly greater in magnitude than approach forces. As a result, cantilever deflections are greater and linear position-sensitive detectors have been employed to capture the amplified signal [68].

AFM studies have confirmed that surface roughness dramatically affects the magnitude of surface and adhesion forces. Therefore interaction theories, although qualitatively useful, have limited use for the quantification of the operative forces. Clearly, the development of interaction theories that explicitly accommodate surface topography is required. Progress in this area has been limited to the work of Czarnecki et al. [69,70] that focuses on the attenuation of the van der Waals attraction and more recently the modified DLVO theory of Suresh and Walz [71]. In the latter case the motivation was explanation of interactions involving latex particles in an aqueous electrolyte, which were modeled as hemispheres on an otherwise smooth surface. Further theoretical development and critical comparison with force measurements are required. The use of the AFM in imaging mode, to provide actual topographical parameters, can facilitate this development.

ACKNOWLEDGMENTS

We are grateful to Dmitri Vezenov for making a preprint of Ref. 43 available.

REFERENCES

1. RG Horn. In: ASM Handbook, Volume 18: Friction, Lubrication and Wear Technology. ASM International, 1992, p 399.
2. JN Israelachvili. Intermolecular and Surface Forces. 2nd ed. London: Academic Press, 1992.
3. DR Dixon. Chem Aust 59:394, 1992.
4. ML Fielden, RA Hayes, J Ralston. Langmuir 12:3721, 1996.
5. I Kashkoush, A Busnaina, F Kern, Jr, R Kunesh. In: KL Mittal, ed. Particles on Surfaces 3: Detection. Adhesion and Removal. New York: Plenum Press, 1991, p 217.
6. BV Derjaguin, L Landau. Acta Physicochim 14:633, 1941.
7. EJW Verwey, J Th. G Overbeek. Theory of Stability of Lyophobic Colloids. Amsterdam: Elsevier, 1948.
8. KL Johnson, K Kendall, AD Roberts. Proc R Soc Lond Ser A 324:301, 1971.
9. RG Horn, JN Israelachvili, F Pribac. J Colloid Interface Sci 115:480, 1987.
10. G Binnig, C Quate, G Gerber. Phys Rev Lett 56:930, 1986.
11. G Binnig, H Rohrer. Helv Phys Acta, 55:726, 1982.
12. JN Israelachvili, GE Adams. J Chem Soc Faraday Trans, I, 74:975, 1978.
13. WA Ducker, TJ Senden, RM Pashley. Nature, 353:239, 1991.
14. H-J Butt. Biophys J 60:1438, 1991.
15. CJ Drummond, TJ Senden. Colloids Surfaces A 87:217, 1994.
16. JE Sader, LR White. J Appl Phys 74:1, 1993.
17. WA Ducker, TJ Senden, RM Pashley. Langmuir 8:1831, 1992.
18. TJ Senden, WA Ducker. Langmuir 10:1003, 1994.
19. JP Cleveland, S Manne, D Bocek, PK Hansma. Rev Sci Instrum 64:403, 1993.
20. P Mulvaney. Personal communication.
21. M Jaschke, H-J Butt. Rev Sci Instrum 66:1258, 1995.
22. G Toikka, RA Hayes. J Colloid Interface Sci 191:102, 1997.
23. LL Springsteen, E Matijevic. Colloid Polym Sci 267:1007, 1989.
24. S Gangolli, RE Partch, E Matijevic. Colloids Surfaces 41:339, 1989.
25. E Matijevic. J Mater Educ 10:177, 1988.
26. R Sprycha, E Matijevic. Colloids Surfaces 47:195, 1990.
27. B Aiken, WP Hsu E Matijevic. J Mater Sci 25:1886, 1990.
28. F Porta, WP Hsu, E Matijevic. Colloids Surfaces 46:63, 1990.

29. E Matijevic, Q Zhong, RE Partch. Aerosol Sci Technol 22:162, 1995.
30. DM Wilhelmy, E Matijevic. J Chem Soc Faraday Trans I 80:563, 1984.
31. B Aiken, WP Hsu, E Matijevic. J Am Ceram Soc 71:845, 1988.
32. E Matijevic, P Scheiner. J Colloid Interface Sci 63:509, 1978.
33. I Larson, CJ Drummond, DYC Chan, F Grieser. J Phys Chem 99:2114, 1995.
34. TH Muster, G Toikka, RA Hayes, CA Prestidge, J Ralston. Colloids Surfaces A 106:203, 1996.
35. G Toikka, RA Hayes, J Ralston. Langmuir 12:3783, 1996.
36. Y Liu, T Wu, DF Evans. Langmuir 10:2241, 1994.
37. E Liu, B Blanpain, JP Celis. Wear 192:141, 1996.
38. G Toikka, RA Hayes, J Ralston. J Adhesion Sci Technol 11:1479, 1997.
39. JM Neumeister, WA Ducker. Rev Sci Instrum 65:2527, 1994.
40. NA Burnham, DD Dominguez, RL Mowery, RJ Colton. Phys Rev Lett 64:1931, 1990.
41. CD Frisbie, LF Rozsnyai, A Noy, MS Wrighton, CM Lieber. Science 265:2071, 1994.
42. A Noy, CD Frisbie, LF Rozsnyai, MS Wrighton, CM Lieber. J Am Chem Soc 117:7943, 1995.
43. A Noy, DV Vezenov, CM Lieber. Annu Rev Mater Sci 27:381, 1997.
44. T Eastman, D-M. Zhu. Langmuir 12:2859, 1996.
45. JM Williams, T Han, TP Beebe Jr. Langmuir 12:1291, 1996.
46. ML Ott. Proceedings of 19th Annual Adhesion Society Meeting, 1996, p 70.
47. H Mizes, K-G Loh, ML Ott, RJD Miller. In: KL Mittal, ed. Particles on Surfaces Detection, Adhesion and Removal. New York: Marcel Dekker, 1995, p 47.
48. ML Ott, HA Mizes. Colloids Surfaces A 87:245, 1994.
49. HA Mizes, KG Loh, RJD Miller, SK Ahuja, EF Grabowski. Appl Phys Lett 59:2901, 1991.
50. HA Mizes. J Adhesion 51:155, 1995.
51. DM Schaefer, M Carpenter, R Reifenberger, LP DeMejo, DS Rimai. J Adhesion Sci Technol 8:197, 1994.
52. L Meagher, VSJ Craig. Langmuir 10:2736, 1994.
53. YI Rabinovich, R-H Yoon. Langmuir 10:1903, 1994.
54. YI Rabinovich, R-H Yoon. Colloids Surfaces A 93:263, 1994.
55. L Meagher. J Colloid Interface Sci 152:293, 1992.
56. BV Derjaguin, VM Muller, YP Toporov. J Colloid Interface Sci 53:314, 1975.
57. VM Muller, VS Yuschenko, BV Derjaguin. J Colloid Interface Sci 77:91, 1980.
58. BD Hughes, LR White. Q J Mech Appl Math 32:445, 1979.
59. G Toikka, RA Hayes, J Ralston. J Colloid Interface Sci 180:329, 1996.
60. F Podczeck, JM Newton, MB James. J Appl Phys 79:1458, 1996.
61. A Milling, P Mulvaney, I Larson. J Colloid Interface Sci 180:460, 1996.
62. H-J Butt. J Colloid Interface Sci 166:109, 1994.
63. WA Ducker, Z Xu, JN Israelachvili. Langmuir 10:3279, 1994.
64. P Mulvaney, JM Perera, S Biggs, F Grieser, GW Stevens. J Colloid Interface Sci 183:614, 1996.
65. S Basu, MM Sharma. J Colloid Interface Sci 181:443, 1996.
66. BA Snyder, DE Aston, JC Berg. Langmuir 13:590, 1997.

67. V Hlady, M Pierce, A Pungor. Langmuir 12:5244, 1996.
68. M Pierce, J Stuart, A Pungor, P Dryden, V Hlady. Langmuir 10:3217, 1994.
69. J Czarnecki, V Itschenskij. J Colloid Interface Sci 98:590, 1984.
70. J Czarnecki, T Dabros. J Colloid Interface Sci 78:25, 1980.
71. L Suresh, JY Walz. J Colloid Interface Sci 183:199, 1996.

5

Plasma Treatment of Polymers to Improve Adhesion

Michael R. Wertheimer, Ludvik Martinu, and Jolanta E. Klemberg-Sapieha
Ecole Polytechnique, Montreal, Quebec, Canada

Gregory Czeremuszkin
Polyplasma, Inc., Montreal, Quebec, Canada

I. INTRODUCTION

A. Background

In recent years, we have witnessed a remarkable growth in the use of the synthetic organic polymers in technology, both for high-technology and for consumer-product applications (see Fig. 1 [1]). Polymers have been able to replace more traditional engineering materials such as metals, because of their many desirable physical and chemical characteristics (high strength-to-weight ratio, resistance to corrosion, etc.) and their relatively low cost. However, fundamental differences between polymers and other engineering solids have also created numerous important technical challenges, which manufacturing operations must overcome. An important example is the characteristic low surface energy of polymers and their resulting intrinsically poor adhesion [2–6]; the term "adhesion," as it is used here and elsewhere in this text, may be briefly defined as the mechanical resistance to separation of a system of bonded materials [7]. Because adhesion is largely a surface property, often governed by a layer of molecular dimensions, it is possible to modify this near-surface region without affecting the desirable bulk properties of the material.

Over the years, several methods have been developed to modify polymer surfaces for improved adhesion, wettability, printability, dye uptake, and other

---

This chapter was submitted to the publisher March, 1997.

139

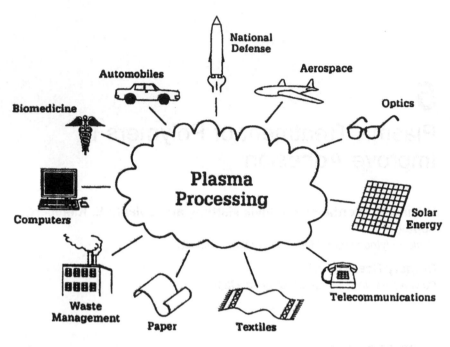

Figure 1 Plasma processing is a critical technology in many vital industries that make extensive use of plastics. (From Ref. 1.)

technologically important characteristics. These methods are discussed in detail in various chapters of this book, and they include mechanical treatments; wet-chemical treatments; and exposure to flames, photons, ion beams and other types of ionizing radiation, corona discharges, and glow discharge plasmas. An important objective of any such treatment is to remove loosely bonded surface contamination and to provide intimate contact between the two interacting materials on a molecular scale, for energies of molecular interactions across an interface decrease drastically with increasing intermolecular distance [4]. Theoretical adhesion models have been proposed by various authors to account for a wide range of related experimental observations, and they are reviewed by Schultz and Nardin in the first chapter of this book. Each of the theoretical models published so far has certain merits but also many weaknesses; at present, none of them taken alone can adequately account for any large subset of all of the experimental observations related to the bond strengths of joined materials. We feel, however, that there is a growing body of evidence that chemical reactions at the interface can play a key role in many cases, a view that is amply documented in various chapters of this book, including the present one.

A concept that has been gaining much support among adhesion scientists in recent years is the existence of an "interphase" (see Fig. 2), loosely defined as a region intermediate to two contacting solids that is distinct in structure and properties from either of the two contacting phases. Sharpe [8] argues very convincingly that interphases exist in many macrosystems such as adhesive joints, coating-substrate systems, and fiber- or particulate-reinforced composites; that they may control the overall mechanical behavior of these systems; and that failure to take them into account is likely to lead to flawed models.

Returning to surface treatments for adhesion enhancement, mechanical roughening alone has limited effectiveness; wet-chemical treatments with solvents, oxidants such as permanganates or chromates, strong acids or bases, or the sodium–liquid ammonia treatment for fluoropolymers [9] are becoming increasingly unacceptable because of environmental and safety considerations. Furthermore, wet-chemical treatments tend to have inherent problems of uniformity and reproducibility, criticisms that are also often leveled against treatments with flames, photons, and ion beams. Modification of polymer surfaces by plasma treatment, both corona and low-pressure glow discharges, presents many important advantages and overcomes the drawbacks of the other processes mentioned above—reasons why these plasma processes have been gaining wide acceptance over the years in diverse industrial applications. Corona treatment, the longest established and most widely used plasma process [10,11], is the subject of a

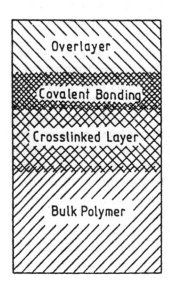

Figure 2 Schematic diagram of the interphase structure between an adhering overlayer and the plasma-treated polymer base.

separate chapter by Uehara; it has the advantage of operating at atmospheric pressure, the reagent gas usually being ambient air. However, this circumstance may also be considered a limitation, in that chemical effects other than surface oxidation, achievable by using reagent gases other than air, tend to be uneconomical or hazardous or both. The same can also be said about atmospheric pressure glow discharge (APGD) processing, in which the reagent gas is heavily diluted with helium [12]. The restrictions just mentioned are largely absent in the case of high-frequency, low-pressure glow discharge treatments, the plasma type that we discuss exclusively in the remainder of this chapter. We and others have compared corona, APGD, and low-pressure plasma surface treatments [13,14] and have reviewed polymer surface modification by glow discharge plasma [15–18].

In the following text we describe basic plasma processes and plasma-polymer interactions; the next section then deals with bonding enhancement through plasma treatment, followed by the description of different plasma sources used for laboratory-scale and industrial-scale treatment of polymers.

B. Low-Pressure Plasma Processes

Most matter in the universe is believed to exist in the plasma state, which can be succinctly defined as partially or fully ionized gas, in which the charged particles have collective interactions [19–21]. Man-made plasmas have been steadily gaining prominence in science and technology since Irving Langmuir's pioneering research in the 1920s and 1930s—it was this Nobel laureate who first proposed the term "plasma" in the current context.

Various natural and man-made plasmas can be classified according to their position on a plot of charged particle density, n, versus electron temperature, T_e (or kinetic energy, u), where both these parameters can vary over many orders of magnitude. In this chapter we are concerned with "glow discharge" plasmas, which exist at reduced gas pressures (typically $0.1 \lesssim p \lesssim 10^3$ Pa); whereas T_e is typically in the 10^4 K (~ 1 eV) range, the temperature of heavy particles (ions, molecules, molecular fragments, . . .) remains near ambient temperature (300 K or 0.025 eV). This state of thermodynamic nonequilibrium results from the relatively low frequency of energy-transfer collisions between the "hot" electrons and the heavier particles, the reason why one often refers to it as "cold" plasma. This characteristic of glow discharge plasmas has rendered them invaluable for the processing of thermally sensitive materials (semiconductors and plastics, the latter of course being the object of this chapter), along with other characteristics such as their "uniqueness" (the absence of competing processes with equal capabilities) and their environmentally benign nature.

Now, the electron "gas" is not monoenergetic but possesses a statistical distribution [the so-called electron energy distribution function, EEDF, $F_0(u)$], with a peak at a relatively low value of u and a tail that can extend to several tens of electron volts or more. Electrons in this tail of the distribution possess enough kinetic energy to break covalent bonds (a few eV) and even to cause further ionization, thereby sustaining the discharge ($10 < u \lesssim 20$ eV, depending on the gas). The chemically reactive species thus created can partake in homogeneous (gas-phase) or heterogeneous reactions with a solid surface (here, a polymer) in contact with the plasma. The rate coefficient C_j for creation of a particular reactive species or state j (radicals, ions, molecular excited states, . . .), via inelastic collisions between "hot" electrons and ground-state gas atoms or molecules, is given by

$$ C_j = \left[\frac{2e}{m} \right]^{1/2} \int_0^\infty \sigma_j(u) F_0(u) u \, du \tag{1} $$

where $\sigma_j(u)$ is the particular process cross section and $F_0(u)$ is the EEDF. It is noteworthy that nearly all processes of importance here possess an energy threshold, u_j, for which

$$ \sigma_j(u) \equiv 0, \qquad u \le u_j \tag{2} $$

The number density $\dot{n}_j$ of species j produced per second in the plasma from ground-state molecules (of number density N cm^{-3}) clearly also depends on the electron density n; that is,

$$ \dot{n}_j = C_j N n \tag{3} $$

The power balance between the applied electromagnetic field (of frequency $f = \omega/2\pi$) and the plasma can be expressed by

$$ P_a = \theta_a n V \tag{4} $$

where P_a is the power absorbed in the volume V of plasma and θ_a is the average power absorbed per electron. Under steady-state conditions $\theta_a = \theta_l \equiv \theta$, where θ_l is the power loss per electron; the parameter θ is readily measurable and can also be considered as the power required to sustain an electron-ion pair in the plasma. Ferreira and Loureiro [22] have calculated θ values for the "simple" case of a low-pressure argon plasma; their model shows that at constant pressure, θ decreases as f is raised from "low frequency" (LF, $\ll 100$ MHz) to microwave (MW, >300 MHz) frequencies. In other words, the efficiency of producing electron-ion pairs is greater at MW than at LF for a given power density absorbed in the plasma. Indeed, the yields of other types of chemically reactive particles such as free radicals, which at 20% or more of all species are the majority constituent, are also found to be higher in MW plasmas. As reviewed by Moisan et al. [23], this is attributed to a significantly higher fraction of energetic electrons in the tail of the EEDF and to a higher n value than for lower frequency plasmas. It is, therefore, not surprising that among recent generations of commercial plasma reactors, an increasing number of these systems operate at the 2.45-GHz MW frequency (see Sec. IV).

An additional important parameter controlling the plasma excitation and plasma-surface interactions is the bombardment of surfaces by energetic ions. The ions are accelerated across the radio frequency (RF)-induced plasma sheath, or "dark space," to their maximal kinetic energy $E_{i,\max}$, by the difference between the plasma potential V_p and the surface potential V_B:

$$E_{i,\max} = e|V_p - V_B| \tag{5}$$

In practical situations, at higher gas pressures, the ions lose part of their energy through inelastic collisions, and they possess a broader ion energy distribution function IEDF [24].

From the aforementioned it is clear that to ensure high quality and reproducibility of a given plasma process, numerous parameters must be controlled with care. Foremost among these parameters that the operator of a modern plasma system must select and control are:

The nature of the reagent gas or gas mixture
Gas pressure and flow rate
The discharge power density
The surface temperature and electrical potential of the workpiece
The excitation frequency(ies) of the power generator(s), which is generally fixed at the time of acquisition of the system

More will be said about this in Sec. IV, where different plasma sources are described.

II. INTERACTION OF PLASMAS WITH POLYMER SURFACES

A. Physicochemical Effects of Plasma Treatment

Materials processing with low-pressure glow discharge plasmas can be divided into three categories:

1. Plasma etching: dry removal of material from the surface of an object being treated, via the formation of volatile reaction products.
2. Surface modification, whereby surface-specific properties, such as adhesion, can be drastically altered; here, little or no material is removed from (or added to) the surface.
3. Deposition of thin films; here, plasma-chemical reaction of one or more

volatile precursor compounds gives rise to the formation of solid reaction products, for example, a so-called plasma polymer [25].

In the plasma treatment of polymers to improve adhesion, the theme of this chapter, all three of these categories can play key roles. For lack of space, however, we have chosen to largely exclude category (3); the reader is referred to an extensive, specialized literature on this subject [25–28].

In the other two process categories, which do not give rise to film deposition, energetic particles and photons generated in the plasma interact strongly with the polymer surface, usually via free radical chemistry [3,15,25]. Four major effects on the surfaces are normally observed. Each is always present to some degree, but one may be favored over the others, depending on the substrate and gas chemistry, the reactor design, and the choice of operating parameters listed at the end of Sec. I.B. The four major effects are [15]:

1. Surface cleaning, that is, removal of organic contamination from the surface
2. Ablation, or etching, of material from the surface, which can remove a weak boundary layer and increase the surface area
3. Cross-linking or branching of near-surface molecules, which can cohesively strengthen the surface layer
4. Modification of surface-chemical structure, which can occur during plasma treatment itself and upon re-exposure of the treated part to air, at which time residual free radicals can react with atmospheric oxygen or water vapor

All these processes, alone or in synergistic combination, affect adhesion and other physico-chemical properties that are governed primarily by the polymer surface. We now briefly examine each of these four effects.

1. Cleaning

This is one of the major reasons for improved adhesion to plasma-treated surfaces. Most other cleaning procedures leave a layer of organic contamination that interferes with adhesion processes; other sources of contamination are additives such as antioxidants, plasticizers, solvents, release agents, and the like, which are deliberately incorporated in polymer formulations to improve their processability and which can "bloom" to the polymer surface. Plasmas of various gases, but especially oxygen-containing plasmas, are capable of removing such surface contamination, the exposure duration depending on the thickness of the contaminant layer. Of course, any clean surface rapidly reacquires a layer of contamination when exposed to ambient atmosphere.

2. Ablation

Ablation, or plasma etching, is distinguished from cleaning only by the amount of material that is removed. Ablation is important for the cleaning of badly contaminated surfaces, for the removal of weak boundary layers (WBLs) formed during the fabrication of a part, and for the treatment of filled or semicrystalline materials. Because amorphous polymer is removed many times faster than either its crystalline counterpart or inorganic filler material, a new surface topography can be generated, with the amorphous zones appearing as valleys. This change in surface morphology can improve mechanical interlocking and can increase the area available for chemical or molecular interactions. Indeed, to obtain a high peel strength of metallic film deposited on polymeric substrates (up to 2 kN/m), surface roughening by a plasma process may even be necessary [29]. Some ablation of reinforcing fibers tends to improve composite properties, but the fibers must not be significantly reduced in diameter by overtreatment because thinner fibers will be weaker [30].

3. Cross-Linking

CASING (cross-linking via activated species of inert gases) was one of the earliest recognized plasma treatment effects on polymer surfaces [31]. As suggested by the acronym, CASING occurs in polymer surfaces exposed to noble gas plasmas (e.g., He or Ar), which are effective at creating free radicals but do not add new chemical functionalities from the gas phase. Ion bombardment or vacuum ultraviolet (VUV) photons (at wavelengths below 175 nm, the absorption edge of oxygen [32]) can break C—C or C—H bonds, and the free radicals resulting under these conditions can react only with other surface radicals or by chain-transfer reactions; therefore, they tend to be quite stable [33]. If the polymer chain is flexible, or if the radical can migrate along it, this can give rise to recombination, thus causing unsaturation, branching, or cross-linking. The latter may improve the heat resistance and bond strength of the surface by forming a very thin, cohesive skin, the effect so dramatically illustrated by the early CASING experiments [31]. VUV photochemistry, a subject currently undergoing an important revival [34–36], is treated in some detail in our earlier review [15] but is not discussed further here, for lack of space. It has even been shown that less energetic UV irradiation of silicone rubber can significantly improve the adhesion to epoxy resins [37].

4. Chemical Modification

The most dramatic and widely reported effect of plasma is the deliberate alteration of the surface region with new chemical functionalities capable of interacting with adhesives or other materials deposited on the polymer. As it is a core subject of the present chapter, much of the remaining text is devoted to chemical modification and its diverse applications.

B. Characterization of Modified Surfaces

All of the processes identified in Sec. II.A can affect adhesion; it is therefore crucial that one be able to characterize a given plasma treatment in terms of the resulting changes in surface chemical composition, structure, and physical or functional properties. Of course, this applies equally to the subject matter of other chapters in this book; because other authors also deal with the topic of surface characterization, and to minimize excessive overlap and duplication, we limit ourselves here to providing only a brief inventory of the most relevant techniques. For more detail, the reader is referred to our earlier review [15] and elsewhere [2,3,17,18,25,27].

A variety of surface-specific techniques are available for the characterization of polymers. Among the most powerful and frequently used are X-ray photoelectron spectroscopy (XPS) or electron spectroscopy for chemical analysis (ESCA) [2,3,6,17,38–40], static secondary ion mass spectroscopy (SSIMS) [2,3,41], Fourier transform infrared spectroscopy (FTIR) [2,6,42], infrared spectroellipsometry [43], and contact angle goniometry [2,3,6,34,44,45]. Also very useful are high-resolution electron energy loss spectroscopy (HREELS) [46], scanning electron microscopy (SEM) and scanning probe microscopy (SPM), and techniques based on ion beam probes such as elastic recoil detection (ERD) analysis [47], ion scattering spectroscopy (ISS) [48], and Rutherford backscattering spectroscopy (RBS) [48]. Of course, a wide variety of ''functional'' test methods have been devised to evaluate the relative merits of a given surface treatment, e.g., mechanical peel or lap-shear tests, electrical property measurements, and others. For example, laser-induced spallation and interferometric detection of transient surface displacement have recently been proposed as a powerful noncontact method for the investigation of adhesion properties of solid coatings [49]. Other methods are too numerous to recite here, and the reader is directed to Ref. 50. The same applies to plasma diagnostic techniques [51], the most common of which are optical emission and mass spectrometries and electric probe measurements.

III. ADHESION ENHANCEMENT THROUGH PLASMA TREATMENT

A. General Considerations

The various effects of the plasma-surface interactions discussed in Sec. II.A contribute in a synergistic manner to the different mechanisms of adhesion. Indeed, adhesion enhancement can be regarded as resulting from the overlap of the following effects: (1) removal of organic contamination and of weak boundary layers by cleaning and ablation, (2) cohesive strengthening of the polymeric surface by the formation of a thin cross-linked layer that mechanically stabilizes the surface and serves as a barrier against the diffusion of low-molecular-weight (LMW) species to the interface, and (3) creation of chemical groups on the stabilized surface that result in acid-base interactions and in covalent linkages believed to yield the strongest bonds (see Fig. 2).

Experience has shown that the plasma treatment conditions necessary to achieve maximal bond strength must be optimized for any given materials combination. This is accomplished by exercising control over the effects mentioned above, a process in which the chemistry of the substrate material, the energetics at the plasma-polymer interface (ions, photons), and the gas-phase plasma chemistry play the major roles. In addition, the dependence on storage time and environment on treated polymer surfaces should be taken into account [17,44,45,52]: "Aging," a time dependence of properties that can be either beneficial or detrimental, can be ascribed to a combination of four effects [53]: (1) thermodynamically driven reorientation of polar moieties away from the surface into the subsurface; (2) the diffusion of mobile additives or oligomers from the polymer bulk to the surface; (3) the formation of LMW species in the subsurface during plasma exposure and their subsequent migration to the surface; and, finally, (4) the reaction of residual free radicals with the ambient. Beside the nature of the polymer surface, the extent of aging effects is directly related to the "degree" of treatment.

In the following we give examples of the effect of plasma treatment on the adhesion for different combinations of materials.

B. Polymer-Polymer Bonding

Plasma treatment can be used with great effect to improve the bond strength to polymers. In these cases, the improved properties result from both increased wettability of the treated polymer by the adhesive and the modification of the surface chemistry of the polymer. The altered surface chemistry facilitates reaction of the adhesive with surface species during curing to form covalent bonds with the plasma-treated surface.

Frequently, measurement of the water contact angle is used to estimate adhesion. In some cases, however, even if the surface is sufficiently hydrophobic to repel water, it may still remain wettable by an epoxy adhesive [15]. Plasma treatment usually changes the surface free energy, γ, of a polymer by affecting its dispersive and, even more, its nondispersive component. For example, surface cross-linking in He or Ar plasmas may increase dispersive interactions. Certain plasma treatments also allow one to decrease the γ value below 20 mJ/m^2, thus producing a surface with relatively poor adhesion to most adhesives; poor adhesion can be a desirable feature in polymeric devices, when antisticking agents may be unwanted. Incorporation of fluorine onto polymer surfaces, for example, by exposure to fluorocarbon plasmas, decreases the γ value and renders the surface hydrophobic, nonwettable by polar solvents, and, in general, weakly adhering to other polymers. Similarly, O_2/CF_4 (20/80) plasma was found to decrease to zero the nondispersive component of the surface free energy of carbon fibers, thus lowering their γ value to 18 mJ/m^2. However, increasing the oxygen content to $\sim$40% in the plasma preferentially accelerated oxidation of the fiber surface, thus strongly increasing the nondispersive part of the γ [54]. However, the broad subject of surface energetics, its correlation with adhesive bonding, and the methods for measuring or estimating γ clearly exceeds the scope of this chapter.

Reference 55 presents what is believed to be the first direct experimental evidence for the formation of covalent bonding between plasma-generated surface functionalities and an epoxy adhesive. The proof of this covalent bonding is extremely important because it would predict greatly improved hot-wet (100% relative humidity, 100°C) stability of the adhesive bonds to plasma-treated polymers. In fact, it has been reported that O_2 plasma treatment of the surface of graphite-polyimide parts improves the hot-wet stability by a factor of 2, as compared with solvent wiping [56]. Shanahan and Bourges-Monnier [57] recently proposed a model including chemical bond formation to explain the improved adhesion between epoxy adhesive and epoxy-graphite composite treated by oxygen and nitrogen plasmas. They used an epoxy gel, in place of epoxy resin, that facilitated examining the adhesion by the peel test technique. Ammonia and hydrogen plasma treatment was found to greatly enhance the adhesion of epoxy resin to Teflon, poly(tetrafluoroethylene) (PTFE), to the extent that cohesive failure within the fluoropolymer bulk was observed in pull-off tests [58]. Cohesive failure may, therefore, be taken as an indication of very good adhesion between polymeric surfaces.

Besides polymeric laminated structures using adhesives, efforts are continuing to achieve good polymer-polymer bonding by mere plasma activation. Interfacial phenomena have been studied for the cases of polyethylene (PE)-PE and poly(ethylene terephthalate) (PET)-PE laminates without adhesives, following treatment in a low-pressure MW air plasma or in an ambient air corona discharge [13]. The adhesion force was found to exhibit a pronounced maximum for a surface concentration of bound oxygen between 11 and 14 at%, independent of the type of treatment. Based on high-resolution XPS analysis, it has been concluded that the maximal adhesion occurs when the concentration of hydroxyl, ether, or epoxide groups is highest and that of carboxyl (acid) groups is lowest. As already indicated, these results suggested that the highest adhesion force appears when the surface is mechanically stabilized by cross-linking and when the effect of a weak boundary layer (WBL) due to an excessive amount of LMW species is minimal (low carboxyl concentration) [13].

The acidity of a strongly oxidized surface suggests that another type of chemical bonding mechanism might be of importance here, namely acid-base interactions [59–61]. Adhesion of dye molecules, an example drawn from our own experience, strongly supports this view: nonpolar polymers such as PE or polypropylene (PP) cannot be treated with inexpensive water-based acid dyes. If, however, basic (Lewis base) moieties are grafted to the surface, the acid-base interactions between these and the dye (Lewis acid) can give rise to strong (chemical) adhesion at the polymer surface. Surface nitrogenation (amination) using a plasma of N_2, NH_3, or volatile amines has been shown to yield the desired results, which are attractive for both economic and environmental reasons [62–65].

Finally, to close this subsection, Tables 1 and 2 give examples of typical bonding improvement data for a number of polymers. It must be stressed that these data are intended only to show the trend and the magnitude of improvement that can be obtained through plasma treatment. These data represent typical values, which will differ from one experimenter to the other depending on the polymer formulation, the type and amount of additives, the adhesive, the cure cycle,

Table 1 Typical Examples of Lap-Shear Bond Strength Improvement After Various Plasma Treatments

Material	Control (MN/m^2)	Plasma-treated (MN/m^2)
Polyimide (PMR-15)/graphite	2.90	17.93
Polyphenylene sulfide (Ryton R-4)	2.00	9.38
Poly(ether sulfone) (Victrex 4100G)	0.90	21.65
Polyethylene/PTFE (Tefzel)	Very low	22.06
HDPE	2.17	21.55
LDPE	2.55	10.00
Polypropylene	2.55	21.24
Polycarbonate (Lexan)	2.83	6.40
Polyamide (Nylon)	5.86	27.58
Polystyrene	3.93	27.58
Poly(ethylene terephthalate) (Mylar A)	3.65	11.45
PVDF (Tedlar)	1.93	8.96
PTFE	0.52	5.17

Source: Ref. 15.

Table 2 Typical Examples of Peel-Strength Improvements After Various
Plasma Treatments

Material	Control (N/m)	Plasma-treated (N/m)
Silicone (red. Durometer 50)	70.1	3330
RTV silicone (D.C. type E)	Very low	403
Perfluoroalkoxy (PFA)	17.5	1450
Fluorinated ethylene propylene (FEP)	17.5	1820
Tefzel	17.5	2770
PTFE	17.5	385
Polyimide (Kapton)	700	2800

Source: Ref. 15.

the time between plasma treatment and bonding, and, of course, the other plasma parameters.

C. Polymer-Matrix Composites

Composites filled or reinforced with fibers or powders, and also laminates, multilayer structures, etc., constitute a broad domain of plasma application. Improvement of the tensile strength, impact resistance, and also of electrical properties and aging characteristics strongly depends on good adhesion between the matrix and the filler and on stability of the interfaces exposed to humidity, solvents, or temperature cycling, for example. Because of the large contribution of interfaces in a typical fiber- or powder-filled composite, any beneficial change in surface characteristics of the filler and the resulting enhanced interaction with the matrix usually have a dramatic effect on the performance of the material.

A most undesirable feature of composites is inherent incompatibility of their components. However, this is difficult to avoid, by definition, when a given matrix is to be improved by incorporating a component of very different characteristics. The two components differ in their mechanical, thermal, chemical, and electrical properties, with the result that stresses and weak points appear in the composite during its formation or use, which can then lead to failure. Good adhesion and stability of an interface (or, better, "interphase" [8]) help to distribute the stress accumulation, to dissipate energy under load, and to prevent permeation of contaminants such as moisture, lubricants, or solvents. Plasma treatment of multifarious fillers has proved to be a very effective method of adhesion improvement in polymeric matrices [66]. However, due to the very wide variety of commercial fillers used, which are often pretreated (with compatibilizers; sizing, coupling, or slip agents; etc.), any process for improving adhesivity or manufacturing process runnability requires a detailed optimization for each composite system.

Graphite, glass, aramid, and polyethylene fibers may be discussed as examples of plasma treatment effectiveness. Graphite fibers of very high strength (or modulus) have a low surface energy value, thus a repellence toward epoxy resins (ERs), the most common matrix for graphite composites. However, a clean and chemically functionalized fiber surface is obtained by ammonia or nitrogen plasma treatment, which has been found to improve the interlaminar and interfacial shear strength [67]. CO_2 [68], Ar, CF_4—O_2, O_2 and N_2—H_2 plasmas also improved carbon fiber–epoxy matrix adhesion, but the latter two treatments proved the most effective [69]. Argon and oxygen plasmas were shown to reinforce graphite composites based on polyphenylene sulfide (PPS) [70,71] and polyimide [56] matrices. Plasma deposition from styrene, acrylonitrile, or acetylene onto graphite fibers also resulted in better adhesion to PPS or ER; however, longer duration treatments were found to decrease the fibers' tensile strength [71,72]. Similarly, glass fibers, treated by oxygen or air plasma, as well as hexamethyldisiloxane (HMDSO) preceded by nitrogen plasma activation [73], displayed excellent adhesion to polymeric resins, thus eliminating the necessity of using primers or compatibilizers.

PE fibers absorb and dissipate shock energy in composites, thus improving their impact resistance. Here, too, good adhesion to the resin is an important issue. Treatments by plasma and silane coupling agents enhanced the interlaminar shear strength in the carbon fiber–ultrahigh modulus polyethylene (UHMPE) fiber hybrid composites [74]. In this case, plastic and elastic deformations of the UHMPE fibers also played a key role in improving the impact properties of the composite. Oxygen plasma oxidizes and cross-links the PE fiber surface region, thus also increasing its cohesive strength [75,76]. Overtreatment may well enhance surface roughness and interlocking, but it reduces the fiber's strength [76]. Both UHMPE and aramid (Kevlar 49) fibers displayed enhanced adhesion to epoxy resins after air plasma treatment; for these materials interlocking and interfacial molecular forces are believed to be responsible for the improvement [77,78]. After NH_3, O_2, or H_2O plasma treatments of aramid fibers (Kevlar 49), the interfacial shear strength of a fiber–epoxy resin composite was markedly improved (43–83%), while the fiber strength was little affected (a loss of less than 10%). The authors explain the improvement by the formation of covalent bonds between the newly created reactive functionalities at the modified fiber surfaces and the epoxy groups of the resin [79]. Ammonia plasma modification of specially prepared Kevlar fibers was used to enhance interfibrillar adhesion by incorporating reactive amine groups on the fibril surfaces [80].

Particulate fillers such as $CaCO_3$ powders [81], either ground or precipitated, mica flakes [82], and cellulosic powders (fibers) [83] have also been shown to improve composite performance markedly after plasma treatment in various gases.

Improvement of laminate peel strength after various plasma treatments is usually related to modification of the roughness, surface chemistry, and viscoelastic properties of subsurface zones of the component(s). Having already discussed this in connection with the modification of polymer surfaces, we mention it only briefly here, even though it has wide industrial implications.

Ammonia plasma has been shown to improve the strength and durability of adhesive bonding of PP, low-density polyethylene (LDPE), and high-density polyethylene (HDPE) with cyanoacrylates under hot and humid conditions [84], the bond strength remaining unchanged for 1 month. The authors proposed that covalent bonding between the cyanoacrylate adhesive and newly formed amino groups was responsible for this adhesion stability. A structure of the interphase in the form of an interpenetrating network, followed by covalent bonding and curing of penetrated adhesive, was also suggested as a contributing factor. Jama et al. [65], who treated PET and polycarbonate (PC) specimens in N_2 and O_2-N_2 plasmas, found the oxygen concentration to be a key parameter for surface modification. The plasma produced both nitrogen- and oxygen-containing groups on the polymers, but with O_2 present in the plasma, no N-containing species were found on the PC. Oxygen-hydrogen plasma treatment of PE, PP, and PET strongly increased their bond strength with various adhesives. Although the polar component of the surface free energy was observed to increase, the dispersive one was only weakly influenced [85]. Polyetheretherketone (PEEK) has been treated with oxygen, air, argon, and ammonia plasmas, which greatly improved its adhesion to an epoxy film adhesive [86]. However, wiping the treated surfaces with acetone reversed the beneficial effects of plasma treatment.

Plasma-induced grafting is also a powerful technique to improve adhesion. Here, the active sites for grafted-chain initiation, either radicals or other reactive groups, are created on polymer surfaces by physical or chemical action of a plasma. Grafted polymer chains (for example, vinylimidazole) may form a charge transfer complex and contribute to improved adhesion between Kapton polyimide and electrodeposited copper metal [87]. Plasma-grafted polymers (as well as plasma polymers), when used as primers for subsequently deposited coatings or adhesives, may be considered as an artificial "interphase" that can improve the adhesion by distributing and equalizing stresses, being compatible with both contacting surfaces. The composition and properties of a plasma-deposited "primer" may change continuously across its thickness, from one polymer to another. Such a film may be prepared by plasma polymerization, using a feed gas with a continuously changing composition [88,89]. Plasma-deposited films may absorb the mismatch of thermal expansion coefficients of adhering materials, thus diminishing the risk of delamination and/or cracking [90].

D. Adhesion of Vacuum-Deposited Films

There is strong evidence in favor of the chemical reaction mechanism (see Sec. II.A.), the one that should, in principle, lead to the strongest bonds, to explain strong film-polymer adhesion. Burkstrand [91] was among the first to show that evaporated metals can react with oxygen-containing polymer surfaces, which can lead to metal-oxygen-carbon (M—O—C) type linkages.

Plasmas can be used with success to provide the necessary surface functionalities that can form strong bonds. For example, in situ XPS studies have revealed the presence of Ag—O—C and Ag—N—C linkages [92] after exposure of PE to oxygen and nitrogen plasmas followed by metallization, respectively. For this particular system, the metal-PE adhesion was found to improve according to the following sequence of plasma gases used: $Ar < O_2 < N_2$. Similar effects were observed in another in situ study of Mg on PP [93]. The highest sticking probability for evaporated Mg atoms was found on a PP surface following exposure to an N_2 plasma or, for comparison, following argon ion bombardment at a dose of 5×10^{15} ions/cm^2.

In practical situations, the plasma conditions needed for treatment of a given polymer surface must be optimized for every metal-polymer combination. Departure from optimal treatment conditions can lead to various effects, often involving the presence of a WBL [94]. An example of the mechanical failure of a WBL is the case of PTFE; this material can readily be plasma treated for cleaning and surface modification to give good wetting and to provide oxygen-containing or nitrogen-containing groups. The adhesion, however, is still found to be poor, because the subsurface structure of PTFE remains weak and not easily stabilized by cross-linking.

Being critically important to the microelectronics industry, copper adhesion to polymers is receiving much attention, mostly to polyimide [95] and to fluoropolymers. Adhesion of copper on Teflon PFA (polytetrafluoroethylene—co-perfluoroalkoxy vinyl ether) has recently been found to be substantially enhanced by plasma treatment [96], and the results are illustrated in Fig. 3 [97]. In this work, the adhesion was evaluated using a microscratch test, whereby a hemispherical diamond stylus is moved along the film surface and the critical load, L_c, at which the metal film starts to delaminate from the interface is determined. For the Cu-PFA system, the highest L_c values were obtained following N_2 plasma treatment.

To obtain insight into the adhesion mechanism, the surface modified by the N_2 plasma at 67 Pa (optimized adhesion) was analyzed by XPS before and after in situ Cu deposition. The F/C, O/C, and N/C atomic ratios were found to change from their initial values of 1.86, 0.01, and 0 to 0.28, 0.19, and 0.56, respectively. Changes in the XPS C($1s$) spectra following plasma treatment and Cu deposition are shown in Fig. 4. Untreated PFA exhibits a single CF_2 peak at 291.8 eV, the small peak near 283 eV being due to X-ray satellites [98]. Plasma

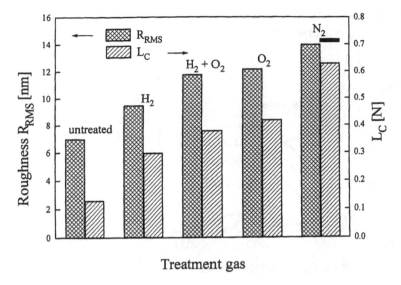

Figure 3 Effect of treatment gas in microwave plasma on the critical load, L_c, of 200-nm-thick evaporated copper films and on the mean surface roughness (RMS) of as-treated Teflon PFA surfaces. The bar indicates an experiment in which the Teflon PFA surface was pretreated in He plasma prior to N_2 plasma exposure. (From Refs. 97 and 98.)

treatment is seen to have caused a substantial fluorine loss, leading to a decrease in the CF_2 peak intensity and to a pronounced increase in the signal intensity at lower binding energies between 289 and 286 eV, due to the formation of nitrogen- and oxygen-containing groups. The deposition of Cu (12 nm) on the untreated PFA surface causes a decrease in the CF_2 peak intensity due to the shielding effect of the metal, combined with a slight loss of fluorine. A small increase in the peak near 285 eV is observed, probably due to the formation of C—C and C—O—Cu structures [96,98]. Compared with the untreated surface, all of the components in the C(1s) spectra of the plasma-treated surface were reduced in intensity after metal deposition, even though the nominal thickness of metal deposited was substantially less (<12 nm). This indicates an enhanced sticking probability of Cu to the plasma-treated surface, as opposed to its untreated counterpart. Interactions of Cu with plasma-induced functionalities result in the formation of C—N—Cu and C—O—Cu groups, which may be responsible for the relative increase of the peak near 285 eV (see Fig. 4).

Further improvement of adhesion of Cu on Teflon PFA was obtained when the N_2 plasma was preceded by a He plasma exposure (see Fig. 3). The strong VUV radiation ($\lambda = 58$ nm) from the He plasma presumably contributed to mechanical stabilization of the interfacial region by cross-linking (CASING; see

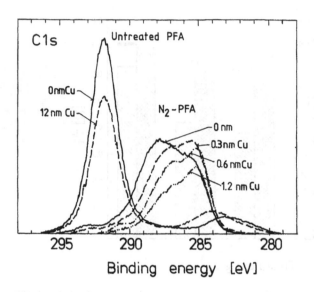

Figure 4 C(1s) spectra for the untreated and N_2 plasma–treated PFA, before and after Cu deposition. (From Ref. 98.)

Sec. II.A), which was then followed by N_2 plasma functionalization. Surprisingly, hydrogen-containing plasma systematically resulted in lower L_c values (see Fig. 3), even though the most pronounced defluorination and surface chemical restructuring was observed [99]. It was proposed that this arises from weaker C—H groups, as suggested by SSIMS measurements [100].

The same plasma-treated Teflon PFA surfaces were also analyzed by AFM for their microroughness, expressed by the RMS value. An example of surface topography is shown in Fig. 5, where RMS values of 6.5 and 14 nm are observed, respectively, for the untreated and N_2 plasma–treated surface [97]. The measured RMS values are also shown in Fig. 3, and their variation with the treatment gas can be compared with the evolution of L_c. One can see that both L_c and RMS values qualitatively follow a similar pattern. It may, therefore, be concluded that surface microroughening can also contribute to adhesion improvement, probably by two effects, namely by enhanced mechanical interlocking and by making more surface area available for subsequent chemical bonding. We believe that these two mechanisms, together with surface mechanical stabilization by cross-linking, synergistically contribute to enhanced film adhesion to plasma-treated polymer surfaces.

Several studies have focused on adhesion enhancement of plasma-deposited optical and protective coatings (such as SiO_2 and $SiN_{1.3}$) on plastic substrates,

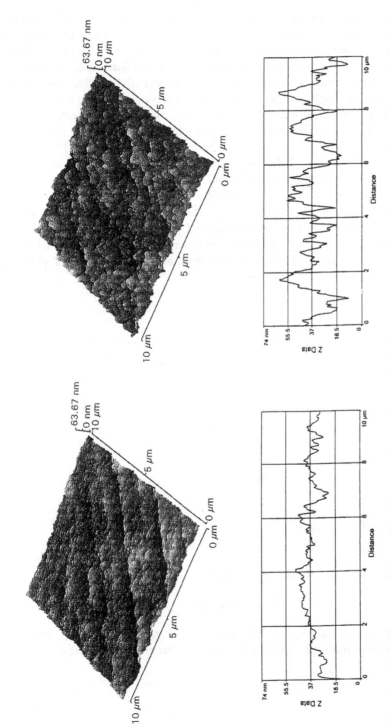

Figure 5 Surface topography and line scan profiles of (left) untreated and (right) N_2 plasma–treated Teflon PFA. (From Ref. 97.)

for example, on optical-grade PC [101,102] and on other polymers such as poly-propylene [103]. As an example, the effect of different treatment gases in MW plasma on the improvement of adhesion of 0.5-μm-thick $SiN_{1.3}$ layers on PC is illustrated in Fig. 6. Even though the $SiN_{1.3}$-PC combination passed the adhesive tape peel test, L_c values could be increased substantially by additional plasma modification, the highest value being attained for N_2 plasma here too.

The adhesion improvement in the $SiN_{1.3}$-PC system has also been correlated with the formation of covalent bonds at the interface, as illustrated in Fig. 7 by the results of XPS analysis. We note the presence of C—Si (283.9 eV) and C—O—Si (286.3 and ~289.0 eV) bonds when $SiN_{1.3}$ is deposited directly on PC (Fig. 7b); after N_2 plasma modification, however, up to 16 at% of N is incorporated (Fig. 7c), and the surface is chemically "activated" by the presence of new C—N and C≡N bonds. Following deposition of a thin (~5 nm) $SiN_{1.3}$ layer, C—N—Si linkages were formed instead of C—Si bonds (see Fig. 7d). Again, the presence of strong covalent C—N-film linkages at the interface can be correlated with the observed adhesion enhancement, as in the case of metal-polymer bonding discussed earlier in this section.

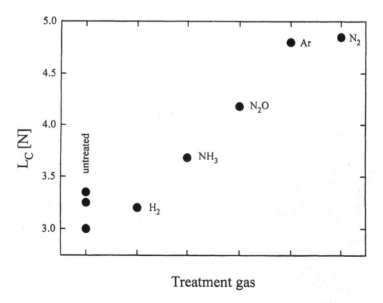

Treatment gas

Figure 6 Critical load, L_c, for $SiN_{1.3}$ films (0.5 μm thick) deposited from SiH_4-NH_3 mixtures onto polycarbonate following MW plasma pretreatment in different gases. (From Ref. 102.)

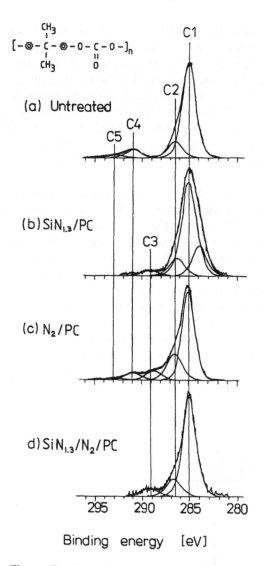

Binding energy [eV]

Figure 7 C1s XPS spectra of polycarbonate: (a) untreated; (b) coated by ~5 nm of plasma-deposited $SiN_{1.3}$; (c) treated for 1 min in microwave plasma using N_2; (d) treated by N_2 plasma followed by a deposition of ~5 nm of $SiN_{1.3}$. (Modified from Ref. 102.)

IV. PLASMA SOURCES AND INDUSTRIAL PROCESSES

A. General Considerations

Plasma technology has been used in the laboratory for at least 50 years, but it was not a practical industrial technology until the last 15 or 20 years. The "spark" to the growth of low-pressure plasma processing of materials came in the late 1960s with the advent of modern integrated circuit technology, now a multi-hundred-billion-dollar annual market worldwide, in which dry etching and plasma deposition processes (see Sec. II.A) play indispensable roles. Commercial equipment development was originally driven by the requirements of the semiconductor market, but more recently equipment is being designed and built specifically for the "industrial" market, applications that do not involve semiconductors and associated products (see Fig. 1).

Plasma equipment usually consists of six modules or functions: pumping system, reactor chamber, power supply and monitor, matching network, process control instrumentation, and process diagnostics (Fig. 8). As already mentioned, low-pressure plasma systems for surface modification generally operate in the pressure range from 10 to 500 Pa (about 0.1–3.5 torr), with a continuous gas flow into the reactor. Therefore, the vacuum system must be able to maintain this pressure-flow regime; this moderate vacuum level does not require sophisticated pumps, so two-stage mechanical pumps are often satisfactory. The pump package is usually sized to allow pumpdown to the operating pressure in a few minutes at most and to maintain an inlet gas flow of several tens of standard cubic centimeters per minute (sccm) for the smallest systems to several thousands of sccm for systems several cubic meters in volume. Pump maintenance and the perfluori-

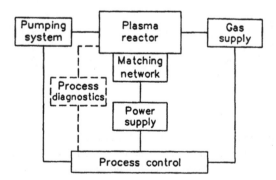

Figure 8   Modules that constitute a typical RF plasma equipment setup. (From Ref. 16.)

nated pump fluid that is required if an oxygen plasma is to be used tend to represent a large part of the total plasma system operating and maintenance costs.

As also mentioned earlier, a major advantage of plasma surface treatments as compared with most other (wet-chemical) treatment processes is the lack of harmful by-products, for there are no toxic or hazardous liquids or gases that must be disposed of. Usually, the main process by-products are CO_2, water vapor, and in some cases also CO, methanol, and others, none of which is present in toxic quantities; in other words, plasma surface treatment is very benign toward the environment.

The excitation power of plasma systems generally ranges from 50 W to several kW, or even as much as 30 kW, depending on the size of the reactor. Plasma reactors have been built utilizing a wide range of excitation frequencies, from DC to MW: DC plasmas are not advantageous, primarily because of the need for a current-limiting resistor to stabilize the plasma and prevent arcing and because of the very large bias effects that are present. Most plasma reactors therefore use AC electrical power supplies, sometimes provided with antiarcing devices, operating at audio, radio, or microwave frequency [16,27,104]. More specifically, international regulatory agencies have allocated certain "ISM" (industrial, scientific, medical) frequencies for use in applications other than telecommunications, and it is these frequencies that equipment builders favor for obvious technical and economic reasons. Commercial plasma systems therefore usually operate in the low-frequency (40–450 kHz), radio frequency (RF, 13.56 or 27.12 MHz), or microwave (MW, 915 MHz or 2.45 GHz) ISM ranges [105].

Most industrial plasma systems are equipped with an automated process control unit (see Fig. 8), which ensures the process reliability. This unit operates with signals fed from other system components; it monitors the pressure, gas flow, power, and discharge stability, but it may also use a feedback from various diagnostic tools, such as optical emission spectroscopy, mass spectrometry, and others [51].

There exist several types of plasma reactor designs: dielectric (e.g., fused quartz) or metal and batch or continuous. Quartz chambers are important in the semiconductor industry because of the requirements for extreme cleanliness and particle-free operation, but they may also have advantages in some industrial applications. In the present context, for plasma treatment for polymer surface modification, extreme cleanliness is rarely required, and the danger of breakage favors the use of metal. Aluminum is the metal of choice for constructing plasma reactors, because it has excellent thermal and electrical conductivity and chemical resistance. Aluminum is not readily attacked by any plasma gas except the heavy halogens (Cl_2, Br_2, or I_2); it is routinely fabricated into cylindrical vessels (for example, "downstream" systems; see further below) and into rectangular reactor chambers with shelf or cage electrodes. It is also possible to fabricate specially shaped electrodes for particular applications. The only size limitation on metal reactors is related to the strength of large vacuum vessels, the largest currently available commercial reactors having dimensions in excess of 2 m (see Sec. IV.B).

In the case of audio or radio frequency plasma sources, the electrode design is of fundamental importance, because this will determine the electric field distribution in the plasma zone and hence such key factors as active species creation and ion bombardment effects. This important topic clearly transcends the scope of the present chapter, and the reader is referred to the specialized literature for more information [20,27,104], including our review of plasma sources [16]. In the latter reference, we also discuss the increasingly important subject of MW plasmas. Beside advantages mentioned in Sec. I.B, MW excitation does not require electrodes; instead, a wide variety of power-coupling schemes are now available, permitting parts of various geometries to be exposed to "direct" (in situ) plasma treatment or to "downstream" treatment, which uses long-lived excited species in the effluent of the MW discharge [106,107].

B. Industrial Plasma Reactors

Most commercially available plasma systems are designed for batch operation, which involves loading a batch of parts, evacuation, plasma processing, purging to atmospheric pressure, and removal of the parts. Although this has been satisfactory for many applications, the use of polymers in ever more sophisticated applications, such as composite structures, packaging films, and many others, increasingly calls for the continuous processing of filaments, yarns, film, and fabric. We begin this section with a discussion of low-pressure plasma treatment of flexible substrate materials, namely webs or filaments, after which we consider solid objects, either large or small.

1.   Treatment of Flexible Substrates

Literature reports on the continuous plasma treatment of flexible polymeric webs or fibers go back to the 1960s and even earlier, plasma systems being based on either "air-vacuum-air" or "cassette-to-cassette" ("batch-continuous") configurations. In air-to-air systems, there are sequentially pumped chambers on each side of the reactor chamber, which are connected by some form of material feed-through system. This feed-through system makes it possible to bring material continuously from atmospheric pressure to the low reactor pressure and then back to ambient atmosphere; a major engineering problem in air-to-air systems is the design of the feed-through.

Ever stricter environmental regulations, for example, the banning of wet-chemical chlorination to improve printing, dyeing, and machine washability characteristics of natural (wool) and synthetic (polyester) fabrics, have encouraged the use of plasma treatment in the textile industry worldwide. An example of a currently used batch-continuous machine is illustrated in Fig. 9. Four of these KPR-180 units, developed in Russia, have been in use for several years in Pavlov-sky, near Moscow, to treat wool fabric. It is claimed that the print definition and color yield are equivalent to those obtained by chlorination. Cloth of 160 cm width is treated in 13.56-MHz RF air plasma at speeds up to 80 m/min, giving rise to batches of 10,000 to 12,000 m per 8-h shift. The firm Technoplasma S.A. in Switzerland is licensed to market this technology worldwide and is now developing the KPR-270 unit, which will have a working width of 270 cm. These and several smaller plasma operations for textiles in Poland, China, Japan, and the United States are outlined in a paper by Chevet [108], which also provides machine and treatment cost figures. Other modern uses of low-pressure plasma for treating flexible polymeric substrates are the following: Eastman-Kodak Company described an N_2 glow discharge pretreatment of PET for the purpose of obtaining superior adhesion of evaporated silver coatings [92,109,110]. Two U.S.-based vendors of industrial plasma equipment (GaSonics International and BOC Coating Technology Inc.) offer RF plasma systems for air-vacuum-air or cassette-to-cassette treatment of multiple parallel strands of fibers or yarn (e.g., PET or aramid tire cord, ultrahigh molecular weight polyethylene (UHMWPE), or carbon fibers) to improve adhesion strength with polymer matrix materials [16].

2. Treatment of Large Objects

Since the mid-1980s there has been a sharp increase in interest in plasma treatment of large, three-dimensional polymeric objects, mostly automobile parts. Newer models of automobiles from North American, European, and Asian manufacturers incorporate many plastic parts, for example, bumpers, grills, dash assemblies and interior panels, fenders, mirror housings, and fuel tanks. These components are light in weight, low in cost, and readily recycled. Most of these molded parts have complex surface features; for protection and for decorative purposes, it is necessary to paint them, but paint adhesion is inadequate on the untreated surfaces. For reasons mentioned earlier, mostly related to the environmental impact of "conventional" surface treatment methods, the automotive industry is adopting plasma treatment methods to achieve the required paint adhesion. Of course, the economic processing of many large parts calls for very large plasma chambers. Two concepts have evolved, namely (1) immersing batches of parts in the direct, primary plasma, and (2) exposing the batches "downstream," that is, to long-lived active species carried into the treatment chamber by the

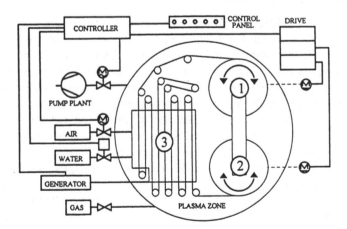

Figure 9 ''Batch-continuous'' machine for plasma treatment of webs up to 160 cm in width, up to 80 m/min. (Model KPR-180, courtesy of Technoplasma S.A.)

effluent from a plasma source outside the chamber. Whereas the latter approach has so far always used MW (2.45 GHz) power to excite the plasma, commercial equipment in the former category includes MW, RF [111], and LF [112] excitations. Table 3 summarizes the available data on large treatment chambers from various vendors [16].

In the case of the two downstream (or "remote") treatment processes using flows of plasma-activated N_2 or O_2, the long-lived active species are believed to be $N(^4S)$ ground state atoms and electronically and vibrationally excited N_2 molecules [106] and O atoms and O_2 ($a^1\Delta$) singlet molecular oxygen [107], respectively.

3. Treatment of Small Parts and Particulates

For at least 15 years, numerous European and North American vendors have been offering plasma reactor systems for treating discrete parts of various geometries, from electronic circuit boards (for desmear or etch back of via holes), to more complex shapes, to batches of thousands of small plastic parts for consumer, biomedical, and other uses. Plasma chambers are generally metallic (aluminum or stainless steel), of rectangular or cylindrical geometry. The former are usually equipped with multiple horizontal tray electrodes, onto which the parts are manually loaded for treatment; Table 4 provides a list of current suppliers and some characteristics of their largest equipment for this purpose. Note that systems are available covering the entire range of excitation frequencies, LF, RF, and MW, in some cases optional from the same vendor.

Batches of many small parts, granules, or powder can be treated in chambers with a rotary drum attachment of the type depicted in Fig. 10; such drums are available in volumes up to 150 liters (0.15 m^3), capable of treating 50- to 60-liter batches of parts at a time. The apparatus of Fig. 10 functions at 2.45 GHz, but rotary drum equipment is also available for LF and RF operations. Probably the largest "drum" treater is the "Phoenix" machine of Polar Materials Inc. (PMI), where the plasma reactor is a 1 m^3 rotating and tilting Pyrex glass vessel.

On the other extreme of size, several vendors offer very small, inexpensive tabletop plasma chambers (Harrick Scientific Corp., March Instruments Inc., Plasmatic Systems) that can be used for treating individual components or small batches.

Finally, one can find at least two system designs in the U.S. patent literature for low-pressure plasma treatment of powder particles, one using MW power [113], the other RF [114]. The latter, also operated by PMI, treats 50 kg/h of 60-μm polymer powder, used as a filler in high-performance polymer matrix composites, for the purpose of enhanced interfacial adhesion with the polymeric matrix.

Table 3 Characteristics of Plasma Chambers for Automotive Parts

Manufacturer (model)	Chamber size (m³)	Direct (D)/ remote (R)	Excitation frequency (power)	Gas	Capacity (throughput)
ATEA (Matis)	3.2	R	2.45 GHz	N₂	22 dash assemblies
Balzers (BPA 2010)	9.0	D	50 Hz	Air	4 bumpers (65/h)
BOC Coating Technology ("Hammer II")	2.4	D	13.56 MHz (5 kW)	—	—
Buck Plasma Electronic GmbH	5.5	D	13.56 MHz (4.8 kW)	O₂	7 bumpers (70/h)
Technics Plasma GmbH (Mammouth)	2.3	D	2.45 GHz (8 sources) 4.8 kW	O₂	—
Toyota Motor Co.	6.2	R	2.45 GHz (1.5 kW)	O₂	—

Table 4 Plasma Treatment Chambers for Discrete Parts

Vendor	Model no.	Chamber or electrode dimensions (cm or m³)	Frequency/power	Remarks
Advanced Plasma Systems Inc.	B Series	71 × 71 × 97 (0.5 m³)	40 kHz or 13.56 MHz	Up to 48 circuit boards
BOC Coating Technology	PS 0500	33 × 50 × 40 (0.066 m³)	13.56 MHz/0.6 kW	Three metal shelves, parts up to 12 cm high
Buck Plasma Electronic GmbH	Domino 80M	37 × 30 × 40 (0.044 m³)	LF/RF/MW (optional) 1.2 kW	Various options: trays, rotary drum, three frequencies
Gasonics International	7100	41 × 41 × 76 (0.125 m³)	13.56 MHz/1 kW	—
Plasma Etch	MKII-3	24 process levels of 61 × 46 cm electrodes	13.56 MHz/5 kW	Optional automatic loading of parts
Technics Plasma GmbH	3067-G	80 × 84 × 108 (0.73 m³)	2.45 GHz/1.2 kW	—
Yield Engineering Systems	YES-R4	2 shelves, 61 × 61 cm electrodes	40 kHz/0.4 kW	—

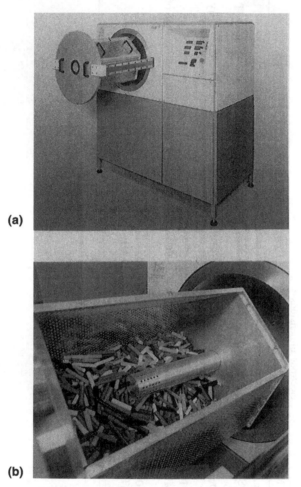

Figure 10 Commercial plasma system with rotating drum for treatment of many small plastic parts (Technics Plasma GmbH, model 3000-5D): (a) overall view; (b) plasma processing of pen casings in the bulk treatment drum; (c) schematic diagram of the microwave plasma reactor chamber.

V. CONCLUSIONS

"Cold," low-pressure plasma treatment can give rise to profound changes in the surface and interfacial properties of materials, particularly of polymers. Unlike corona treatment, a similar process in several respects, which has been in indus-

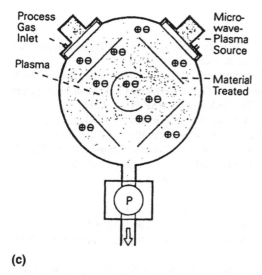

Process Gas Inlet

Micro-wave-Plasma Source

Plasma

Material Treated

P

(c)

Figure 10   Continued

trial use for several decades, in which the reactive gas is usually air, and which is generally restricted to simple surface geometries such as flexible webs, plasma treatment is extremely versatile in its capabilities. Its four main effects—cleaning, ablation or etching, cross-linking, and surface chemical modification—occur together in a complex synergy, which depends on many parameters controlled by the operator. All of these contribute to enhancing the surface characteristics of the treated polymer, particularly its adhesion.

Plasmas can be readily applied to objects of all possible geometries, ranging from webs or films to large solid objects with complex shapes and to small discrete parts in large quantities (even including fine particles such as powders, fibers, or flakes). Industrial plasma systems are available for treating webs, either in a continuous air-vacuum-air or in a batch-continuous (cassette-to-cassette) mode of operation, while other reactor designs are readily adapted for treating discrete parts with the characteristics listed above.

ACKNOWLEDGMENTS

The authors are grateful to several collaborators, especially to Dr. E. M. Liston, for many fruitful discussions. Our plasma research is supported by grants from the Natural Sciences and Engineering Research Council of Canada (NSERC) and by the fonds "Formation de Chercheurs et Aide à la Recherche" of Québec.

REFERENCES

1. Plasma Processing of Materials: Scientific Opportunities and Technological Challenges. Washington, DC: National Academy Press, 1991.
2. DT Clark, WJ Feast, eds. Polymer Surfaces. New York: John Wiley & Sons, 1978.
3. M Strobel, CS Lyons, KL Mittal, eds. Plasma Surface Modification of Polymers: Relevance to Adhesion. Zeist, The Netherlands: VSP Press, 1994.
4. S Wu. Polymer Interface and Adhesion. New York: Marcel Dekker, 1982.
5. LH Lee, ed. Fundamentals of Adhesion. New York: Plenum Press, 1991.
6. G Akovali, ed. The Interfacial Interactions in Polymeric Composites. NATO-ASI Series E: Applied Sciences, Vol 230. Dordrecht, The Netherlands: Kluwer, 1993.
7. LH Sharpe, H Schonhorn. Adv Chem Ser 43:189, 1964.
8. LH Sharpe. Chapter 1 in Ref. 6.
9. LM Siperko, RR Thomas. J Adhesion Sci Technol 3:157, 1989.
10. M Goldman, A Goldman, RS Sigmond. Pure Appl Chem 57:1353, 1985.
11. DA Markgraf. Proceedings Coextrusion Conference. Atlanta: TAPPI Press, 1986, p 85.
12. S Kanazawa, M Kogoma, T Moriwaki, S Okazaki. J Phys D Appl Phys 21:838, 1988.
13. S Sapieha, J Cerny, JE Klemberg-Sapieha, L Martinu. J Adhesion 42:91, 1993.
14. M Strobel, MJ Walzak, JM Hill, A Lin, E Kerbashewski, CS Lyons. J Adhesion Sci Technol 9:365, 1995.
15. EM Liston, L Martinu, MR Wertheimer. Chapter 1 in Ref. 3; also J Adhesion Sci Technol 7: 1091, 1993.
16. MR Wertheimer, L Martinu, EM Liston. In: DA Glocker, SI Shah, eds. Handbook of Thin Film Process Technology. Bristol, IOP Publishing, UK: 1996, chap E3.0.
17. OM Küttel, S Novak. In: J Pouch, SA Alterowitz, eds. Plasma Properties, Deposition and Etching. Aedermansdorf, Switzerland: Trans Tech Publications, 1993, p 705.
18. FD Egitto, LJ Matienzo. IBM J Res Dev 38:423, 1994.
19. SC Brown. Basic Data of Plasma Physics. Cambridge, MA: MIT Press, 1967; re-edited, Woodbury, NY: AIP Press, 1994.
20. B Chapman. Glow Discharge Processes. New York: John Wiley & Sons, 1980.
21. FF Chen. Introduction to Plasma Physics and Controlled Fusion. New York: Plenum Press, 1983.
22. CM Ferreira, J Loureiro. J Phys D Appl Phys. 17:1175, 1984.
23. M Moisan, C Barbeau, R Claude, CM Ferreira, J Margot, J Paraszczak, AB Sá, G Sauvé, MR Wertheimer. J Vac Sci Technol B 9:8, 1991.
24. P Reinke, S Bureau, JE Klemberg-Sapieha, L Martinu. J Appl Phys 78:4855, 1995.
25. R d'Agostino, ed. Plasma Deposition, Treatment and Etching of Polymers. Boston: Academic Press, 1990.
26. J Mort, F Jansen. Plasma Deposited Thin Films. Boca Raton, FL: CRC Press, 1986.
27. SM Rossnagel JJ Cuomo, WD Westwood, eds. Handbook of Plasma Processing Technology. Park Ridge, NJ: Noyes Publications, 1990.
28. RE Clausing, LL Horton, JC Angus, P Koidl, eds. Diamond and Diamond-like

Films and Coatings. NATO-ASI Series B Physics. Vol 266. New York: Plenum Press, 1991.
29. G Hecht, H Kupfer. Proceedings International Symposium on Trends and New Applications of Thin Films, TATF. Paris: Société Française du Vide, 1996, p 119.
30. C Jones, E Sammann. Carbon 28:509, 1990; Carbon 28:515, 1990.
31. RH Hansen, H Schonhorn. J Polym Sci Polym Lett Ed B4:203, 1966.
32. JAR Samson. Techniques of Vacuum Ultraviolet Spectroscopy. New York: John Wiley & Sons, 1967.
33. H Yasuda. J Macromol Sci Chem 10:383, 1976.
34. FD Egitto, LJ Matienzo. Polym Degrad Stabil 30:293, 1990.
35. A Höllander, MR Wertheimer. J Vac Sci Technol A 12:879, 1994.
36. A Höllander, JE Klemberg-Sapieha, MR Wertheimer. J Polym Sci A Polym Chem 33:2013 (1995); Macromolecules 27:2893, 1994.
37. A Kondyurin, Y Klyachkin. J Appl Polym Sci 62:1, 1996.
38. DT Clark, A Dilks. J Polym Sci Polym Chem Ed 17:957, 1979.
39. DT Clark, A Dilks. J Polym Sci Polym Chem Ed 15:2321, 1977.
40. JE Klemberg-Sapieha, OM Küttel, L Martinu, MR Wertheimer. J Vac Sci Technol A 9:2975, 1991.
41. J Lub, FCBM van Vroonhoven, E Bruninx, A Beninghoven. Polymer 30:40, 1989.
42. F Poncin-Epaillard, B Chevet, J-C Brosse. Makromol Chem 192:1589, 1991.
43. A Canillas, E Pascual, B Drévillon. Rev Sci Instrum 64:2153, 1993.
44. M Morra, E Occhiello, F Garbassi. Surf Interface Anal 16:412, 1990.
45. JE Klemberg-Sapieha, L Martinu, OM Küttel, MR. Wertheimer. In: KL Mittal, ed. Metallized Plastics 2: Fundamental and Applied Aspects. New York: Plenum, 1991, p 315.
46. JJ Pireaux, C Grégoire, M Vermeersch, PA Thiry, M Rei Vilar, R Caudano. In: Metallization of Polymers. E Sacher, JJ Pireaux, S Kowalczyk, eds. ACS Symposium Series No 440. Washington, DC: American Chemical Society, 1990, p 47.
47. SC Gujrathi. In Ref. 46, p 88.
48. Y De Puydt, P Bertrand, P Lutgen. Surf Interface Anal 12:486, 1988.
49. MW Sigrist, Opt Eng 34:1916, 1996.
50. KL Mittal, ed. Adhesion Aspects of Polymeric Coatings. New York: Plenum, 1983.
51. O Auciello, DL Flamm, eds. Plasma Diagnostics. Boston: Academic Press, 1989.
52. E Ochiello, M Morra, F Garbassi, D Johnson, P Humphrey. Appl Surf Sci 47:235, 1991.
53. HL Spell, CP Christenson. Tappi J 62:77, 1979.
54. K Tsutsumi, K Ban, K Shibata, S Okazaki, M Kogoma. J Adhesion 57:45, 1966.
55. HF Webster, JP Wightman. J Adhesion Sci Technol 5:93, 1991.
56. JD Moyer, JP Wightman. Surf Interface Anal 17:457, 1991.
57. MER Shanahan, C Bourges-Monnier. Int J Adhesion Adhesives 16:129, 1996.
58. B Chabert, Tran Minh Duc. Int J Adhesion Adhesives 16:173, 1996.
59. FM Fowkes. J Adhesion Sci Technol 1:7, 1987.
60. MB Kaczinski, DW Dwight. J Adhesion Sci Technol 7:165, 1993.
61. KL Mittal, HR Anderson Jr, eds. Acid-Base Interactions: Relevance to Adhesion Science and Technology. Zeist, The Netherlands: VSP, 1991.

62. L Cop, J Jordaan, HP Schreiber, MR Wertheimer. US Patent 4, 744, 860, 1987.
63. N Shahidzadeh-Ahmadi, MM Chehimi, F Arefi-Khonsari, N Foulon-Belkacemi, J Amouroux, M Delamar. Colloids Surfaces A Physicochem Eng Aspects 105:277, 1995.
64. N Shahidzadeh, F Arefi-Khonsari, MM Chehimi, J Amouroux. Surf Sci 352:888, 1996; N Shahidzadeh, F Arefi-Khonsari, MM Chehimi, J Amouroux, M Delamar. Plasmas Polym 1:27, 1996.
65. C Jama, O Dessaux, P Goudmand, B Mutel, L Gengembre, B Drévillon, S Vallon, J Grimblot. Surf Sci. 352–354:490, 1996.
66. JE Klemberg-Sapieha, L Martinu, S Sapieha, MR Wertheimer. In Ref. 6, p 201; EM Liston. In Ref. 6, p 223.
67. C Jones. Composites Sci Technol 42:275, 1991.
68. RE Allred, WC Schimpf. J Adhesion Sci Technol 8:383, 1994.
69. XS Bian, L Ambrosio, JM Kenny, L Nicolais, E Occhiello, M Morra, F Garbassi, AT DiBenedetto. J Adhesion Sci Technol 5:377, 1991.
70. N Chand, E Schultz, G Hinrichsen. J Mater Sci Lett 15:1374, 1996.
71. LY Yuan, SS Shyu, JY Lai. Composites Sci Technol 45:9, 1992.
72. W Weisweiler, K Schittler. Thin Solid Films 207:158, 1992.
73. W Michaeli, M Stollenwerk, F Steinkampf. Proc SPE Annu Tech Conf ANTEC 94(2):2386, 1994.
74. J Jang, S-I Moon. Polym Composites 16:325, 1995.
75. U Plawky, M Londschein, W Michaeli. Acta Polym 47:112, 1996.
76. B Tissington, G Pollard, IM Ward. Composites Sci Technol 44:185, 1992.
77. DA Biro, G Pleizier, Y Deslandes. J Appl Polym Sci 47:883, 1993.
78. Y Chaoting, S Gao, Q Mu. J Mater Sci 28:4883, 1993.
79. GS Sheu, SS Shyu. Composites Sci Technol 52:489, 1994.
80. A Mathur, AN Netravali. J Mater Sci 31:1265, 1996.
81. HP Schreiber, MR Wertheimer, M Lambla. J Appl Polym Sci 27:2269, 1982.
82. HP Schreiber, YB Tewari, MR Wertheimer. J Appl Polym Sci 20:2663, 1976.
83. A Bialski, SJ Manley, MR Wertheimer, HP Schreiber. J Macromol Sci Chem A10: 609, 1976.
84. DW Wu, W (Voytek), S Gutowski, S Li, HJ Griesser. J Adhesion Sci Technol 9: 501, 1995.
85. W Petasch, E Raeuchle, M Walker, P Elsner. Surface Coatings Technol 75:682, 1995.
86. J Comyn, L Mascia, G Xiao, BM Parker. Int J Adhesion Adhesives 16:97, 1996.
87. N Inagaki, S Tasaka, M Matsumoto. Macromolecules 29:1642, 1996.
88. A Tasaka, A Komura, Y Uchimoto, M Inaba, Z Ogumi. J Polym Sci A Polym Chem 34:193, 1996.
89. M Cheaib, M Brassier, R Bosmans, P Leprince. Le Vide Sci Technol Appl Suppl 275:96, 1995.
90. K Singh, DR Gilbert, J Fitz-Gerald, S Harkness, DG Lee. Science 272:396, 1996.
91. JM Burkstrand. J Vac Sci Technol 15:223, 1978.
92. LJ Gerenser. J Vac Sci Technol A 6:2897, 1988.
93. S Nowak, R Mauron, G Dietler, L Schlapbach. In: KL Mittal, ed. Metallized Plastics 2: Fundamental and Applied Aspects. New York: Plenum, 1991, p 233.

94. A Nihlstrand, T Hjertberg, K Johansson. J Adhesion Sci Technol 10:123, 1996.
95. G Rozovskis, J Vinkevicius, J Jaciauskiene. J Adhesion Sci Technol 10:399, 1996.
96. MK Shi, A Selmani, L Martinu, E Sacher, MR Wertheimer, A Yelon. J Adhesion Sci Technol 8:1119, 1994.
97. JE Klemberg-Sapieha. In R d'Agostino, ed. Plasma Treatment and Deposition of Polymers. Dordrecht, The Netherlands: Kluwer Academic Publishers, 1997, p 233.
98. JE Klemberg-Sapieha, MK Shi, L Martinu, MR Wertheimer. Proceedings SPE Annual Technical Conference ANTEC 95, Boston, 1995.
99. MK Shi, L Martinu, E Sacher, A Selmani, MR Wertheimer, A Yelon. Surf Interface Anal 23:99, 1995.
100. MK Shi, JE Klemberg-Sapieha, G Czeremuszkin, L Martinu, E Sacher. Proceedings 12th International Symposium on Plasma Chemistry ISPC 12. In: JV Heberlein, DW Ernie, JT Roberts, eds. Minneapolis: University of Minnesota, 1995 p 51.
101. S Vallon, B Drévillon, F Poncin-Epaillard, JE Klemberg-Sapieha, L Martinu. J Vac Sci Technol A 14:3194, 1996.
102. JE Klemberg-Sapieha, D Poitras, L Martinu, NLS Yamasaki, CW Lantman. J Vac Sci Technol A 15(3):985, 1997.
103. S Vallon, R Brenot, A Hofrichter, B Drévillon, A Gheorghiu, C Sénémaud, JE Klemberg-Sapieha, L Martinu, F Poncin-Epaillard. J Adhesion Sci Technol 10: 1313, 1996.
104. CM Ferreira, M Moisan, eds. Microwave Discharges: Fundamentals and Applications. NATO ASI Series B: Physics, 302. New York: Plenum, 1993.
105. M Moisan, J Pelletier. In Ref. 104, p 8.
106. B Mutel, O Dessaux, P Goudmand, F Luchier. Rev Phys Appl 25:1019, 1990.
107. F Normand, A Granier, P Leprince, J Marec, MK Shi, F Clouet. Plasma Chem Plasma Process 15:173, 1995.
108. B Chevet. Le Vide Sci Technol Appl Suppl 275:104, 1995.
109. RG Spahn, LJ Gerenser. US Patent 5,324,414, 1994.
110. JM Grace, DR Freeman, R Corts, W Kosel. J Vac Sci Technol A 14:727, 1996.
111. TA Wilde, H Grünwald, G Stipan, K Naunenburg. In: JE Harry, ed. Proceedings 11th International Symposium on Plasma Chemistry ISPC 11, Loughborough, UK, 1993, p 1198.
112. JE Steinmann. Int Plastics Mag, May 1993.
113. M Hirose, K Nishida. US Patent 4,423,303, 1983.
114. RJ Babacz. US Patent 5,234,723, 1993.

6

Flame Treatment of Polymers to Improve Adhesion

Derek McHardy Brewis and Isla Mathieson
Loughborough University, Loughborough, Leicestershire, United Kingdom

I. INTRODUCTION

Good adhesion to polymers is required in a number of important technologies including adhesive bonding, printing, and painting. To achieve a satisfactory level of adhesion it is often necessary to pretreat the polymer by one of a wide range of methods. Two books are of particular interest [1,2]. In the case of polar polymers such as nylon 66 and epoxide thermosets, a treatment may not be necessary, or if the surfaces are contaminated, a physical method such as solvent degreasing or grit blasting to remove the contaminants may be all that is required. On the other hand, if a polymer lacks suitable functionality, it will be necessary to modify its surface chemically. Polymers with no active functionality include low-density polyethylene (LDPE), high-density polyethylene (HDPE), and polypropylene (PP). A wide variety of methods for introducing new groups is available, including the use of low-pressure plasmas, corona discharges, flames, etchants, and active gases.

Flame treatment to enhance adhesion to polymers has been used since the early 1950s, one of the first applications being to enhance print adhesion to low-density polyethylene. Since that time, flame treatment has been used with many other polymers in a variety of applications. Flame treatment has a number of advantages over the other main method of treating large areas of polymers, i.e., the corona treatment. These include no reverse-side treatments, no creation of pinholes, no ozone production, and better aging characteristics.

This chapter will discuss the applications of flame treatment and give de-

175

tails of equipment used, the mechanisms of combustion, the mechanism of oxidation, and details of individual research studies.

II. COMBUSTION PROCESS

Flame treatment involves the combustion of air and natural gas or a specific alkane, e.g., propane. The hydrocarbons and air are well mixed and then reacted as shown schematically in Fig. 1.

The ratio of air to hydrocarbon for complete combustion is known as the stoichiometric ratio. For example, the complete combustion of 1 volume of methane requires 9.55 volumes of air, so the stoichiometric ratio for an air-methane flame is 9.55:1. In addition to the hydrocarbon/air ratio, the other variables in the treatment are the total flow of the gases, the treatment time, and the position of the polymer in relation to the flame.

An excellent account of the complex combustion process has been provided by Strobel et al. [3]. Some of the main points made are now given.

For all hydrocarbons except methane, there is an initial hydrogen abstraction reaction

$$RH + X \rightarrow R\cdot + HX$$

where X is a radical such as H, O, or OH. The alkyl radicals then thermally decompose.

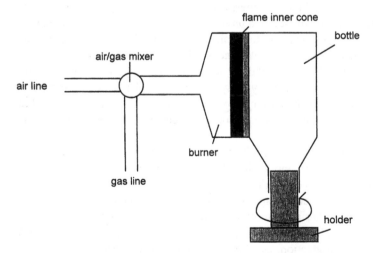

Figure 1 Schematic of the flame treatment of a plastic bottle.

$$R \cdot \rightarrow \text{alkene} + R' \cdot$$

The alkyl radicals and alkenes are oxidized to carbon monoxide.

If methane is the fuel, the first step of combustion is reaction with a radical, OH being especially important:

$$CH_4 + X \rightarrow CH_3 \cdot + HX$$

The methyl radical then either (1) oxidizes to carbon monoxide by reactions with oxygen and other radicals or (2) recombines to form ethane, ethyl radicals, or ethene.

During the combustion of hydrocarbons, water is formed in a number of reactions, including

$$CH_x + OH \rightarrow CH_{x-1} + H_2O$$
$$CH_2O + OH \rightarrow CHO + H_2O$$
$$H_2 + OH \rightarrow H + H_2O$$

The final stage of hydrocarbon combustion is the conversion of CO to CO_2, mainly in the luminous portion of the flame. The most important reaction is

$$CO + OH \rightarrow CO_2 + H$$

and it is this reaction that causes the major heat evolution in hydrocarbon oxidation.

III. INDIVIDUAL STUDIES

In one of the first detailed studies of the flame treatment, Ayres and Shofner [4] examined most of the key variables, namely the nature of the gas, the air-to-gas ratio, the effect of contact time, and the distance of the polymer from the flame. The nature of the polyolefin used was not stated. The effectiveness of the treatments was assessed using a tape peel test and the adsorption of a dye. The authors found that higher levels of treatment were achieved with an excess of air over the stoichiometric requirements to burn all the propane (Fig. 2); this was also true with methane and butane. The authors concluded that the optimal treatment time was about 0.02 s and that the optimal distance between the polymer and the top of the inner cone was about 10 mm.

The study of Ayres and Shofner [4] was carried out without access to X-ray photoelectron spectroscopy (XPS), a technique that enables a quantitative elemental analysis of the surface regions of a polymer to be carried out. Briggs et al. [5] utilized this technique to study the flame treatment of LDPE. It was found that high levels of oxygen and significant levels of nitrogen were intro-

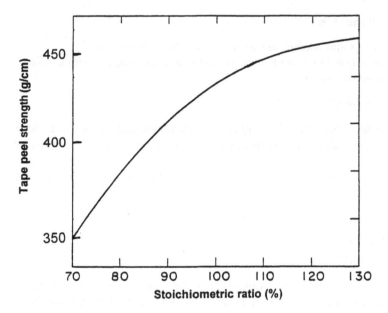

Figure 2 Effect of air/propane ratio in terms of % of stoichiometric ratio on the peel strength. (From Ref. 4.)

duced into the surface. It was calculated that the depth of chemical modification was only about 9 nm, which is much less than for chromic acid treatment, for example. There were no significant changes in the XP spectra or the adhesion levels after 12 months storage of pretreated LDPE.

Garbassi et al. [6] found that the flame treatment of polypropylene resulted in large increases in the adhesion of polyurethane and acrylic paints to the polymer. Curve fitting of the C1s spectra indicated that hydroxyl and carbonyl were the predominant groups formed, although some carboxyl groups were formed after repeated flame treatment (Fig. 3).

Sutherland et al. [7–11] carried out a detailed study of the pretreatment of various PP types. Improvements in adhesion were assessed by means of a butt test. In some cases the adhesion to an epoxide adhesive was assessed, but in many cases the PPs were first coated with a polyurethane paint so that improvements in paint adhesion could be determined (Fig. 4).

Changes in adhesion to a PP homopolymer after flame treatment under a variety of conditions are given in Table 1 [7]. It can be seen that good adhesion is obtained under a broad range of conditions. The lack of sensitivity to the air/ gas ratio is shown in Fig. 5, where a broad peak for surface oxygen and a broad trough for water contact angle can be seen. Likewise, the oxygen concentration

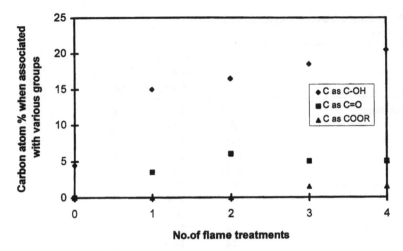

Figure 3   Flame treatment of polypropylene showing concentration of different groups introduced. The speed of the polymer was 15 m min^{-1} and the distance from the flame was 40 mm. (From Ref. 6.)

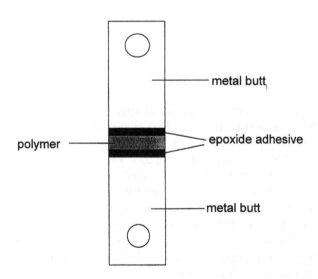

Figure 4   Schematic representation of butt joint test.

Table 1 Adhesion Tests of Polyurethane-Painted, Flame-Treated Polypropylene

Property		Tensile strength[a] (MPa)	Locus of failure[b]
Air-to-gas ratio[c]	8:1	24.7	C
	9:1	25.4	C
	11:1	26.4	C
	13:1	26.7	C
	14:1	25.8	C
Total flow rate[d] (L/min)	12	26.0	I
	18	25.6	C
	24	26.4	C
	36	27.2	C
	48	24.0	C
Distance from inner cone tip[e] (mm)	2.5	22.8	C
	10	26.4	C
	20	22.1	C
	40	6.5	I
	60	4.2	I

[a] For comparison, the tensile strength involving PP wiped with trichloroethane was 2 MPa.
[b] I, interfacial between paint and polymer; C, complex failure including cohesive within PP.
[c] Total flow rate 24 L min^{-1}; distance 10 mm.
[d] Air-to-gas ratio 11:1; distance 10 mm.
[e] Air-to-gas ratio 11:1; total flow rate 24 L/min.
Source: Ref. 7.

and water contact angles are insensitive to changes in total flow rate above a value of about 24 liters per minute (Fig. 6). The effect of the distance between the flame and polymer was more pronounced (Fig. 7).

A similar lack of sensitivity to air/gas ratio and total gas flow rate (above a certain minimum) was observed with a rubber-modified propylene-ethylene block copolymer [8] and a polypropylene blended with EPDM [9].

Vapor-phase derivatization of treated PP homopolymer with trifluoroacetic anhydride (TFAA) showed that for a wide variety of treatment conditions about 20% of the oxygen introduced was in the form of hydroxyl groups. This compared with about 30% for both the block copolymer and the EPDM terpolymer [9]. TFAA reacts with hydroxyl groups as follows [10].

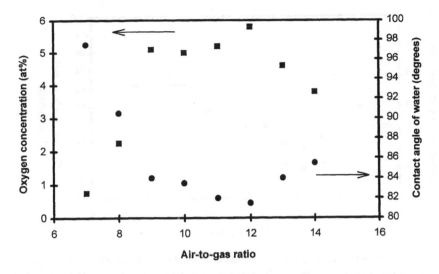

Figure 5 Effect of gas composition (air and natural gas) on the surface chemistry and water contact angle of polypropylene. (From Ref. 7.)

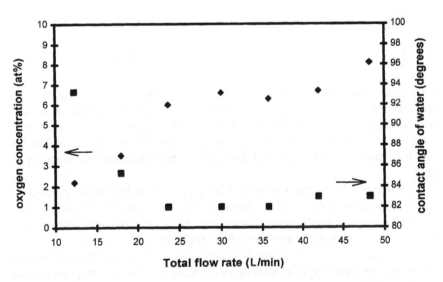

Figure 6 Effect of total gas (air and natural gas) flow rate on the surface chemistry and water contact angle of polypropylene. The air-to-gas ratio was 11:1 and the distance from the inner cone tip was 10 mm. (From Ref. 7.)

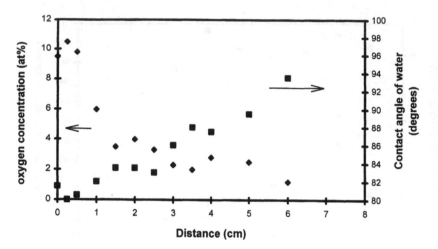

Figure 7 Effect of the distance between the flame and the polymer on the surface chemistry and water contact angle of polypropylene. The total flow rate was 24 L min^{-1} and the air-to-gas ratio was 11:1. (From Ref. 7.)

$$-CH_2-\underset{\overset{|}{OH}}{CH}- \quad + \quad CF_3-\overset{\overset{O}{\|}}{C}-O-\overset{\overset{O}{\|}}{C}-CF_3 \quad \longrightarrow \quad -CH_2-\underset{\overset{|}{O}}{CH}- \quad + \quad CF_3-\overset{\overset{O}{\|}}{C}-OH$$

Variations in composition as a function of depth were studied using different take-off angles in XPS [11]. The results indicated an enrichment of oxygen concentration near the surface at a total gas flow rate of 33.9 L/min; with more intense flames (67.8 L/min) the maximal oxygen concentration was below the outermost surface. This difference has been discussed by Sheng et al. [11].

Wu et al. [12] treated sheets of HDPE, a polypropylene homopolymer PP, and a propylene-ethylene rubber. Subsequently, these treated polyolefins were assembled with styrene-butadiene copolymer SBR sheets containing benzoyl peroxide and the two sheets were then heated together under pressure.

The authors determined the dispersion and polar contributions to surface energy for the three polymers after up to 10 flame treatments. The polar component for HDPE increased much more than for the other two polymers, indicating a greater concentration of O-containing groups with HDPE.

The authors also determined the peel strengths for the treated polyolefin-SBR laminates in air and also when the failure zone was in contact with liquid polydimethylsiloxane (PDMS); the results for PP and HDPE are shown in Figs. 8 and 9, respectively. The higher peel strength in air with treated HDPE is in line with a greater concentration of functional groups. Wu et al. attributed the much better resistance to PDMS in the case of HDPE to the existence of chemical bonds. It was concluded that substantial unsaturation was produced by flame treatment of HDPE and this could result in cross-linking to the SBR via radicals produced by the benzoyl peroxide.

Strobel et al. [3] made a detailed study of a biaxially oriented PP homopolymer. In particular, they examined the effects of gas stoichiometry and the flame-film distance on wettability and surface chemistry. They expressed their results in terms of the equivalence ratio Φ, which is defined as the stoichiometric air/fuel ratio divided by the actual air/fuel ratio. Thus a fuel-lean mixture will have a value of $\Phi < 1$, whereas with a fuel-rich mixture, $\Phi > 1$. Using scanning electron microscopy at a resolution of less than 50 nm, corresponding to a magnification of $\times 30,000$, they did not detect any changes in topography even after severe treatments.

To assess the effect of flame treatment under various conditions, Strobel

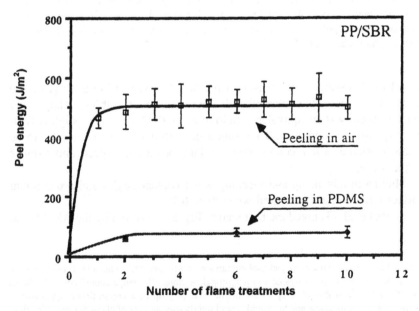

Figure 8 Effect of flame treatment on the peel strengths between PP and SBR in air or in polydimethylsiloxane. The total flow rate (air and methane) was 36 L min⁻¹ and the distance between the polymer and the upper planar flame front was 8 mm. A stoichiometric mixture of gases was used. (From Ref. 12.)

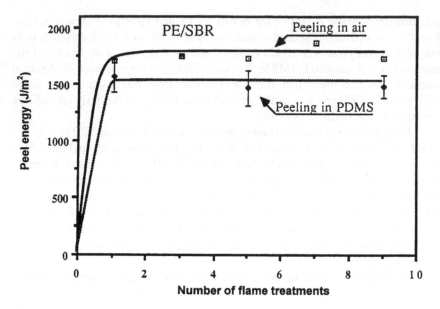

Figure 9 Effect of flame treatment on the peel strengths between PE and SBR in air or in polydimethylsiloxane. The total flow rate (air and methane) was 36 L min^{-1} and the distance between the polymer and the upper planar flame front was 8 mm. A stoichiometric mixture of gases was used. (From Ref. 12.)

et al. [3] used contact angle measurements, the ASTM wetting test* [13], and XPS. Figure 10 shows the effect of equivalence ratio on wettability. They found optimal wettability at an equivalence ratio of about 0.93, i.e., under slightly fuel-lean (oxygen-rich) conditions. They found the optimal equivalence ratio to be the same for all the various combinations of flame power, film speed, and distance from the flame.

Minima in advancing and receding water contact angles and a maximum in surface oxidation also occurred when $\Phi \approx 0.93$.

Strobel et al. [3] noted the best wettability, as measured by the ASTM test,

---

* A range of formamide and 2-ethoxyethanol mixtures with different surface tensions is used to study the wetting behavior of polyethylene and polypropylene films. The range should cover 30–46 in steps of 1 mN/m and thereafter up to 56 mN/m. The very tip of a cotton-tipped applicator is wet with one of the mixtures and the liquid spread lightly over an area of about 6.5 cm^2. The time required for the continuous film of liquid to break up into droplets is noted. If this takes longer than 2 s, a mixture with a higher surface tension should be taken and the procedure repeated. The mixture that comes closest to wetting the film for 2 s is taken as the wetting tension of the plastic film.

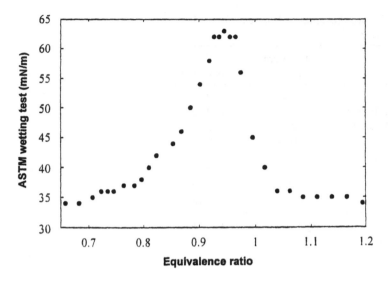

Figure 10 Effect of equivalence ratio on the wetting tension of flame (air and natural gas) treated polypropylene. (From Ref. 3.)

when the distance between the tips of the flame cones and the PP was about 2 mm. However, some improvement in wettability was still evident at a distance of 20 mm.

Strobel et al. [14] compared five gas-phase methods for treating polypropylene and poly(ethylene terephthalate), namely flame, corona, remote air plasma, ozone, and ultraviolet (UV)-ozone. Whereas the first three methods caused major chemical modification in less than 1 s, the ozone treatments required orders of magnitude greater exposure time to cause similar modifications. Some of the results for polypropylene are given in Table 2.

Most studies of flame treatments have involved polyolefins. However, Mathieson et al. [15] showed that the treatment was very effective with partially fluorinated polymers (Table 3). With poly(tetrafluoroethylene) (PTFE), there was no significant change in surface chemistry and the adhesion level actually fell. With the other two polymers large increases in adhesion and substantial changes in surface chemistry occurred. Treatment of poly(vinyl fluoride) resulted in the introduction of oxygen but no significant defluorination. However, with ECTFE copolymer, introduction of oxygen was accompanied by substantial dehalogenation.

Cope [16] reported up to a three-fold increase in lap shear strength after flame treatment of nylon 11. Dillard et al. [17] studied the flame treatment of sheet molded composite (SMC), which is a glass fiber–reinforced polyester containing

Table 2 Surface Properties of Variously Treated
Polypropylene Films

Treatment	Exposure time (s)	XPS O/C atomic ratio[a]
None	—	0.0
Corona (1.7 J/cm^2)	0.5	0.12
Corona (0.17 J/cm^2)	0.05	0.07
Flame	0.04	0.12
Remote air plasma	0.1	0.12
Ozone	1800	0.13
UV-air	600	0.08
UV-air plus O$_3$	600	0.14

[a] At a 38° electron take-off angle with respect to the surface.
Source: Ref. 14.

Table 3 Flame Treatment of Poly(vinyl fluoride) (PVF),
Poly(tetrafluoroethylene) (PTFE), and Ethylene-Chlorotrifluoroethylene
(ECTFE) Copolymer[a]

Treatment time (s)	XPS (atom %)				Failure load (N)
	C	Cl	F	O	
PVF none	70.4	—	28.8	0.8	360
PVF 0.06	67.6	—	28.0	4.4	3240
PTFE none	38.4	—	61.6	—	420
PTFE 0.04	34.0	—	66.0	—	80
ECTFE none	53.2	14.3	32.5	—	240
ECTFE 0.06	68.8	8.0	17.2	6.0	2980

[a] Composite lap shear specimens were used for the bond strength measurements.
  The joints consisted of a steel strip/adhesive/fluoropolymer/adhesive/steel strip.
  The overlap area was 20 mm (wide) × 10 mm (long).
Source: Ref. 15.

several other components. XPS showed large increases in oxygen-containing
functional groups but the increases in lap shear strength were modest. The poor
adhesion was attributed to a cohesively weak layer.

IV. GENERAL DISCUSSION

Flame treatment is a highly effective method of treating regular shapes such as
film, sheet, and bottles with cylindrical sides. The treatment time is short and

good results can be obtained with a wide range of the other variables. An oxygen-rich mixture is usually recommended [3–5,18]. A wide range of total gas flows, above a certain minimal level, will give effective treatments. The polymer should be a few millimeters from the inner cone tip. Polymers that have been success-fully treated include LDPE, HDPE, PP, nylon 11, poly(ethylene terephthalate), poly(vinyl fluoride), and ethylene-chlorotrifluoroethylene copolymer.

Strobel et al. [3] found optimal oxidation and wetting at an equivalence ratio of about 0.93. This did not correspond to the maximal temperature, which they calculated to occur when $\Phi \approx 1.03$. However, they calculated that the con-centrations of hydroxyl radicals and oxygen atoms were at a maximum when $\Phi \approx 0.95$. Furthermore, the decline in [OH] and [O] is more pronounced on the hydrocarbon-rich side (right-hand side) of the optimal value (Fig. 11). The asym-metry closely follows that which they observed with wettability (Fig. 10). There-fore, these authors seem justified in concluding that the concentrations of OH radicals and oxygen atoms are of paramount importance in flame treatment of polymers. In contrast, they concluded that UV radiation and oxygen molecules (see Fig. 11) and ions (calculated to be at a concentration of 10^{-6} or less relative to the unionized species) were not important.

The initial step in the chemical modification of polymers by flame treatment is probably polymer radical formation by hydrogen abstraction [3]. The most likely processes are

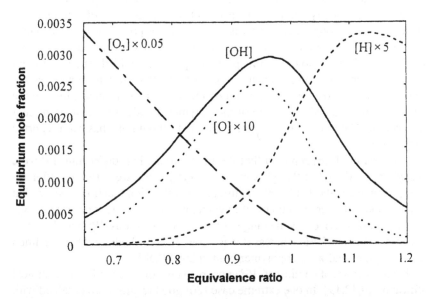

Figure 11 Calculated concentrations of O_2 molecules and H, O, and OH radicals as a function of equivalence ratio for air-methane flames. (From Ref. 3.)

$$RH + O\cdot \rightarrow R\cdot + OH\cdot$$

$$RH + OH\cdot \rightarrow R\cdot + H_2O$$

The alkyl radicals will then react with oxygen atoms and molecules to form alkoxy and peroxy radicals, respectively.

$$R\cdot + O \rightarrow RO\cdot$$

$$R\cdot + O_2 \rightarrow RO_2\cdot$$

The peroxy radicals then react by abstracting hydrogen from other polymer chains to produce hydroperoxides (ROOH). Decomposition of the hydroperoxides leads to the formation of a variety of functional groups. Alkoxy radicals decompose via β-scission to produce ketones [3]:

$$-C-C-C \rightarrow -C=O + -C\cdot$$
$$|$$
$$O\cdot$$

Alcohols will be formed by the reaction between hydroxyl and alkyl radicals.

$$R\cdot + OH\cdot \rightarrow ROH$$

The depth of chemical modification of flame treatment is just a few nanometers. Until fairly recently, reflection infrared techniques have not been useful in providing information on these modifications. However, with modern equipment evidence of chemical modification has been obtained. For example, Sheng et al. [19] found that carbonyl groups had been introduced into LDPE although no evidence for these groups was found with PP [9]. This is probably due to a combination of greater oxidation within a given layer and a greater treatment depth [19]. However, X-ray photoelectron spectroscopy has provided much important information even with polypropylene. Typically, flame treatment introduces between 5 and 15 at% oxygen into polymers, the actual amount depending on the treatment conditions and the nature of the polymer. In addition, a small amount of nitrogen may be introduced into polyethylene [19] but none has been reported in treated PP.

Garbassi et al. [6] reported that the main groups introduced into PP were hydroxyl and carbonyl, although some carboxyl groups were formed after repeated treatment. Sutherland et al. [9] used a derivatization reaction to show that for a particular polymer the amount of oxygen introduced in the form of hydroxyl groups was constant over a wide range of treatment conditions. With a PP homopolymer it was about 20%, whereas it was about 30% for both a rubber-modified block copolymer and a homopolymer containing EPDM.

For a given set of conditions, PE appears to be more susceptible to chemical modification [9,12,19]. In one extreme case [20] good results were obtained with PE but no significant improvement was obtained with PP.

Various studies of the corona discharge treatment have shown that it is subject to aging. This is manifested, in particular, by a decrease in surface energy with time. The flame treatment seems to produce a more stable surface, but further work is required to confirm this.

By introducing various functional groups, the flame treatment will increase the surface energy of the treated polymer. This will improve wetting by the liquid phase, i.e., adhesive, paint, or printing ink. In addition, the interaction between the liquid phase and the treated polymer will be increased. In the extreme, chemical bonding may occur [12].

V. CONCLUSIONS

The flame treatment is a rapid and effective method for treating many polymers.

A variety of functional groups is introduced by the treatment, including hydroxyl and carbonyl.

The depth of chemical modification is only a few nanometers.

The treated surface is relatively stable.

REFERENCES

1. M Strobel, CS Lyons, KL Mittal, eds. Plasma Surface Modification of Polymers: Relevance to Adhesion. Zeist, The Netherlands: VSP, 1994.
2. KL Mittal, ed. Polymer Surface Modification: Relevance to Adhesion. Zeist, The Netherlands: VSP, 1996.
3. M Strobel, MC Branch, M Ulsh, RS Kapaun, S Kirk, CS Lyons. J Adhesion Sci Technol 10:515, 1996.
4. RL Ayres, DL Shofner. SPE J 28(12):51, 1972.
5. D Briggs, DM Brewis, MB Konieczko. J Mater Sci 14:1344, 1979.
6. F Garbassi, E Occhiello, F Polato. J Mater Sci 22:207, 1987.
7. I Sutherland, DM Brewis, RJ Heath, E Sheng. Surf Interface Anal 17:507, 1991.
8. E Sheng, I Sutherland, DM Brewis, RJ Heath. Surf Interface Anal 19:151, 1992.
9. I Sutherland, E Sheng, DM Brewis, RJ Heath. J Adhesion 44:17, 1994.
10. I Sutherland, E Sheng, DM Brewis, RJ Heath. J Mater Chem 4(5):683, 1994.
11. E Sheng, I Sutherland, DM Brewis, RJ Heath. Appl Surf Sci 78:249, 1994.
12. DY Wu, E Papirer, J Schultz. J Adhesion Sci Technol 7:343, 1993.
13. ASTM D-2578-84.
14. M Strobel, MJ Walzak, JM Hill, A Lin, E Karbashewski, CS Lyons. In: KL Mittal, ed. Polymer Surface Modification: Relevance to Adhesion. Zeist, The Netherlands: VSP, 1996, p 233.
15. I Mathieson, DM Brewis, I Sutherland, RA Cayless. J Adhesion 46:49, 1994.

16. BC Cope. Proceedings, Adhesive Sealants and Encapsulants Conference, London, 1985, pp 355–359.
17. JG Dillard, TF Cromer, CE Burtoff, AJ Cosentino, RL Cline, GM MacIver. J Adhesion 26:181, 1988.
18. CW Hurst, RE Schanzle. Mod Packaging 40:163, October 1966.
19. E Sheng, I Sutherland, DM Brewis, RJ Heath, RH Bradley. J Mater Chem 4(3):487, 1994.
20. AR Carter. J Adhesion 12:37, 1981.

7

Corona Discharge Treatment of Polymers

Tohru Uehara
Shimane University, Shimane, Japan

I. INTRODUCTION

The three states of matter are solid, liquid, and gas. A plasma state exists as its fourth state. A plasma consists of positively charged particles and negatively charged electrons existing at almost the same electrical density, it is overall electrically neutral, and it was named plasma by Langmuir in 1928. The easiest way to obtain a plasma state is to induce an electrical discharge in a gas.

A corona discharge treatment is a kind of plasma treatment. Plasmas are classified roughly into two categories: equilibrium plasmas and nonequilibrium plasmas. In equilibrium plasmas, the temperatures of electrons and of the gas are the same. Mainly equilibrium plasmas have been studied, and temperatures of approximately 10,000°C have been reported. In nonequilibrium plasmas the gas is at ambient temperature, but the temperature of electrons is very high (about 10,000°C). These nonequilibrium plasmas are used in chemical applications and are called low-temperature plasmas or cold plasmas. The low-temperature plasmas are classified roughly into two categories: (1) ordinary low-temperature plasmas at low pressure and (2) corona discharges at atmospheric pressure. Ordinary low-temperature plasmas are widely used in chemical modification of the surfaces of materials, especially in semiconductor industries [1] as well as for polymers [2].

Glow discharges are the most widely used technique for low-temperature plasmas. Because the mean free path of activated gas molecules is longer in a vacuum, a greater distance between the electrodes and the samples treated can be used. The typical example of this is the CASING (cross-linking by activated species of inert gases) technique [3]. In this case, activated gas species produced by the discharge react with the polymer surfaces to be treated and induce cross-linking at such surfaces. If samples are placed between the electrodes, the sample surfaces are subjected to both the bombardments of high-temperature electrons and the attack of activated gas species.

The types of electrodes used can be classified into two categories, i.e., capacity-type (capacitor) and induction-type (inductance coil) electrodes. These correspond to the kinds of electrical sources used (direct current, commercial alternating current, and high frequency). In connection with this, many types of discharging cells exist.

All electrical sources for corona discharge treatments generate 15–30 kV, and their frequencies are 50 Hz, 60 Hz, or 1–50 kHz. Electrical sources that can alter their frequencies continuously are rare. Therefore, effects of frequency on treatment results were not examined for the same sample under the same voltage and current conditions.

II. EXPERIMENTAL RESULTS

Because the main advantage of corona discharge treatments is that no evacuation system is needed, the equipment investment necessary is much lower than for ordinary low-temperature plasmas installations. Therefore, corona discharge treatments were used in earlier times to improve the wettability [4] and printability [5] of polymer surfaces.

For laboratory use, a commercial neon transformer (15 kV, 60 Hz, 165 W) is useful as an electrical source for corona discharge treatment [6]. Two classes of electrodes can be used: batch-type electrodes (Fig. 1) [6] and continuous-type electrodes [7]. In the latter case, one of the two electrodes used consists of a cylinder that is covered with an insulation material, and when such an electrode is used the corona discharge treatment is known as the covered-roll method (Fig. 2).

Corona discharge treatments have been used mainly for the surface treatment of polyolefins, i.e., for improvement of the autohesion of polyethylene (Table 1) [6,8,9]. Furthermore, morphological changes (the formation of dendrites to increase the roughness of the treated surface) were found with corona treatment in oxygen or oxygen-containing gases [6,9,10], and the extent of surface roughness and the size of the dendrites increased with the time of treatment. Furthermore, the thermally induced bond between wood and polyethylene and polystyrene was improved by treatment of the wood surface in a corona discharge prior to bonding [11].

In corona discharge treatments, an optimal degree of treatment must be adopted. It is not true that the more a sample is corona treated, the more its bond

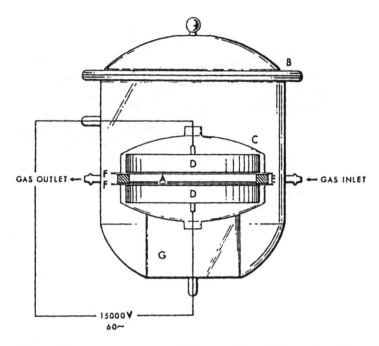

Figure 1 A batch-type corona discharge cell. A, Polymer sheet; B, reaction kettle; C, truncated glass bottle; D, stainless steel electrodes; E, glass spacer; F, 0.76-mm glass plates; G, glass cylinder. (From Ref. 6.)

strength and printability are improved [12]. The bond strength decreases with overtreatment [13]; i.e., a treated material fails cohesively, because of the degradation caused by the overtreatment.

Water contact angles decreased with increasing treatment time (Fig. 3) [14], and there was an increase in the wettability of the corona-treated low-density polyethylene [15]. Wettability is frequently used as a parameter to evaluate the effectiveness of the treatment and improves up to certain level with an increase in the degree of treatment [16]. However, the extent of wettability cannot be observed if the material has been overtreated. In the earlier papers on this subject the wettability was not determined, and the effects of corona discharge treatments were estimated by determining the autohesion strengths [6,10]. Later, the surface free energy of treated surfaces was estimated from contact angle measurements.

The polar component of surface energy, γ_s^p, is the key to understanding the changes in the adhesion behavior of the films after treatment [17]. The dispersion component of surface energy, γ_s^d, changed little (Fig. 4), as was observed with cellophane [18]. The strength of bonds between sheets of cellophane treated in

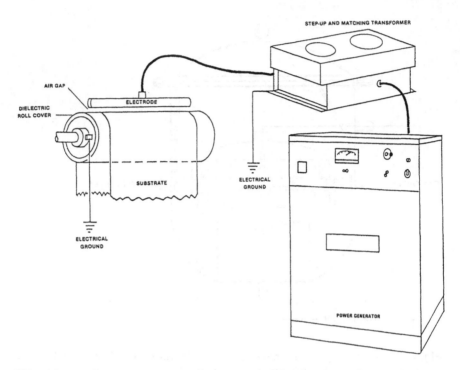

Figure 2   Continuous-type corona treatment equipment. (From Ref. 7.)

Table 1   Effect of Temperature of Lamination and Sample Density on Autohesion of Untreated and Corona-Treated Polyethylene

		Bond strength (kg/cm^2) at temperature of lamination					
		16°C		36°C		55°C	
Sample	Density (g/cm^3)	Untreated	Treated	Untreated	Treated	Untreated	Treated
A	0.919	0.3	12	0.8	>20	2.0	>20
B	0.928	0	1.5	0.5	7.7	0.9	>20
C	0.935	0	0	0.3	1.3	0.7	3.7
D	0.948	0	0	0	0	0	1.8

Source: Ref. 9.

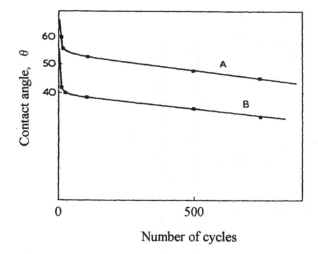

Figure 3 Water contact angles (A, θ_A advancing; B, θ_r receding) versus treatment time (pass times between electrodes). Poly(ethylene terephthalate) (PET), 16 kV, 20 kHz; the rotation speed of the roller was 24 cycles/min. (From Ref. 14.)

a corona discharge was related to changes in solid surface free energy accompanying the treatment. Significant increases in the polarity of the corona-treated surfaces were found (Fig. 5).

From reactions of the corona-treated surfaces with diphenyl-picrylhydrazyl (DPPH), a compound that readily transfers its radical to other species, it was determined that corona treatments produced fairly stable peroxide structures of the forms RO_2R and RO_3R on polyethylene surfaces [17]. A new infrared (IR) band was also noticeable at 1375 cm^{-1} and was related to the C—CH_3 deformation band [19]. Such experimental results support a mechanism of dipole orientation for the origin of piezoelectricity in poly(vinylidene fluoride) [20].

The effect of corona discharge treatment decayed with time after treatment for polyethylene films. The characteristics of the decay of the surface potential of corona-charged polyethylene film in air were investigated, and local differences in the decay rates were observed [21,22].

Corona discharge treatment is also effective for metal surfaces. Aluminum and titanium surfaces have been treated by corona discharge in air, which resulted in bonds of strength similar to those obtained by conventional chemical treatments [23].

The selection of corona treatment equipment [24] and process parameters [25], e.g., cooling rate, influences the effectiveness of corona discharge treatment. Such influence has even been shown at industrial levels.

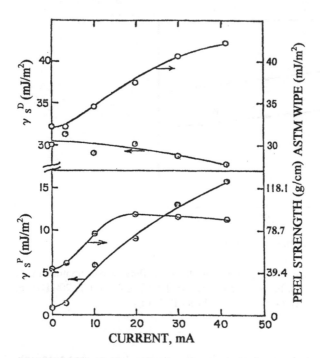

Figure 4 Wetting and adhesion results. Conditions: polyethylene; speed 30.5 m/min; frequency 1900 Hz; thickness of dielectric layer 0.361 mm; gap 1.52 mm; peel strength, ASTM D-2141-65T. γ_s^D the dispersion component and γ_s^P, the polar component of the solid surface energy. (From Ref. 16.)

All the basic experimental facts reported here about the improvement of bond strength by corona discharge treatment were found in the 1970s.

III. WHAT HAPPENS IN CORONA DISCHARGE?

Two theories have been advanced for the effects of corona discharge treatments: electret formation [26] and chemical changes [27–29].

Electrically neutral molecules have negatively charged and positively charged sites on the same molecule. All molecules near the surfaces of materials are oriented by external effects (for example, electrical and/or magnetic fields), the surface being charged positively or negatively. This state is called the electret. Autohesion by the electrical forces of this electret constitutes the theory of electret

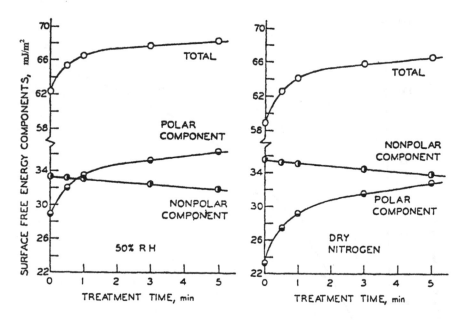

Figure 5 Surface free energy components calculated from contact angle data on the "air side" of corona-treated cellophane samples. (From Ref. 18.)

formation. This theory is especially important for explaining the effect of nitrogen corona [26] on poly(vinylidene fluoride) [20].

The theory of electret formation was popular in earlier times, because of the lack at that time of adequate techniques for surface analysis. Since the development of new surface analysis techniques, a theory involving surface chemical changes has become more accepted. However, the differences in the real effects induced by corona discharge treatment are not established, and these two theories still compete with each other.

Nuclear magnetic resonance (NMR), which is widely used for determining chemical structures of organic compounds, is not useful for surface analysis of plasma-treated materials. The analytical methods for treated surfaces are contact angle measurements [16], adsorption of dyes [30], infrared spectroscopy [10,27], X-ray photoelectron spectroscopy (XPS) (Fig. 6) [31,32], high-resolution electron energy loss spectroscopy (HREELS) [33], and microscopic observations (atomic force microscopy (AFM) [34], scanning electron microscopy (SEM) [6,10,30] etc.). Increments of the C═O absorption band from IR and of the O/C ratio from XPS analysis are reported in many papers.

Even if oxygen is introduced on the surface, the changes in chemical structure on the treated surfaces have not been clarified. Which carbon site is attacked

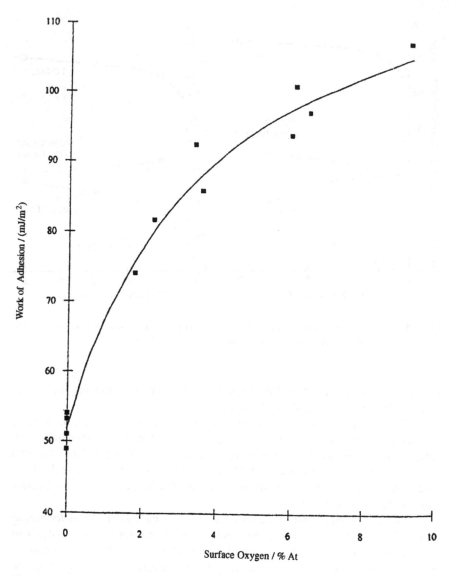

Figure 6 The work of adhesion with water of corona-treated polyolefins versus surface oxygen concentration as determined by XPS. The biaxially drawn films (thickness ~25 µm) consisted of coextruded polypropylene cores with melt coats (thickness ~5 µm) of polyethylene (PE, ~935 kg/m³) and polypropylene (PP, ~906–908 kg/m³). (From Ref. 31.)

by oxygen? By use of time-of-flight secondary ion mass spectrometry (TOF-SIMS), the chemical structure changes in the surface region can be clarified, but the literature on the application of this technique deals only with ordinary low-temperature plasma treatments and not very extensively [35].

IV. MORE RECENT TRENDS

The more recent trends in the study and application of corona discharge treatments are fairly varied. A large number of reports on the treatment of polyolefins have been published (Fig. 7) [13,31,36–38], but the number of investigations reported has decreased. The effects of corona treatment on fluoropolymers [39],

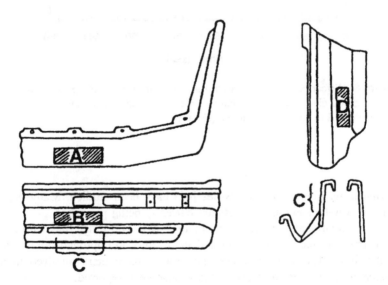

PPCP Treatment	Peel Strength (g/cm)			
	A	B	C	D
With	1,000	1,150	1,050	did not peel
Without	0	0	0	0

Figure 7 Effect of pulse corona–induced plasma chemical process (PPCP) surface treatment for polypropylene automobile bumpers on peel strength between polypropylene (PP) and urethane paint. Pulse peak voltage 50 kV; rise time 50 ns; half-tail 100 ns; 180° peel test. A–D, positions of treatment. (From Ref. 38.)

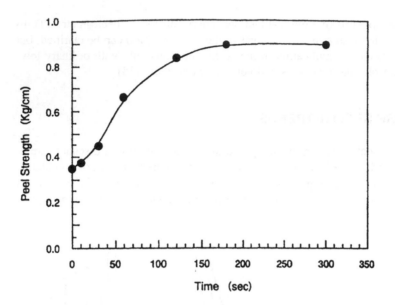

Figure 8 Effect of the hydrogen plasma treatment time on the peel strength between the copper foil clad sheets and partially-cured epoxy impregnated glass cloth. The sheets were preheated to 150°C for 3 hours. (From Ref. 48.)

thermoplastic fiber [40], and textiles [41] have been studied. Many papers on ordinary low-temperature plasmas are currently being published.

The treatment of lubricant-containing plastic film materials with a corona discharge in air has also been reported and resulted in improved storage stability of the films [42]. The principles and process parameters involved in pulsed plasma oil quenching of steels [43] have also been reported.

The reported applications of corona discharge treatments are less numerous than applications reported for ordinary low-temperature plasmas. The reason is that the variety of treatments that exist for ordinary low-temperature plasmas is much greater than that for the corona effect. Corona discharge treatments generally proceed under atmospheric air conditions and are aimed at the oxidation of treated surfaces. Whereas only for ordinary low-temperature plasmas can one select the kinds of gases and the gas pressure to use, selection of a range of voltages, currents, frequencies, and wave shapes is possible for both treatments. As already mentioned, the mean free path of the activated species is very short under atmospheric pressure, and hence the samples must be placed between the

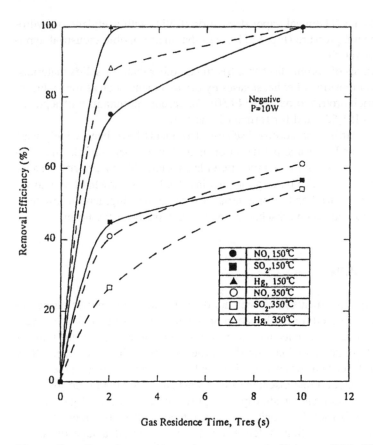

Figure 9   Effect of gas residence time on removal efficiency of NO, SO$_2$, and mercury vapors by the pulse corona–induced plasma chemical process (PPCP). (From Ref. 52.)

electrodes of a corona discharge apparatus. Thus, the very useful remote plasma technique cannot be adopted for corona discharge treatments, and the shape of samples that can be treated is limited to films, plates, or powder.

Only a few fundamental research papers on the more theoretical aspects of corona discharge treatments, for example, the kinetics [44] and the determination of the electron temperature [45], have been published.

Progress in electronic circuitry technology has rendered possible the control of the current wave shape. High-voltage multipulse electrical sources are now available [46], and as a consequence some new applications are expected.

Corona discharge treatments in selected gases are also in the process of being studied—i.e., chemical vapor deposition (CVD) with acetylene or hydrogen at atmospheric pressure (Fig. 8) [47,48]—but their possible industrial applications are not yet clear.

Applications of corona discharge treatment, other than for solids materials, are for the decomposition of exhaust gases by dry treatment: such as for trichloroethylene and trichlorotrifluoroethane [49,50], for gaseous pollutants with a pulsed corona (Fig. 9) [51,52], and for formaldehyde [53].

As stated earlier, the corona discharge treatments have been widely used for applications, and the same is true for ordinary low-temperature plasmas. Experiments with all kinds of low-temperature plasmas have been a priority because of their industrial uses. The basic principles involved are somewhat less known, and hence empirical trial-and-error approaches for the development of low-temperature plasmas and corona discharge treatments have been the norm.

V. CONCLUSION

Corona discharge treatments have been conducted under atmospheric air conditions and have been used for modification of polymer surfaces by oxidative reactions. Because of developments of ordinary low-temperature plasmas, the number of reported investigations of corona discharge treatments has decreased. Now that control of the current wave shape is possible with the progress in electronic circuitry technology, corona discharge treatments can be converted to low-temperature plasma treatments at atmospheric pressure using selected gases. This plasma technique is useful not only for surface treatments of polymers but also for CVD. That no evacuation system needed is the main advantage of corona treatments.

REFERENCES

1. M Sarmadi, F Denes. Tappi J. 79(8):189–204, 1996.
2. M Strobel, CS Lyons, KL Mittal, eds. Plasma Surface Modification of Polymers: Relevance to Adhesion. Zeist: The Netherlands, VSP, 1994.
3. H Schonhorn, RH Hansen. J Appl Polym Sci 11:1461–1474, 1967.
4. JA Coffman, WR Browne. Sci Am 212:90–96, June 1965.
5. K Rossmann. J Polym Sci 19:141, 1956.
6. CY Kim, G Suranyi, DAI Goring. J Polym Sci Part C 30:533–542, 1970.
7. RH Cramm, DV Bibee. TAPPI Proceedings, 1981 Paper Synthetics Conference, 1981, pp 1–11.
8. CY Kim, DAI Goring. J Appl Polym Sci 15:1365–1375, 1971.
9. M Stradal, DAI Goring. J Adhesion 8:57–64, 1976.

10. CY Kim, DAI Goring. J Appl Polym Sci. 15:1357–1364, 1971.
11. CY Kim, DAI Goring. Pulp Paper Mag Can 72(11):93–96, 1971.
12. HL Spell, CP Christenson. Tappi J. 62(6):77–81, 1979.
13. J Friedrich, L Wigant, W Unger, A Lippitz, J Erdmann, H-V Gorsler. Coating 28(4): 123–127, 1995.
14. B Leclercq, M Sotton. Polymer 18:675–680, 1977.
15. A Honkanen, E Laiho, C Bergström. Tappi J. 61(11):93–95, 1978.
16. JF Carley, PT Kitze. Polym Eng Sci 18:326–334, 1978.
17. JF Carley, PT Kitze. Polym Eng Sci 20:330–338, 1980.
18. PF Brown, JW Swanson. Tappi J. 54(12):2012–2018, 1971.
19. B Benaissa, C Mayoux. Angew Makromol Chem 66:155–167, 1978.
20. K Doughty, P Pantelis. J Mater Sci 15:974–978, 1980.
21. EA Baum, TJ Lewis, R Toomer. J Phys D Appl Phys 10:487–497, 1977.
22. EA Baum, TJ Lewis, R Toomer. J Phys D Appl Phys 10:2525–2531, 1977.
23. KW Allen, L Greenwood, TC Siwela. J Adhesion 16:127–132, 1983.
24. LM Salmen. Paper Film Foil Converter 52(10):94–96, 1978.
25. R Krüger, H Potente. J Adhesion 11:113–124, 1980.
26. JM Evans. J Adhesion 5:1–7, 1973.
27. JM Evans. J Adhesion 5:9–16, 1973.
28. DK Owens. J Appl Polym Sci 19:265–271, 1975.
29. DK Owens. J Appl Polym Sci 19:3315–3326, 1975.
30. GJ Courval, DG Gray, DAI Goring. J Polym Sci Polym Lett Ed 14:231–235, 1976.
31. I Sutherland, RP Popat, DM Brewis, R Calder. J Adhesion 46:79–88, 1994.
32. JF Friedrich, W Unger, A Lippitz, T Gross, P Rohrer, W Sauer, J Erdmann, H-V Gorsler, J Adhesion Sci Technol 9:575–598, 1995.
33. CA Fleischer, WP McKenna, G Apai. J Adhesion Sci Technol 10:79–91, 1996.
34. OD Greenwood, RD Boyd, J Hopkins, JPS Badyal. J Adhesion Sci Technol 9:311–326, 1995.
35. D Leonard, P Bertrand, Y Khairallah-Abdelnour, F Arefi-Khonsari, J Amouroux. Surf Interface Anal 23:467–476, 1995.
36. P Rohrer, W Unger, A Lippitz. Gummi Fasern Kunstst 40:382–388, 1994.
37. S O'Kell, T Henshaw, G Farrow, M Aindow, C Jones. Surf Interface Anal 23:319–327, 1995.
38. S Masuda, S Hosokawa, I Tochizawa, K Akutsu, K Kuwano, A Iwata. IEEE Trans Ind Appl 30:377–380, 1994.
39. Y Kusano, M Yoshikawa, I Tanuma, K Naito, M Kogoma, S Okazaki. Symposium on Plasma Science for Materials, 6th Symposium Proceedings, Tokyo, 1993, pp 139–144.
40. BRK Blackman, AJ Kinloch, JF Watts, Composites 25:332–341, 1994.
41. A Nakamura, A Okuno, M Hayashi, H Kawano. R&D Kobe Steel Eng Rep 45(2): 46–51, 1995.
42. KW Gerstenberg. Coating 26(8):260, 262, 1993.
43. F Schnatbaum, A Melber, F Preisser. Stahl 2:48–49, 1994.
44. N Peyraud-Cuenca, P Faucher. J Plasma Phys 54:309–332, 1995.
45. A-M Gomes, J-P Sarrette, L Madon, A Epifanie. J Anal Atomic Spectrom 10:923–928, 1995.

46. T Yamamoto, JR Newsome, DS Ensor. IEEE Trans Ind Appl 31:494–499, 1995.
47. J Salge. Surf Coatings Technol 80(1–2):1–7, 1996.
48. Y Sawada, S Ogawa, M Kogoma. J Adhesion 53:173–182, 1995.
49. T Oda, R Yamashita, T Takahashi, S Masuda. IEEE Trans Ind Appl 32:227–232, 1996.
50. MC Hsiao, BT Merritt, BM Penetrante, GE Vogtlin, PH Wallman. J Appl Phys 78: 3451–3456, 1995.
51. A Mizuno, K Shimizu, A Chakrabarti, L Dascalescu, S Furuta. IEEE Trans Ind Appl 31:957–963, 1995.
52. S Masuda, S Hosokawa, X Tu, Z Wang. J Electrostat 34:415–438, 1995.
53. MB Chang, CC Lee. Environ Sci Technol 29:181–186, 1995.

8

Laser Surface Treatment to Improve Adhesion

Alisa Buchman and Hanna Dodiuk-Kenig
RAFAEL, Haifa, Israel

I. INTRODUCTION

The quality and quantity of adhesion are dependent upon the ability to apply proper surface treatment to the adherents. Both chemical modification and mechanical interlocking induced by surface treatment affect the strength and durability of the adhering system.

Various methods of surface treatment are applied for plastic, metal, composite, and ceramic adherents, among them being abrasive and chemical treatments, thermal welding, and plasma etching.

Excimer laser ultraviolet (UV) irradiation presents a new technology for preadhesion surface treatment and surface modification for various materials and adherends. This technology presents an alternative to use of ecology-unfriendly chemicals involved in conventional etching and abrasive treatments.

The effect of UV radiation on polymers, composites, metals, and ceramics was tested using chemical, physical, and mechanical analytical methods. The effect on adhesional strength and durability was tested on various joint geometries following exposure to a hot, humid environment and open time.

Experimental results, presented in this review, indicated that UV laser surface treatment and modification improved significantly the adhesional shear, tensile, and peel strength as well as other properties such as hardness, wear resistance, conductivity, and appearance.

Optimal UV laser treatment parameters (intensity, repetition rate, and number of pulses) depended on the substrate material and its chemical nature. The mode of failure changed from interfacial to cohesive as the number of UV laser pulses or energy increased during treatment until specific threshold values were reached that are typical of the substrates.

The latter phenomenon resulted in changes in morphology (uniform roughness) as indicated by scanning electron microscopy (SEM) and in chemical modification and removal of contamination as indicated by electron spectroscopy for chemical analysis (ESCA), Auger, Fourier transform infrared (FTIR) spectroscopies and contact angle measurements. A few models tried to predict the radiation effect but none, so far, has been successful. The UV laser affects the material mainly through a photochemical mechanism, but at higher energies a thermal effect is also observed.

It can be concluded that the UV laser excimer has potential as a precise, clean, and simple surface modification technique for a nearly unlimited range of materials. It is restricted to the upper surface and does not cause any damage to the bulk of the material.

II. SURFACE PRETREATMENT FOR ADHESIVE BONDING

A. General

Surface preparation as pretreatment for adhesive bonding is required to attain joint durability as well as joint strength with the various adherends. Surface treatment removes weak boundary layers, cleans the surface, alters the surface energy (primarily through oxidation), and improves microtopographical characteristics. The net effects of these changes are enhanced interfacial bonding, mechanical interlocking between adhesive and adherend, and greater resistance to degradation by moisture or humidity.

Five different mechanisms of adhesion exist that can be enhanced by appropriate surface treatment [1]:

1. Mechanical interlocking—mechanical keying of the adhesive into the substrate.
2. Diffusion—mutual interdiffusion of polymer molecules across the interface (only for polymeric substrate).
3. Electrical—the joint behaves as a capacitor that charges due to contact of two different materials.
4. Adsorption—strength of the joint results from primary and secondary chemical forces across the interface.
5. Acid-base—based on donor-acceptor interaction, occurs widely and follows theories of surface interactions relevant to adhesion [2].

Different adherends require different pretreatment methods. Plastics are the most difficult adherends to treat for the following reasons:

1. The number of different types of plastics available is very high and they differ widely in their behavior and characteristics.
2. The temperature dependence of the mechanical properties of plastics and adhesives is much higher than that of metals or ceramics.
3. Most plastics have a low surface energy. This necessitates severe pretreatment to make plastics wettable by adhesive.
4. Plastics contain numerous components, these can vary considerably in a single group of plastics, and some of them, especially lubricants and plasticizers, hinder adhesion severely. The effect often depends on the migration of the components from the bulk to the surface and also on the temperature influence on their mobility.

B. Pretreatment Processes

To obtain optimal strength of adhesive bonds to adherends, it is necessary to increase the surface energy of the adherend by specific pretreatment processes. These processes can be divided into three groups [3]:

1. Mechanical processes:
   a. Sandblasting
   b. SiC or SACO blasting (using silicate-coated sand blasting material, where the coating is transferred during blasting and adheres to the treated surface)
   c. Grinding, brushing
   d. Abrading
   e. Peel ply
2. Chemical processes:
   a. CSA (chromic–sulfuric acid) treatment—pickling
   b. Ozone treatment
   c. Organic solvent treatment—etching
   d. Coating with chemically active substances—priming
   e. Deposition of active metals
3. Physicochemical processes:
   a. Low-pressure plasma treatment
   b. Corona discharge treatment
   c. Thermal treatment
   d. Flame treatment
   e. Ion etching
   f. Laser irradiation
   g. UV light irradiation

This review concentrates on UV laser treatment as a replacement for many of the surface treatments specified in the preceding list.

III. TYPES OF LASERS

A variety of commercially available laser systems are used to process surfaces
of materials, including carbon dioxide (CO_2), neodymium-doped yttrium–aluminum oxide (ruby) continuous and pulsed, and excimer systems [4] (Fig. 1). The
main types used in surface modification include CO_2, Nd: YAG, and excimer
lasers in various wavelengths. There are notable differences between processing
with CO_2, Nd: YAG lasers operating in the infrared (IR), and excimer lasers
operating in the UV. Much more incident laser energy is needed to provide the
same thermal effect with IR lasers than with excimers, and this results in a large
scale of structural and compositional changes. Because the melt depth is much
smaller for UV processing (micrometers) than for IR processing (hundreds of
micrometers), the resultant surface finish in case of the latter is much smoother.
     A number of important processing factors, including wavelength, power
density, beam diameter, beam speed, beam focusing, beam overlapping, output
mode, pulse length, and scan and repetition rates, need to be considered when
choosing a laser source and its operating parameters. Increasing the beam speed
and decreasing the transverse speed enhance the laser effects. The exposure time
(dwell time or pulse length) strongly influences the depth that will be affected.
Longer exposure times lead to deeper melting or ablation and to more surface
alloys in metals. Excimer lasers can overcome some of the problems inherent
with IR lasers in that shallower melt depths and shorter pulse duration result in
negligible surface topography.

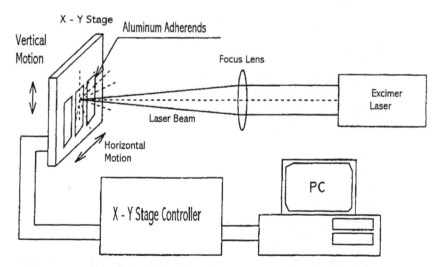

Figure 1 Excimer laser system for surface treatment.

Excimer lasers are chemical gas lasers based on molecular emissions from strongly bound excited state molecules to a weakly bound unstable ground state. Lasers utilizing rare gas halides such as argon-fluoride (ArF) are an important subclass of excimer lasers.

The mixture of gases used in these lasers has a halogen content of about 1% and a few percent of one of the rare gases Xe, Kr, or Ar. The remainder is a buffer gas (Ne or He). The mixture is excited with an electrical arc. The molecule (e.g., ArF) is split into two excimers, Ar and F, which form an excited molecule (ArF*). While returning to the dissociative ground state these molecules emit laser radiation at 193 nm. The wavelength of the laser is determined by the composition of the rare gas halide (Table 1) [5–7].

Excimer lasers operate in a pulse mode. The pulse length is between 10 and 60 ns. To obtain high pulse energy, pressures of 1.47–3.67×10^3 torr are used. A perfectly homogeneous laser beam can be obtained by preionizing with X-rays.

The 193-nm UV laser pulse creates a blast wave followed by gas molecules and monomer vapor from the irradiated substrate. Solid polymer material is also ejected as finely divided particles. No melting is observed. At 284 nm the particles turn into droplets of liquid spray. The surface melts and resolidifies. At 308 nm opaque material is produced very close to the substrate [8].

Many materials irradiated with an excimer laser have a high absorption coefficient in the UV range so that the energy of the laser is absorbed in a thin surface layer typically 0.1–0.5 μm thick. Because of the high photon energy of the UV laser, especially with ArF (6.6 eV), direct bond breaking in nearly all organic materials (organic bond energy is mostly less than 6.6 eV) (Fig. 2) occurs.

The section of the electromagnetic spectrum that extends from 200 to 150 nm is formally defined as far-ultraviolet radiation. The narrow band that lies between 200 and 180 nm is a practical and highly effective region for the photochemical modification of polymers; the photoenergy exceeds the strength of most of the bonds in a typical organic polymer. It is, therefore, quite efficient in bring-

Table 1 Wavelength of Some Excimer Lasers

Excimer gas	Wavelength (nm)
ArCl	170
ArF	193
KrCl	222
KrF	248
XeCl	308
XeF	351

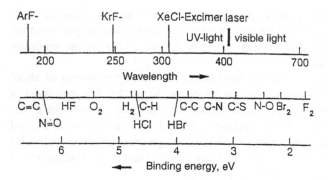

Figure 2 Chart relating the energy of radiation at various wavelengths to the dissociation energy of various chemical bonds.

ing about photochemical reactions. Nearly all organic compounds (with the exception of saturated aliphatic hydrocarbons and fluorocarbons) absorb intensely in this region. As a result, this radiation penetrates organic polymers to only ≈ 300 nm before 95% of its intensity is absorbed [7]. The absorption of the photons by the polymer is governed by Beer's law. As a result, there will be numerous bond breaks in the irradiated volume. Typically, radiative lifetimes of organic molecules in this wavelength region are on the order of picoseconds. The breaking of the bonds in the polymer chain is rapidly followed by recombination processes, so the net reactions are loss of the small gaseous molecules (CO, CO_2, H_2) and degradation of the polymer by the disruption of its linear structure. In the presence of air, oxygen traps the radical ends to give initial oxidation products that can undergo further photolysis to give smaller fragments. Prolonged irradiation results in controlled etching of the polymer. When the photons are delivered in an intense pulse of short duration, the concentration of fragments in the irradiated volume presumably reaches a high value, which, when it exceeds a threshold level, results in the spontaneous ejection of the fragments into the gas phase (ablative photodecomposition) [9].

Because of the short duration of the laser pulse, there is virtually no heat flow across the boundary of the irradiated zone. Thus, high precision is achieved without thermal damage.

It is important to distinguish between the excitation of a molecule into a binding state with the possibility of further reaction (for example, with atmospheric oxygen, photooxidation), which usually occurs at low energy and a low number of pulses, and the excitation of a molecule into a nonbinding state with immediate dissociation (photolysis), a process occurring in solid polymers at high energies and a high number of pulses. These two processes occur simultaneously

Table 2 Ablation Behavior of Some Polymers
Irradiated with Excimer Laser (XeCl)

Material[a]	Threshold energy (J/cm^2)	Ablation rate (μm/p)
PI	0.12	0.5
PEEK	0.2–0.28	0.5–1.0
PES	0.24	0.8
PET	0.18–0.29	1.7–1.9
PA	2.0	—
PMMA	2.0	—
PTFE	2.0	—
PVDF	1.3	1.0
PVC	0.5	1.0
PE	3.3	2.0

[a] PI, polyimide; PEEK, poly(ether ether ketone); PES, polyester;
PET, poly(ethylene terephthalate); PA, polyalcohol; PMMA,
poly(methyl methacrylate); PTFE, poly(tetrafluoroethylene);
PVDF, poly(vinylidene fluoride); PVC, poly(vinyl chloride);
PE, polyethylene.

with one process dominating. As a result, two competing mechanisms occur during laser irradiation: at low energy, photochemical reactions dominate, causing chemical modification, radical formation, cross-linking, and oxidation; at high laser energies, thermal reactions dominate, and melting, amorphization, chain scission, and recombination occur.

Excimer laser processing is affected by the laser parameters as well as the material properties, ablation rate, and threshold energy (Table 2 and Fig. 3). The performance achieved is a complex function of all these parameters.

IV. USES OF EXCIMER LASERS

Since the invention of the excimer laser in 1975, technical development allowed its use for a wide range of industrial purposes. Lasers are used to modify surface characteristics in order to improve properties for a wide variety of industrial applications in addition to improving adhesion. Lasers are well established in cutting, drilling, welding, and machining, but the number of publications indicates that the main interest is in surface modification and the level of interest has been increasing rapidly (Fig. 4) [10]. Lasers are an attractive technology for a number of reasons: surface modification offers the possibility of tailoring the surface

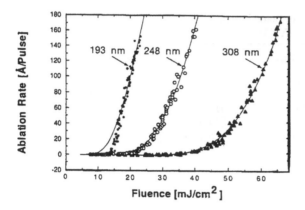

Figure 3   Ablation rate as function of excimer laser wavelength.

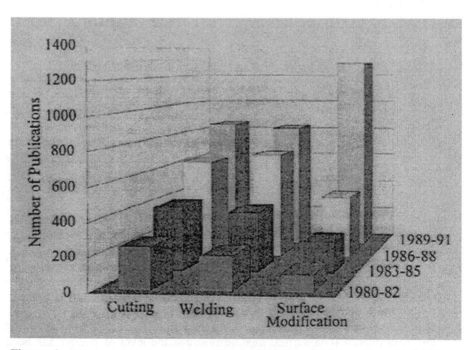

Figure 4   The increase in interest in surface modification with lasers. (From Ref. 10.)

properties without changing the bulk mechanical or physical properties, surface alloying can be produced, and various kinds of materials can be treated, ranging from biological tissues to ceramics and diamonds. This method presents an alternative to other limited and polluting conventional surface treatment methods, such as sandblasting, etching, or welding, which cause occupational health, safety, and environmental risks.

Some examples of laser uses are listed below:

Medical:

Laser surface modification creating pillar structures on polymers to improve biocompatibility to stimulate movement of neutrophils [11].

CO_2 laser treatment of Ti implements for dental adhesives [12].

ArF laser treatment at 260 mJ/p · cm^2, 450 p of dental enamel creating a rough surface and good bond strength (6–10 N/mm) [13].

ArF laser mask imaging on polyolefin artificial blood vessels to introduce hydrophobic and hydrophilic alternating zones on the inner wall of the blood vessel to maintain anticlotting properties [14].

Microelectronics:

Improvement of adhesion of gold to alumina substrate [15,16].

Etching, stripping, cleaning, bonding, and ablation processes in manufacturing and repairing [17].

Metallization and deposition:

Metallization of Al_2O_3 by laser deposition of metal films [18]. Metallization of ethylene acrylic (EAA), poly(ethylene terephthalate) (PET), poly(ethylene naphthanate) (PEN) fibers using a UV laser. The treatment produces an amorphous layer (a few nm thick) that is easy to metallize [19].

Improvement of sputtered coating for hot corrosion resistance by laser treatment; elimination of defects and bond improvement with the substrate by remelting the coating with laser energy and changing the surface morphology [20,21]. Laser treatment of coatings also produces dendrites, solid solution, fine crystal growth at low energy, and amorphic phases at high energy. The durability and the impact resistance are improved [22].

Deposition of Al_2O_3 and Cu on sapphire for electronic packaging, information processing, and thin-film technology [16]. The coating is treated with an XeCl excimer laser at 0.75 J/p · cm^2. A modified layer is obtained with a new chemical bond, Cu—Al—O, as well as crystalline Al_2O_3 and aggregate clusters of particles.

Deposition of carbon on fibers using a CO_2 laser to toughen the fibers, creating higher density and better adhesion to the matrix, resulting in improved mechanical strength [23].

Dyeing:

Effective dyeing of fabric fibers by laser fixation by incorporating polar groups on the surface of the fibers using photoirradiation with an Hg lamp and UV laser at 366 nm. Amino groups and double bonds are created, enhancing surface activity and improving wetting and dyeing [24].

Cleaning the surfaces of textiles and composites from release agents (IR laser) [25].

Textile fibers PET, Nylon 66, and carbon fibers were irradiated with ArF and KrF excimer lasers at 57 mJ/p · cm^2, 10 p for the nylon and 114 mJ/p · cm^2, 30 p for PET at 2 Hz. The laser caused roughness of the surface as well as chemical changes. Both effects enhanced dyeing, adhesion, and wetting and improved optical properties [26,27].

Treatment of carbon and glass fibers for improvement of matrix-fiber adhesion in composites [28,29].

Printing:

Use of lasers for scratching, dotting, and lithography [30].

An F$_2$ laser at 157 nm was used for lithography on polyimide (PI), PET, and polyethylene (PE) films. PI showed the best results in printing at 5 mJ/p · cm^2 [31].

An ArF excimer laser was used for patterning and drilling microholes in flexible sheet materials (plastics and fabrics) [32].

Metal machining:

Laser treatment is used to improve hardness of steel by producing a martensitic phase on the surface [33].

Excimer laser is used to modify the surfaces of ceramics and metals by alloying to obtain surfaces with reduced friction [34].

Ablation:

Lasers are used to ablate the surfaces of polymers [35].

Polymerization:

Laser energy is also used to polymerize monomers by initiation of the reaction. Photopolymerization of monomethyl acrylate (MMA) monomers was done with an Xe F$_2$ laser excimer (351 nm), with 200 p. Pentacosadiynoic acid (PDA) monomer was cured with a KrF excimer at 8 mJ/p · cm^2 [36–38]. Adsorbed acrylic monomers on wafers were polymerized using an ArF excimer laser to obtain a uniform film with no defects [39].

Filtering:

Laser treatment can be used for particle filtration (the high roughness traps the particles), used as an effective substrate for chemical catalysis reactions, an antireflection surface, and damping of hypersonic sound waves [40].

V. ADHESIVES AND ADHERENDS

Laser treatment has been used to improve adhesion between various joint materials. The polymeric adherends can be divided into two groups: strong absorbers and weak absorbers. Strong absorbing materials such as PI, PET, polyetheretherketone (PEEK) and polyester (PES) show a low threshold energy and good interaction with the laser. The ablation rate increases with energy logarithmically at low intensity and linearly at high intensity. Weak absorbers such as poly(methyl methacrylate) (PMMA), polyalcohol (PA), PE, poly(vinylidene fluoride) (PVDF), and Poly(tetrafluoroethylene) (PTFE) have high threshold energies. Ablation stops after a certain number of laser pulses as the UV light scatters within the material and plasma masks the surface.

Some examples of polymers treated with excimer lasers follow:

Bonding of Teflon to stainless steel (SS) using epoxy adhesive. The ArF excimer laser pulls out F atoms from the C—F chain, replacing them with —OH. This generates a hydrophilic surface. Single lap shear (SLS) joint strength was 110 kg/cm^2, compared with 2 kg/cm^2 of the nontreated material [41].

Bonding of polypropylene (PP) with epoxy resin. Bond strength increased by 500% when treated with an excimer laser at 12.5 mJ/p · cm^2, 10k p, and reached 46 kg/cm^2. Hydrogen atoms were pulled out from the irradiated surface and replaced by OH and C=O groups, producing more hydrophilic behavior [42,43]. The same system was irradiated with an excimer laser at 248 and 308 nm, 0.1–1 J/p · cm^2, 1–200 p. Adhesional strength was 36 kg/cm^2; all failures were cohesive in the PP adherend near the interface [44]. A blend of PP and ethylene propylene diene monomer (EPDM) rubber was bonded to SS with polyurethane adhesive, after treating the polymer surface with a KrF excimer laser at 0.1 J/p · cm^2, 100 p, and 3 Hz and with an XeCl laser at 0.3 J/p · cm^2, 20 p, 5 Hz. The XeCl laser gave better results. Both C=O and OH groups were found on the surface and the adhesion strength was 40 kg/cm^2.

Ethylene tetrafluoroethylene (ETFE) and PP were irradiated with ArF and KrF excimer lasers at 0.1–0.5 J/p · cm^2, 1 Hz, 15 pulses. PP showed decomposition of alkylphenol and antioxidant. ETFE created C=O and C=C bonds and induced oxidation [45].

PTFE bonded with epoxy adhesive after treatment with a KrF excimer laser at 60 mJ/p · cm^2, 10 Hz in a glycidylmethacrylate atmosphere, showed increased peel strength of 1.2 kg/cm. Improvement of adhesional strength occurred only after aromatic polymers were added to the adhesive (20 wt%), such as PI, PEEK, or aromatic polyester (APE). Peel strength was 3 kg/cm. XPS showed a decrease in F concentration and increase in C/O ratio. SEM showed creation of fine roughness of 0.1 μm caused by the laser treatment [46,47].

High-density polyethylene (HDPE) bonded with cyanoacrylate adhesive (Loctite 401), treated with a KrF excimer laser at 250 mJ/p · cm², was tested by X-ray photoelectron spectroscopy (XPS), SEM, and ATR and showed increased lap shear strength (60–70% more compared with untreated material) [48].

HDPE (6706) bonded with cyanoacrylate adhesive showed good lap shear strength after laser treatment [28]. Low-density polyethylene (LDPE), HDPE, and PP were treated with ArF, KrF, or XeCl excimer lasers at 120 mJ/p · cm², 50 Hz, resulting in SLS adhesional strength of 3–4 MPa with cohesive failure compared with 0.5–1 MPa for the untreated adherends. After treatment, the typical effects were cleaning, roughening and generation of new functional groups (C=O) on the surface [49].

PS foam was irradiated with 308 nm XeCl excimer laser at 4 J/p · cm², 2 Hz, 60 p for drilling purposes. Only 2 layers of cells deep were ablated [50].

PI was treated with a 248-nm excimer laser at 135 mJ/p · cm², 1 p for adhesion pretreatment. A decrease in N and O was found on the surface. Only the upper 10 nm was affected. Imide bonds were broken, recombined, and rearranged, modifying the surface for better bonding [51]. A similar effect was found in treating PI (Kapton) with an ArF excimer laser as pretreatment for bonding with acrylic adhesives. Improved peel strength was achieved [5].

Polyetherimide (PEI) and polycarbonate (PC) were treated with an ArF excimer laser to improve adhesion with polyurethane adhesives. The optimal laser parameters are summarized in Table 4. The increase in adhesional shear and tensile and peel strength was between 200 and 300% with cohesive or mixed failure compared with untreated adherends. Cleaning of the surface, morphological changes, and chemical modification contributed to this effect [52].

Carbon-filled rubber was also treated with a laser to improve adhesion to metal [53].

Composite materials such as poly(phenylene sulfone) (PPS) or PEEK reinforced with carbon fibers were treated with an excimer laser to improve adhesion. These thermoplastic aromatic materials are known to be inert and tend not to react with adhesive resin. The ArF excimer laser treatment improved shear strength by 400%, as well as tensile and flexural strength and fracture toughness. Laser treatment induced cleaning of the surface, roughening of the surface uniformly, and generation of a tough cross-linked surface for adhesion [54]. Some adhesional strength results after laser treatment compared with conventional treatments are presented in Table 4 [5].

Liquid crystalline polymer (LCP) polyester reinforced with glass fibers was pretreated with a KrF excimer laser at 248 nm, 430 mJ/p · cm^2, 150 p, 10 Hz. As a result, shear and tensile strengths improved by 350% compared with untreated LCP adherend. The failure changed from interfacial to mixed mode. Laser treatment caused fiber exposure and surface cleaning [55].

Some examples of metals treated with excimer lasers follow:

Excimer laser radiation couples strongly with metals, which alters the morphology and the oxidation state and causes annealing, alloy formation, and segregation.

Aluminum 2024 adherends were surface treated with a KrF laser at 250 mJ/p · cm^2, 15–20 p, 2 Hz and bonded with epoxy structural adhesive. Results showed changes in morphology, oxidation of the surface, and adhesional strength improved by 24% compared with nontreated controls and 15% compared with anodized Al. Laser-treated joints exhibited higher plastic deformation, imparting greater fracture toughness [56]. Aluminum 2024 adherends were also surface pretreated with an ArF laser at 190 mJ/p · cm^2, 2000 p, 30 Hz and bonded with epoxy structural adhesive or epoxy-toughened adhesive. Results showed changes in morphology, chemical modification of the surface, and adhesional strength improved by 700% compared with nontreated controls and 150% compared with anodized Al. Laser-treated joints also exhibited higher hardness and higher surface wetting [57].

Stainless steel showed a $\gamma \rightarrow \alpha$ phase transformation after CO_2 laser treatment with formation of carbides and hardened zones [58].

Ti bonded to PMMA was pretreated with a CO_2 laser at 10.6 μm, 3–10 Hz, 120–180 mJ/p · cm^2 in two atmospheres (N_2, O_2). Adhesional strength improved by a factor of 40 [10]. With high-intensity laser radiation a high electrical field was produced and metastable, nonstoichiometric Ti oxides were created.

Deposited Au bonded to Al_2O_3 using a UV laser in an O_2 atmosphere and postannealing to 300°C resulted in improvement of adhesional strength compared with nontreated controls [13]. The pronounced increase in adhesion of the gold to the alumina was obtained when the alumina was laser treated in an O_2-rich atmosphere. The adhesion strength reached 50 MPa, 500% that of the nontreated system. This adhesion enhancement was attributed to the formation of high-energy disordered alumina at the surface with rapid solidification followed by melting [59].

Excimer laser irradiation of sealed chromic acid anodized Al 2024 adherends improved adhesional strength with epoxy by 150% compared with untreated specimens. This was attributed to water removal and dehydration of the bohemite structure of the alumina surface, increased roughness, and crystallization. This treatment enables bonding to hard anodized Al at discrete locations without affecting the corrosion protection of the surrounding areas [60].

Copper, brass, Al, and SS adherends treated using a 248 nm excimer laser (20-ns pulse, 1 Hz) resulted in ablation, increased roughness, cleaning, and improved bond strength [61].

Deposited Cu bonded to Al_2O_3 was laser treated. The bond strength of the treated adherends increased by 200–400% compared with nontreated ones [18].

Titanium and Al were bonded with epoxy resin. The adherends were treated with a KrF 248-nm excimer laser at 0.5 J/p · cm^2, 5–20 p, 10–100 Hz. The Al was more affected than the Ti as more oxides were formed. Adhesion improved by 180% [62,63].

The following are other materials treated with excimer lasers:

Carbon fibers and aramide fibers treated with lasers were bonded to thermoplastic and thermosetting materials to be used as composites. Polyamide and PETF (poly ethylene terephthalate) fibers were also laser treated for textile dyeing, resulting in surface structuring, improved wetting properties, and improved optical appearance [28,29,64].

Ceramics are usually transparent to UV radiation and thus were not expected to be potential substrates for surface treatment. Surprisingly, the ceramic adherends were also susceptible to laser treatment and showed improved adhesion after laser irradiation.

Both fused silica and Piezoelectric $PbZrT:O_3$ (PZT) wafers were laser treated at very high energy (2.2 J/p · cm^2, 100 pulses) and showed improved adhesion to metals (200% higher tensile strength and mixed failure) [5] and better coating ability with Cr deposition [54,65].

Some mechanical results for laser-treated adherends compared with classical treatments are given in Table 3.

VI. SURFACE TESTING

A. Analytical Methods

Among analytical techniques used in characterizing surfaces after laser treatment or bond failure the following techniques are included [4]. The laser microprobe

Table 3 Adhesional Shear Strength and Failure Modes of Various Studied Bonded Joints

	Surface treatments		
Adherends/adhesives	Untreated SLS (failure mode)[b] (MPa)[a]	Conventional SLS (failure mode)[b] (MPa)[a]	Laser-treated SLS (failure mode)[b] (MPa)[a]
Polycarbonate/PU	3.5 (A)	5.0 (M) SiC blast	7.5 (C)
Polyetherimide/PU	2.5 (A)	5.0 (M) SiC blast	5.5 (C)
PEEK composite (APC2/ AS4)			
Structural epoxy FM 3002 K	6.1 (A)	14.7 (M) SiC blast	27.8 (M)
Structural epoxy AF 163-2	21.7 (A)	34.0 (M) SiC blast 20.0 (M) plasma	45.4 (C)
Aluminum alloy/epoxy	2.0 (A)	10.2 (C) Unsealed anod.	14.3 (C)
Aluminum alloy/structural epoxy	12.8 (C)	42.9 (C) Unsealed anod.	34.4 (C)
Sealed anodized aluminum/ epoxy	4.5 (A)	10.2 (C) Unsealed anod.	11.0 (C)
Copper/modified epoxy	6.1 (A)	14.9 (C) sand blast	14.3 (C)
Fiberglass epoxy/(rubber-Modified epoxy)/copper-coated fiberglass epoxy	—	1.7 (C) sand blast	25.6 (A/M)
Fiberglass epoxy/(acrylic adhesive)/copper-coated fiberglass epoxy	—	8.5 (C) sand blast	15.6 (C)
Copper-coated fiberglass/epoxy (acrylic adhesive)/polyimide	Peel test lb/in	21.0 (M) sand blast	25.0 (A)
Invar/fused silica/RTV	—	3.6 (A) alumina	5.6 (C)

[a] ±5% standard derivation (five samples for each test).
[b] A- interfacial; M, mixed; C, cohesive mode of failure.
Source: Ref. 5.

mass analyzer (LAMMA) is capable of characterizing the atomic or molecular weight of surface and bulk constituents. Electron spectroscopy for chemical analysis (ESCA) is a surface-sensitive technique providing elemental and chemical bonding information from the top 4–5 nm of both conductive and insulting materials. Auger electron spectroscopy (AES) is a highly spatially resolved technique that provides the elemental composition of the top 2–5 nm of a surface. Fourier transform infrared (FTIR) spectroscopy provides molecular bonding information on organic materials. X-ray fluorescence (XRF) provides a spectroscopic measurement of the elemental composition of bulk or thin-film inorganic structures. The scanning electron microscopy with energy-dispersive X-ray (SEM/EDX) detector provides the ability to observe changes in morphology of the surface and its atomic composition. Secondary ion mass spectrometry (SIMS) provides the molecular weight of fragmented molecules on the surface. Surface tension, wetting, and surface energy are determined by contact angle measurements. Surface roughness is determined by atomic force microscopy (AFM).

B. Testing Procedure

The various substrates were treated with UV lasers using various parameters: wavelength, intensity, repetition rate, beam area, and number of pulses. The optimal laser treatment for each material was defined in terms of the maximal shear strength of the corresponding bonded joint. Joint properties were determined using single lap shear (SLS) tests according to ASTM D-1002-72. The mode of failure was determined visually to be either interfacial (0% coverage of adherends), cohesive (full coverage of adherends), or mixed. Surface morphologies following laser treatment and following shear failure were analyzed by means of SEM. Surfaces were Au-Pd sputtered prior to analysis. Surface roughness and microstructure after laser treatment were analyzed using AFM or scanning tunneling microscopy (STM). Chemical changes at laser-irradiated surfaces and at failure surfaces were investigated by means of FTIR spectroscopy in an external specular mode, equipped with a horizontal stage in near to normal incidence and a gold-coated mirror as reference. Surfaces of the laser-treated compared with untreated adherends were also investigated for chemical changes due to laser irradiation by ESCA with an Al K_α X-ray source, at 10 kV, 40 mA, and pressure of 3×10^{-8} torr. Microhardness tests as well as contact angle measurements were also carried out for the adherends following laser treatment.

Two types of references were usually used for each set of experiments: a nontreated substrate or a conventionally treated adherend. Some examples are as follows. For the thermoplastics and composite thermoplastics the adherends were untreated and/or were conventionally treated by abrasive treatment with SiC (36 mesh). For the aluminum and sealed anodized aluminum adherends the references used were untreated aluminum and/or unsealed chromic acid anodized according to MIL B 8625. For copper the references were black conventional treatment (abonite) and sandblasting; for the fiberglass epoxy the reference was sandblasting; and for the invar and alumina the reference was alumina blasting and silane priming (see Table 3) [57].

Table 4 Optimal Laser Parameters for the Treatment of Various Adherends

Adherend	Laser energy (J/p · cm²)	Repetition rate (Hz)	No. of pulses
Polycarbonate	0.08	10	12
Polyetherimide	0.08	10	200
PEEK composite/carbon fibers (APC2/AS4)	0.18	5	100
	1.0	5	10
Aluminum alloy	0.19	30	2000
Sealed anodized aluminum alloy	0.8–1.9	30	100–1000
Copper	2.7	10	50
Fiberglass epoxy	0.18	30	100–1000
Copper-coated fiberglass epoxy	2.1	30	50–100
Polyimide	0.18 (Kapton)	30	1000
Fused silica	2.2	30	10

VII. OPTIMAL IRRADIATION PARAMETERS

As indicated in Sec. III, there are many laser parameters to consider in order to achieve the optimal surface modification. The optimal condition is obtained when the surface is activated but not totally ablated and the adhesional properties are superior.

Table 4 summarizes the optimal laser treatment parameters for some materials tested [52]. As can be seen from the results, different conditions are required for different adherend materials. Pure polymers, which are easy to activate, require the least intensity for laser treatment (0.1 J/p · cm²); composites, which are more inert, require higher energy (0.2–1 J/p · cm²); and metals and ceramics, which have strong chemical bonds and high thermal conductivity, require high energy (>2 J/p · cm²). A model predicting optimal laser conditions and their correlation with the material properties has not yet been established and will be studied in the future.

VIII. MORPHOLOGY OF TREATED SURFACES

Laser irradiation causes changes in the surface morphology of the adherends treated. These changes depend on the adherend material, the laser parameters, and the treatment atmosphere. The main effect is to provide increased roughness, which generates a larger surface area and enables keying of the adhesive into the adherend. Another advantage is the uniformity of the roughness that is created, resulting in a regular homogeneous microstructure of the surface for improved adhesion, which presents an advantage over abrasive treatment. Some example of specific morphologies created are given next.

The most common features occurring on the surface after laser irradiation are globules. In some cases these globules show poor adhesion and are easy to remove (e.g., on Zn and Sb or Al) [66], especially when the surface is already ablated, but in most cases the globules are tough and enable keying of the adhesive to the adherend (PC, PEI, PEEK, Cu, fiberglass) [56].

Scanning electron micrographs of laser-treated adherends revealed different morphological changes depending on the adherend material, laser energy, and number of pulses. The PC and PEI thermoplastic adherend surfaces exhibited conical and rounded granules spread all over the area (Fig. 5 and 6). The carbon-reinforced PEEK exhibited the same granules accompanied by partial exposure of the fibers (Fig. 7) at low laser energies of $0.1-0.2$ J/p $\cdot$ cm^2, whereas at higher laser energies (above 1 J/p $\cdot$ cm^2) the laser-irradiated surface was smooth with randomly spread cracks (Fig. 8).

Scanning electron micrographs of the Al adherend after laser treatment showed no morphological changes at low laser energies ($0.18-0.2$ J/p $\cdot$ cm^2). Increasing the laser energy resulted in a fine microstructure of the treated surface demonstrating arrays of cracks about 1 μm wide and small holes at energies of 0.7 J/p $\cdot$ cm^2 (Fig. 9).

Irradiation of the sealed anodized Al specimens at low energy densities

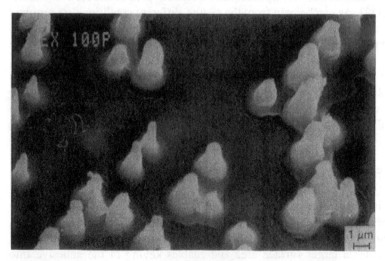

Figure 5 Surface morphology of polycarbonate after optimal laser treatment.

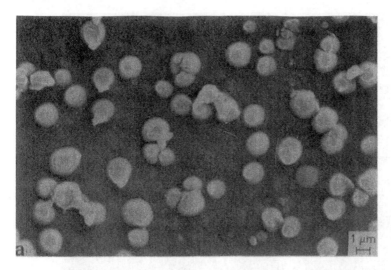

Figure 6 Surface morphology of polyetherimide after optimal laser treatment.

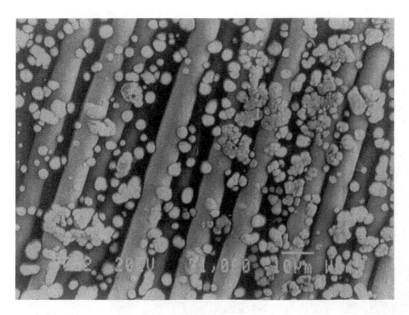

Figure 7 Surface morphology of PEEK-carbon composite (low energy) after optimal laser treatment.

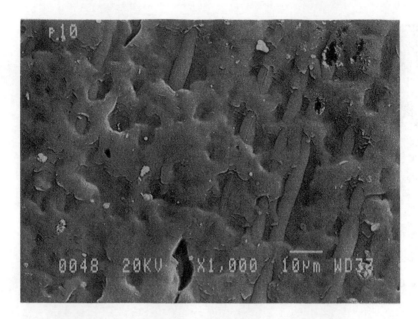

Figure 8 Surface morphology of PEEK-carbon composite (high energy) after optimal laser treatment.

Figure 9 Surface morphology of Al 2024 after optimal laser treatment.

(0.2 J/p · cm^2) resulted in no change in surface morphology even after 1000 pulses. At 0.7 J/p · cm^2, changes in morphology included open bubbles, probably resulting from water removal, some spherical droplets of Al_2O_3 due to splashing (caused by laser ablation of Al_2O_3), and cracks (Fig. 10). Irradiation of copper at low energy led to only color changes due to oxidation. At higher energies morphological changes were observed, with round spheres spread evenly on the entire surface (Fig. 11).

Irradiated fiberglass showed the same morphology as PEEK composite with the difference that the globules were surrounded by round circles that seemed like diffraction patterns. Invar irradiated at low energies showed a change of color and at higher energies changes in morphology described as "moonlike" ridges and canals. At very high energies the ridges formed into smooth shapes (Fig. 12). The fused silica showed no morphological change at low energies and the same morphology as sealed anodized Al at high energy.

Kapton showed features similar to cones sticking out from the surface (Fig. 13). All these morphologies increase the roughness of the surface, enabling mechanical interlocking of the adhesive.

PTFE shows a nodular structure ~0.3 μm wide, The nodules are loosely attached to the surface [47].

PMMA shows an irregular wavelike structure, the depth increasing with increasing number of pulses [8].

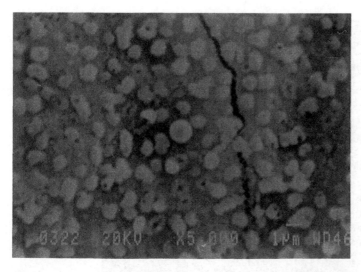

Figure 10 Surface morphology of Al_2O_3 (anodized Al 2024) after optimal laser treatment.

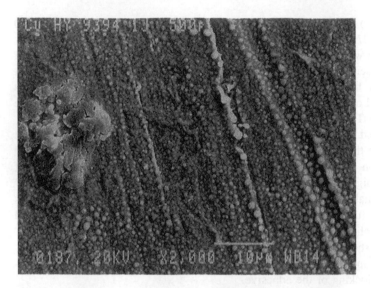

Figure 11   Surface morphology of copper after laser treatment.

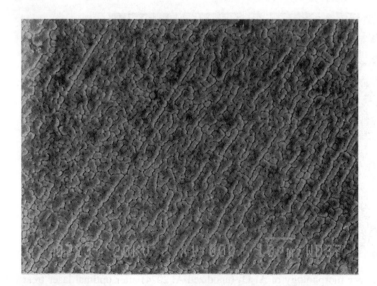

Figure 12   Surface morphology of invar after laser treatment.

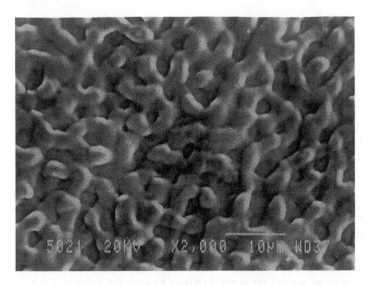

Figure 13 Surface morphology of polyimide (Kapton) after laser treatment.

Various polymer films, PEEK, PI, PES, and PET, showed ripple-like patterns after irradiation with an excimer laser (XeCl and KrF). Between the main ripples uniform and finer secondary ripples were observed. The reason for the ripples is probably a scattering and reflecting mechanism. The "valleys" erode less than the "hills," up to 1 μm. After a finite number of pulses there is saturation as the plasma plume absorbs the laser energy (a model is presented); see Fig. 14 [67,68]. Morphology is also influenced by stretch direction during production or by flow marks [69]. AFM and STM studies showed a periodic modification of submicrometer size aligned parallel that is formed on smooth surfaces, resulting from incident and surface-scattered waves [70].

Textile fibers irradiated with a KrF excimer laser at 29 mJ/p · cm^2, 10 p, showed a morphology of hills and valleys and higher magnification showed aggregates. Higher energy caused an increase in distance between the hills and sharper features. Among the three types of fibers tested, PET was the most sensitive. Carbon fibers were not affected at all. PET fibers showed a ripple-like structure perpendicular to the fiber axis. A higher pulse number caused a coarser structure. This structure probably resulted from frozen-in stresses [26,27].

Mylar, PC, polybutyleneterephthalate (PBT), and PET films showed ripple-like structures and some nodules. The structures depended on the homogeneity of the polymer material, surface stresses, and flow. Morphological changes occur only at $I < I_{critical}$ (ablation) and depend on the chemical composition of the substrate, wavelength, and angle of irradiation of the laser. Mylar showed cross-

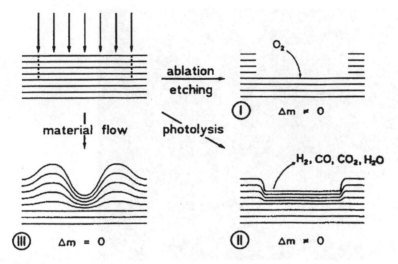

Figure 14 Different mechanisms for ripple formation by laser irradiation.

linking on the surface. Heating the film to 160°C caused all the created morphology to disappear [71].

PEEK thermoplastic polymer at low energy or a low number of pulses shows smooth craters and microfusion spots, independent of the repetition rate or laser wavelength. At high energy or a high number of pulses an irregular structure is created with redeposited debris on the surface. This shields the surface from further etching [72].

PVC irradiated with an ArF excimer laser at 0.11 J/p · cm^2 shows conical structures; at 1 J/p · cm^2 the surface is smoothly ablated. The same phenomenon is also typical for PEEK, PET, PVDF, and PEI. PC shows a conelike structure on top of a wavy background. PVDC has an ice cream cone–like structure resulting from collapsing of the cones built during irradiation. PS shows fluffy cones because material jets up from the surface [40].

The concentration and size of the features created on the surface by laser irradiation change with time of irradiation and/or laser energy. The higher the energy or the number of pulses, the denser are the features until ablation occurs at the threshold point. At this point the morphology changes and becomes similar to that obtained with IR (CO$_2$ laser) treatment.

Ceramics are also affected by laser treatment and exhibit morphological changes:

Al$_2$O$_3$ becomes smoother after treatment due to changes in composition and porosity (at 6 J energy) [18]. The same phenomenon was also found on other ceramics such as fused silica or PZT wafers. The roughness of the original wafer is smoothed by creating very shallow globules, which become denser as the energy of the laser increases [60]. Both Al_2O_3 and Si_3N_4 showed cone- and lavalike structures [40] and highly disordered features on the surface during rapid solidification [15].

Metals irradiated by laser show features different from those shown by polymers. The plasma produced by the laser generates shock waves, which deform the material and form striations through liquid flow, leading to concentric rings with lips. The roughened surface is then uniformly oxidized [68].

The surface topography of Ti after laser treatment becomes similar to that after freeze melting [12] or shows periodic ridges 3 μm wide at 20 p irradiation and 1.6–1.8 μm wide at 5 p [62].

Aluminum 2024 shows concentric circular wavelike relatively smooth structures 8–12 μm wide [56].

The enhanced adhesion due to mechanical interlocking is clearly revealed in the SEM micrographs for nearly all UV laser treatments and adherend materials. The cohesive mode of the failure is indicated by the adhesive layer covering each adherend and by showing a replica of the granules or bubbles formed during laser treatment. Cohesive failure indicates far better adhesion than the interfacial failure of the untreated adherend. Figures 15 and 16 are micrographs showing

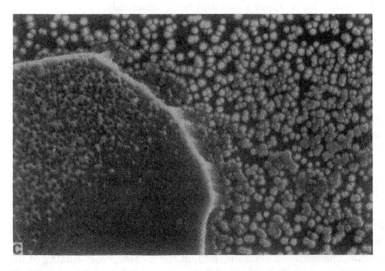

Figure 15   Failure surfaces of laser-treated, bonded, and shear-tested adherends. Replica of the globules is observed: PEI.

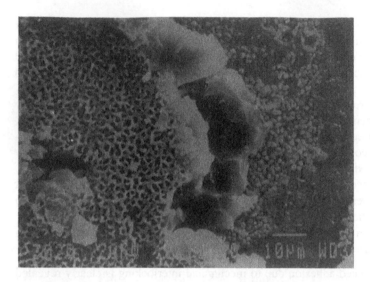

Figure 16   Failure surfaces of laser-treated, bonded, and shear-tested adherends. Replica of the globules is observed: Al_2O_3 (anodized Al 2024).

the replica surface formed during failure of an anodized Al joint and a thermoplastic joint, respectively. The same phenomenon is also observed in PEEK composite, Cu, figerglass, invar, fused silica and other adherends.

It should be noted that even if visually the failure sometimes seems partly interfacial, high magnification shows that a thin layer of adhesive is always present on the laser-treated adherend.

At high laser intensity or high number of pulses there is a weakening of the treated layer of the adherend due to ablation, resulting in loosely connected granules at thermoplastic surfaces, deformed exposed fibers of composites, molten dendritic areas of the Al and anodized Al surfaces, loosely adhering spheres on the Cu surface, and molten smooth areas in the ceramic irradiated areas [52].

IX.   EFFECTS OF LASER TREATMENT

The main effect presented in this chapter is the improvement of adhesion to the treated materials. This effect is achieved through surface roughening, removal of contamination, chemical modification, and, in some cases, cross-linking of the surface layers and creation of a hydrophobic layer capable of repelling moisture. Laser irradiation has some other specific effects in addition to improved adhesion that can be utilized for applications other than adhesion or coating, as will be specified further in this section.

Improvements in adhesive shear strength by 200–600% are achieved for laser-treated adherends compared with nontreated ones and of 100–200% compared with conventionally treated ones. The best results are achieved for thermoplastic composite the materials, as expected, because the coefficient of absorption of these materials is the highest.

Visual inspection of the failure surfaces shows clearly that laser treatment causes the mode of failure to change from interfacial in non–laser-treated adherends to mostly cohesive under optimal laser operating conditions, which indicates that the interfacial adhesion was significantly improved [54].

In addition to the pronounced morphological modification, which was covered in the previous section, chemical changes, especially oxygen formation or elimination, take place following UV laser treatment of the adherends. These changes are depicted by FTIR, XPS (or ESCA). An increase in the surface polarity of polycarbonate occurs due to the scission of carbonate bonds to form hydroxyls and carboxyls [52]. These changes probably increase the surface reactivity, followed by increased adhesional strength. More energetic treatment causes massive degradation, even in the aromatic part of the skeletal chain. This degradation causes the weakening of the outer layer as observed by SEM and concomitantly a reduction in the adhesional strength.

FTIR analysis of polyetherimide treated with an excimer laser reveals smaller chemical changes occurring mainly at the amide ring, to form polyamic acid, which improves the surface chemical activity [52].

FTIR analysis of the PEEK composite reinforced with carbon fibers reveals formation of hydroxyls and aldehydic groups from the scissioned carbonate bonds, which enhance adhesion. It also reveals massive cross-linking of the outer surface layer, resulting in a tougher surface with no weak boundary layers. High laser energies result in a total spectral peak reduction, observed by FTIR, due to degradation and ablation (Fig. 17). ESCA analysis confirms the FTIR results and in addition reveals another important advantage of laser treatment, namely cleaning of the surface by evaporating the contaminants and the weak boundary layers from the adherend surface at the initial stage of the treatment (Fig. 18). This effect was observed for all the adherend materials tested.

FTIR spectra of anodized Al treated with a laser reveal removal of adsorbed and structural water molecules from the anodic film and dehydration of the hydroxide to Al oxide [52]. A Growth of a new Al oxide layer on the laser-treated surface is observed by FTIR for the pure Al alloy.

PET treated with a KrF laser shows a decrease in the O/C ratio [73]. There is also oxygen loss in PEEK resulting in a less polar surface (loss of CO and CO_2) and formation of a hydrophobic surface [69]. This phenomenon is an advantage in drying moisture from the surface before adhesion.

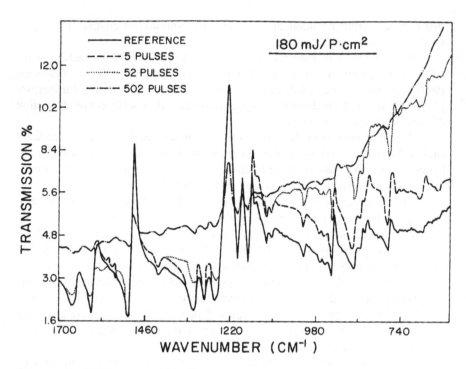

Figure 17 FTIR spectra of PEEK/carbon composite laser treated at 0.18 J/p · cm² at various numbers of pulses.

The ceramic adherends showed an excess of oxygen on the surface. This oxygen can create strong metal-ceramic bonds [74]. In laser irradiation of ceramics (fused Si and sapphire) there are two stages in the radiation mechanism—a photochemical ablation, which is a mild process, and a high-rate thermal ablation, which is a more aggressive process [75]. The oxygen that is adsorbed in the second stage creates new kinds of oxides that are stronger and do not contain water molecules [60].

PI irradiated by an excimer laser (18 mJ/p · cm²) resulted in a reduction of the O concentration due to loss of $C=O$, $N-O$, C_6H_6O, and $C_2H_2ON_2O_5$ [76].

Measurement of the microhardness of metal adherends that were polished prior to laser treatment shows a slight increase in hardness of both Cu and Al metals (of about 4–8%). This increase in hardness is typical of other laser processes as well (welding, cutting, printing, etc.) and probably results from oxidation and repeated melting and condensation on the surface.

Wetting characteristics were assessed by contact angle measurements on Al adherends and PEEK composite adherend following laser treatment. The Al adherend shows a decrease in contact angle (better wetting) after laser treatment

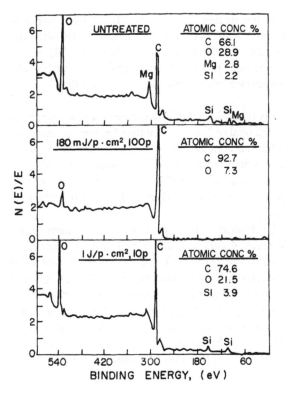

Figure 18 Comparison of C_{1S} and O_{1S} ESCA spectra for nontreated and laser-treated PEEK/carbon fiber composite.

Table 5 Contact Angles on Nontreated and Laser-Treated Adherends and Comparison with Adhesional Shear Strengths

Adherend	Surface treatment	Contact angle θ ($°$)		SLS (MPa)
		Water	MeI	
APC-2/AS-4	None	34	52	5.9
PEEK composite	Laser 0.18 J/p · cm^2, 100 p	110	43	27.2
	Laser 1 J/p · cm^2, 10 p	59	8	27.8
Al-2024	None	90	—	12.8
	Anodized	—[a]	—	43.0
	Laser 0.18 J/p · cm^2, 2000 p	41	—	34.0

[a] Anodized surface reacts with water.

due to chemical modification of the surface (oxidation). The PEEK composite shows an increase in contact angle with water due to increased cross-linking, which lowers the polarity and increases the hydrophobic character [54]. Using a hydrophilic liquid (methylene iodide) proved this assumption (Table 5). The hydrophobic property enhances surface drying and increases durability in a wet environment.

Another interesting effect is found in ceramic materials after excimer laser treatment: AlN turns into Al on removal of N atoms and thus becomes conductive [77]. The same phenomenon of better conductivity was also observed by us for PZT wafers, where pure metal was detected on the surface. Both Al_2O_3 and sapphire showed metallic Al formed on the surface [72]. This phenomenon can be used to enhance surface conductivity, to enable soldering to these materials, or increase electrochemical reactivity.

Kapton irradiated with a UV laser at 1 J/p $\cdot$ cm^2, 1 p, and an IR laser shows charring of the surface due to residual debris of scissioned aromatic rings and creation of CN, HCN, C_6H_6, and C$=$O groups. The depth of the effect is 20 nm [57,78]. Another effect observed in Kapton irradiated with KrF at 300 mJ/p $\cdot$ cm^2, 50 p, and a YAG laser was increased conductivity of the surface layer due to the charring and creation of polar groups (NH_2, C$=$O, and OH) and graphite [79,80]. A similar phenomenon was observed for PMMA irradiated by a UV laser (0.16 J/p $\cdot$ cm^2, 100–200 p, 5 Hz) and an IR laser, which created a charred surface while forming CO and CO_2 [8]. This phenomenon can be utilized to create conductive polymers.

Deposited coatings irradiated with a CO_2 laser at 1.4 kW exhibit increased hardness, improved wear resistance, improved intercoat strength and bonding, sealing and reduced porosity, elimination of cracks on the surface, and creation of a branched dendritic crystal structure growth [81–83]. These effects can be utilized to improve existing coating properties.

Metal can be melted and evaporated and thus purified. Segregation of active metals and oxides also occurs [75]. This phenomenon can create more durable metal surfaces.

One of the main advantages of the excimer laser treatments is that all the effects mentioned above are very localized and no heat penetrates the bulk material; only the surface is affected (within 30 nm). This results from the fact that the reflection in the vibration band is low and the exposure time is extremely short [5,30,73].

Ablation is another phenomenon caused by the laser. It involves material loss through vaporization and melt flow and chemical reaction between the plasma generated by the laser and the surface of the material. Ablated surfaces have wide use in space technology for applications such as heat shields [56].

PET irradiated by a laser shows increased crystallinity. At high laser energy crystallization was reduced as the surface ablated by chain scission (UV laser) or radical formation (IR laser) [84]. As laser irradiation is localized, it could be employed to form hybrid crystalline structures. A similar phenomenon was observed for PEEK [54].

Cross-linking is another phenomenon observed in some irradiated surfaces. It is caused by chain breaking and network formation at the surface layer. The branching and subsequent cross-linking of the surface result in a thermally stable, harder, and more wear resistant surface structure [85]. Such a surface can better endure fatigue, erosion, and cracking.

Another effect is the annealing of thermoplastic materials such as PEEK. The annealing occurs through melting and re-quenching. This can reduce the time and cost of the annealing processes. The annealing reduces residual stresses at the surface and creates a more durable material [85].

X. COMPARISON TO OTHER TREATMENT METHODS

Historically, treatments to improve the adhesion of coatings to plastics consisted of mechanical abrasion, solvent wipe, solvent swell followed by acidic or caustic etching, and corona treatment. Each of these treatments has limitations, which provided a strong driving force to seek alternative treatments. Often what works for one material will not be effective for another, and specific treatments need to be developed for each. Mechanical abrasion or sandblasting is operator sensitive, dirty, and difficult to do on small parts or for high production volumes. Abrasive treatment creates mechanical keying of the adhesive to the substrate, but the durability in a humid environment is poor. Grit blasting also presents both occupational health and safety and environmental risks. Solvent wipe and swell treatments pollute, present safety and expensive hazard disposal problems, and often do not work. Acid etching, although more effective than solvent swelling, usually aggregates handling problems. Also, it is easy to overtreat and damage parts [3].

Corona treatment is limited to materials that are responsive to the treatment and have a simple part configuration. Often the treatment is marginal in effectiveness and has a short open time. Because a corona relies on ambient air, the process can change from day to day and from location to location. Complex shapes cannot be easily treated, as the treatment quality is a function of the distance from the electrodes. Arcing is prevalent in atmospheric discharge systems (corona) and can damage the coated material either thermally or electrically. Components of a corona line must be treated to eliminate the ozone generated during the process.

Coating with a primer is usually insufficient as the sole treatment, as it activates the surface but does not contribute to surface roughening or cleaning.

It is also very sensitive to the concentration and amount of primer applied (thickness of the layer). Each adherend-adhesive couple needs a specific primer.

Plasma treatment has advantages similar to those of laser treatment, but the method requires a special environment and vacuum. The treated material needs thorough cleaning and the surface accumulates electrostatic charges, which interfere with production [86].

Far-UV mercury lamps, can introduce functional groups only in a special atmosphere, such as liquids or gases, and no ablation occurs [88]. The main effect is production of a thin amorphous layer on the surface [89].

The CO_2 laser is based mostly on a heat (thermal) effect [52,87]. The result is an increase of only 22% in the adhesional strength of laser-treated Al 2024 bonded with epoxy, compared with untreated material, as only a small amount of surface modification occurs. The surface created is smooth and molten dendrites are observed after treatment, so no mechanical interlocking can occur [90,91].

XI. MODELS FOR LASER SURFACE TREATMENT

Many models try to explain the interaction of laser radiation with materials and the appearance of conical features and periodic wavelike structures on the surface. Up to now, there is no leading theory to explain the various phenomena. There is also no theory that can predict the various laser parameters needed to achieve optimal surface treatment for a specific material. Some models are presented in this section.

The laser interaction with a surface is wavelength dependent as different physical mechanisms are involved in each case: the shorter the wavelength, the more pronounced the interaction and its effects [61].

A wavelike structure is created during laser irradiation by shock waves induced by the surface plasma that is generated when the surface is vaporized by incident and reflected heat radiation. Threshold power densities of the laser are required to melt the surface and create this plasma [12]. Impurity or surface imperfections can also cause a wavelike structure, as certain parts of the irradiated surface absorb the laser radiation more than other parts. Island-like structures form gradually layer by layer, creating peaks and valleys, which break into cone-like structures [61].

An excimer laser at 193 nm has predominantly a photochemical ablation mechanism due to electronic excitation motion. This behavior is different from the thermal ablation process that results from intense local heating of the material by longer wavelength lasers [92].

Infrared lasers show predominantly a thermal mechanism. The heating rate of the adherend depends on the laser parameters (energy and number of pulses) and the material properties (absorptivity, heat capacitance, density). The surface temperature calculated is $T = 6\text{--}12 \times 10^3$ K [93].

A UV laser has a different mechanism than an IR laser. Most models predict two competing mechanisms occurring with UV lasers: a photochemical and a thermal mechanism. A photothermal model based on a zero-order ablation reaction has been proposed for UV laser irradiation [94]. The main factors for ablation are the repetition rate of the laser and the absorption coefficient of the treated material. Ablation begins much before the ablation threshold, during the photochemical process. Materials with low thermal diffusivity show pure ablation. The model predicts an optimal pulse duration that depends on the properties of the material irradiated.

The temperature gradient developed at the surface was modeled for laser treatment. Heating depends on the thickness of the sample, density ρ, specific heat C_p, thermal conductivity K, laser intensity I, and repetition rate. Results show a maximal temperature of 450 K at the surface, decreasing to room temperature at a depth of 460 nm [95].

The ablation temperature for polymers irradiated with a laser excimer was calculated on the basis of density, heat capacity, thermal conductivity of the material, and pulse duration of the laser [96]. An ArF laser displays a very sharp ablation point resulting from a transition from photochemical to thermal reaction. The ablation thresholds of various materials are summarized in Table 6.

There are two critical stages for ablation: one ablation point is due to contamination and nonuniformity of the surface, and the second ablation point is due to scission of bonds and buildup of plasma. The ablation points depend on the structure of the adherend material and its ability to form plasma [97]. There is an incubation period before the ablation process, but the removal of material begins even before the threshold [98].

The ablation rate increases exponentially with laser wavelength. The tem-

Table 6 Ablation Temperature and Ablation Threshold of Various Polymeric Materials (ArF Laser)

Material	Ablation temperature (K)	Ablation threshold (mJ/p · cm²)
PMMA	1100	1600
PI	2700	
PET	1500	30
PC	2500	
PS	2200	800
PE	800	1780
PP	1000	

perature of the surface obtained by ablation is modeled with the Arrhenius equation [99].

XII. DISADVANTAGES OF LASER TREATMENT

Some problems concerning surface modification by excimer lasers are briefly summarized here:

1. High temperature fails to provide protection in coating and improvement in adhesion [18]. This limits the laser treatment to relatively low temperature curing adhesives up to about 150°C.
2. The photochemical process is non-linear and therefore difficult to control. The extent of ablation can be controlled by adding dopants to the polymer surface (chromophors) [100].
3. Small size laser excimers are not yet industrially available. This limits the use of lasers in field repairs.
4. Scanning speed for high absorbance materials is relatively low which causes the process to be rather long and costly for some adherends.

XIII. ADVANTAGES OF LASER TREATMENT

Excimer laser surface treatment has many advantages:

1. The laser treatment is highly effective for a variety of materials because of a short penetration depth (30 nm) and high quantum yield for bond breaking [88].
2. Its control is very accurate and only the surface is affected, with no adverse effect on bulk properties [77].
3. It cleans the surfaces from all contaminants, adsorbed water, and natural oxides [101].
4. It protects the environment from pollution by introduction of solvent-free surface treatment [57].
5. It is highly localized and very precise and can be easily automated and applied to any surface geometry.
6. There is a long open time between treatment and application of adhesive or coating (14 days or more).
7. Treatment is performed at room temperature and in air, so no special environment is needed.

Table 7   Adhesional Shear Strength (MPa) of APC-2/AS-4 Adherends Bonded with FM 300-2K with Various Surface Treatments Exposed to Heat and a Wet Environment

Time (days) at 60°C, 95% RH	No treatment	Abrasive (SiC)	Laser treatment 0.18 J/p · cm², 100 p
0	6.1 ± 0.8ᵃ (A)ᵇ	14.7 ± 1.9 (A)	27.2 ± 0.8 (M)
10	2.6 ± 0.1 (A)	16.6 ± 1.5 (M)	38.6 ± 2.2 (M)
30	3.9 ± 0.4 (A)	13.2 ± 0.6 (A)	32.9 ± 6.0 (M)
60	4.5 ± 0.2 (A)	12.8 ± 1.5 (A)	29.2 ± 4.3 (M)

ᵃ ±5% standard deviation.
ᵇ A, interfacial; M, mixed mode of failure.

XIV. DURABILITY OF JOINTS AFTER LASER TREATMENT

Durability tests were conducted on Al 2024-T4 and on PEEK composite adherends treated with a UV laser, bonded, and then exposed to 60°C and 95% relative humidity (RH) for up to 60 days. For both joints no degradation was observed during this period. An example of durability results is given in Table 7.

The open time of laser-irradiated surfaces is 4–15 days [49,57].

XV. CONCLUSIONS

The use of the excimer laser as a preadhesion surface treatment for various substrates (plastics, ceramics, composites, and metals) has been reviewed in this chapter.

The UV laser irradiation as a surface treatment and modification proved to be most efficient compared with other conventional treatments such as chemical etching, abrasive blasting, plasma treatments, and UV lamp irradiation.

The general phenomena observed with UV laser treatment were surface cleaning by removal of contamination and weak boundary layers through evaporation, modification of surface chemistry by introducing polar groups such as oxide derivatives and hydroxides, surface cross-linking and hardening (as observed by Auger, ESCA, and FTIR spectroscopies), and change of surface morphology by introduction of uniform roughness (as observed by SEM). These phenomena were also reflected in contact angle measurements, a decrease of contact angle for metal adherends indicating an enhancement in surface energy (γ_s), and an increase in water contact angle for PEEK composite because of a hydrophobic effect, i.e., reduction of polar groups on the surface and removal of adsorbed water. An increase in hardness due to cross-linking (recombination) and removal of weak boundary layers were also observed.

As a result of these effects, adhesional strength was improved by 200–400% compared with conventional or no treatment. In addition, a change in failure mode from interfacial to cohesive was observed as a result of increased chemical and physical interactions and mechanical interlocking.

It can be concluded that all four factors affecting adhesional strength were met using UV laser treatment: cleaning, mechanical interlocking, chemical modification, and increased wetting. The potential of using UV lasers for surface treatment was shown for a wide variety of substrates, each requiring its optimal laser parameters for successful treatment.

Most of the phenomena observed on the surface after UV laser irradiation are similar for most adherends. It might be concluded that UV laser treatment has many advantages: it enhances adhesional strength and durability; is restricted to a surface effect without damaging the bulk properties of the adherend treated; can be applied in an air atmosphere and at room temperature; and is an effective, clean, environmentally friendly, precise, and safe process.

It should be noted that deviating from optimal laser conditions may result, in most cases, in decomposition, melting, and ablation, which should be avoided.

A model for predicting optimal laser treatment for a specific material needs further investigation and study.

REFERENCES

1. DE Packham. In Handbook of Adhesion. Avon, UK: Longman Scientific & Technical, 1992, chap A.
2. KL Mittal, HR Anderson Jr. Acid-Base Interactions: Relevance to Adhesion Science and Technology, Zeist, The Netherlands: VSP, 1991.
3. WJ Prame. Polym News 19(9):281, 1995.
4. JW Dini. SAMPE J 32(3):58, 1996.
5. A Buchman, H Dodiuk, M Rotel, J Zahavi. Int J Adhesion Adhesives 11:144, 1991.
6. L Dorn, W Wahono. Proc. 6th International Symposium of Swiss Bonding, Basel, Switzerland, 1992, p 87.
7. PW Milloni, JH Eberly. Lasers. New York: John Wiley & Sons, 1988, chap 13.
8. R Srinivasan. J Appl Phys 73:2743, 1993.
9. S Kawanishi, Y Shimizu, S Sugimoto, N Suzuki. Polymer 32:979, 1991.
10. MU Islam. Adv Mater 3:215, 1996.
11. JA Hunt, RL Williams, SM Tavakoli, ST Riches. J Mater Sci Mater Med 6:813, 1995.
12. WD Mueller, K Seliger, J Meyer, JA Gilbert. J Mater Sci Mater Med 5:692, 1994.
13. U Stratmann, K Schaarschmidt, M Schurenberg, U Ehmer. Scanning Microsc 9: 469, 1995.

14. M Murahara. Lambda Highlights 43:3, 1994.
15. AJ Pedraza, RA Kumar, DH Lowndes. Appl Phys Lett 66:1065, 1995.
16. MJ Godbole, AJ Pedraza, DH Lowndes, EA Kenik. J Mater Res 4:1202, 1989.
17. B Braren. SPIE Proc 1598:240, 1991.
18. DH Lowndes, M DeSilva, MJ Godbole, AJ Pedraza, T Thundat, RJ Warmack. Appl Phys Lett 64:1791, 1994.
19. A Hagemeyer, H Hibst, J Heitz, D Bauerle. J Adhesion Sci Technol 8:29, 1994.
20. HJ Sun, HC Jang, E Chang. Surf Coatings Technol 64:195, 1994.
21. A Heilmann, J Werner, F Hamilius, F Muller. J Adhesion Sci Technol 9:1181, 1995.
22. OI Balits'kii, VI Pokhmurs'kii, MO Tikhan. Sov Mater Sci (USA) 27(1):51, 1991.
23. MD Digilov, G Neiman. Carbon 29(2):284, 1991.
24. K Kearney, A Maki. Text Chem Colo 26(1):24, 1994.
25. J Newbould, KJ Schroeder. Adv Composites 365:51, 1987.
26. T Bahners, E Schollmeyer. Angew Makromol Chem 151(2509):19, 1987.
27. T Bahners, E Schollmeyer. Angew Makromol Chem 151(2513):39, 1987.
28. JG Dillard. ASM, Proc. Surface Preparation of Composites Conference, Metals Park, OH, 1990, p 281.
29. D Knittel. Development of scientific fundamentals for excimer laser surface structuring of synthetic fibers. NTIS No. TIB/A96-01598/XAB, 1995.
30. P Holzer, F Bachmann. Kunstst Ger Plast 79:485, 1989.
31. PE Dyer, J Sidhu. J Opt Soc Am B 3:792, 1986.
32. Z Kollia, E Hotzopoulos. SPIE Proc 1503:215, 1991.
33. JC Ion, T Moisio, TF Pedersen, B Sorensen, CM Hansson. J Mater Sci 26:43, 1991.
34. TR Jervis, JP Hirvonen, M Nastasi. Adv Mater 25:4, 1993.
35. RF Cozzens, P Walter, AW Snow. Proceedings of the 17th Adhesion Society Conference, Hilton Head, SC, 1994, p 35.
36. CE Hoyle, MA Trapp, CH Chang. J Polym Sci A 27:1609, 1989.
37. K Ogawa. J Phys Chem 93:5305, 1989.
38. CE Hoyle, MA Trapp, CH Chang. Polym Mater Sci Eng 57:579, 1987.
39. Y Kizaki, T Taguchi, S Yamasaki. Jpn J Appl Phys 29:1561, 1990.
40. DW Thomas, CF Williams, PT Rumsby, MC Gower. Proc. Laser Ablation of Electronic Materials Conference, Gironde, France, 1992, p 221.
41. MM Murahara, M Okoshi, K Toyoda. Gas Flow and Chemical Lasers: 10th International Symposium, Friedrichshafen, Germany, 1994.
42. J Breuer, S Metev, G Sepold. Mater Manuf Process 10(2):229, 1995.
43. M Murahara, M Okoshi. J Adhesion Sci Technol 9:1593, 1995.
44. J Breuer, S Metev, G Sepold, OD Hennemann, H Kollek, G Kruger. Appl Surf Sci 46:336, 1990.
45. W Kesting, T Bahners, E Schollmeyer. Appl Surf Sci 46:326, 1990.
46. M Nishii, S Sugimoto, Y Shimizu, N Suzuki, S Kawanishi, T Nagase, M Endo, Y Eguchi. Proceedings of the 17th Adhesion Society Conference, Hilton Head, SC, 1994, p 165.
47. S Yahaura, H Kurokawa, T Fujimoto, F Baba, T Ando. Polym Mater Sci Eng 69:357, 1993.
48. DJ Doyle. SPIE Proc 1990:260, 1993.

49. ST Tavakoli, ST Riches. Proc. SPE-ANTEC'96, 1996, p 1219.
50. WW Duley, G Allan. Appl Phys Lett 55:1701, 1989.
51. F Kokai, H Saito, T Fujioko. J Appl Phys 66:3252, 1989.
52. E Wurzberg, A Buchmann, E Zylberstein, Y Holdengraber, H Dodiuk. Int J Adhesion Adhesives 10:254, 1990.
53. PL Baldeck, SY Yang, RR Alfano, RH Callender, C Bennett Jr, SK Mowdood, WH Waddell. Opt Eng 30:312, 1991.
54. M Rotel, J Zahavi, A Buchman, H Dodiuk. J Adhesion 55(1–2):77, 1995.
55. HC Man, M Li, M Yu. J Adhesion Sci Technol 11:183, 1997.
56. E Sancaktar, SV Babu, E Zhang, GC D'Couto. J Adhesion 50:103, 1995.
57. A Buchman, H Dodiuk, S Kenig, M Rotel, J Zahavi, TJ Reinhart. J Adhesion 41: 93, 1993.
58. SS Gurkovsky. J Mater Sci Lett 9:1338, 1990.
59. AJ Pedraza. Appl Phys Lett 66:1065, 1995.
60. Z Gendler, A Rosen, M Bamberger, M Rotel, J Zahavi, A Buchman, H Dodiuk. J Mater Sci 29:1521, 1994.
61. F Henari, W Blau. Appl Opt 34:581, 1995.
62. E Sancaktar, SV Babu, W Ma, GC D'Couto, H Lipshitz. Proceedings of the 19th Adhesion Society Conference, Myrtle Beach, SC, 1996, p 203.
63. E Sancaktar, E Zhang. ASME 79:65, 1994.
64. T Bahners, W Kesting, E Schollmeyer. SPIE Proc 1503:206, 1991.
65. AJ Pedraza. Proc. International Conference on Beam Processing of Advanced Materials, Chicago, 1992, p 69.
66. J Kuldeep, K Aninesh. Nucl Instrum Methods Phys Res B B15(1–6):250, 1986.
67. PE Dyer, RJ Farley. Appl Phys Lett 57:765, 1990.
68. WW Duley, K Dunphy, G Kinsman, S Mihailov. J Laser Appl 23:9, 1990.
69. T Lippert, F Zimmermann, A Wokaun. Appl Spectrosc 47:1931, 1993.
70. S Lazare, P Benet, M Bolle, P De Donato, E Bernardy. SPIE Proc 1810:546, 1992.
71. M Bolle, S Lazare. J Appl Phys 73:3516, 1993.
72. O Occhiello, F Garbassi, V Malatesta. Angew Makromol Chem 169:143, 1987.
73. F Kokai. Jpn J Appl Phys P1 29:158, 1990.
74. AJ Pedraza, JW Park, HM Mayer, DN Braski. J Mater Res 9:2251, 1994.
75. DJ Doyle. Proceedings PRI Third International Conference on Structural Adhesives in Engineering, Bristol, UK, 1992, p 15/1.
76. A Brezini, N Zekri. J Appl Phys 75:2015, 1994.
77. KJ Schmatjko. Ind Anz 111(99):39, 1989.
78. R Srinivasan. Appl Phys A56:417, 1993.
79. LH Dao, XF Zhong, A Menikh, R Paynter, F Martin. Proc. SPE-ANTEC'91, p 783.
80. PV Nagarkar, EK Sichel. J Electrochem Soc 136:2979, 1989.
81. S Weeramareddy, S Bahadur. Wear 157:245, 1992.
82. JM Rigsbee. Laser Processing of Metals and Coatings for Enhanced Surface Properties. NATO ASI Ser E Appl Sci 85:300, 1984.
83. G Christodoulou, WM Steen. ASM, Proc. Lasers in Material Processing Conference, Metals Park, OH, 1983, p 116.
84. MF Sonnenschein, CM Roland. Appl Phys Lett 57:425, 1990.

85. VV Korshak, IA Gribova, EE Said-Galiev, AP Krasnov, TM Babchinitser, LN Nikitin, MM Tepliakov, RA Dvorikova. Mekh Kompoz Mater 6:985, 1987.
86. P Davis, C Courty, N Xanthopoulus, HJ Mathieu. J Mater Sci Lett 10:335, 1991.
87. SL Kaplan, PW Rose. Int J. Adhesion Adhesives 11:109, 1991.
88. R Srinivasan, S Lazare. Polymer 26:1297, 1985.
89. DS Dunn, AJ Oudekirk. SPIE Proc. 2374:42, 1995.
90. GW Critchlow, DM Brewis, DC Emmony, CA Cottam. Int J Adhesion Adhesives 15(4):233, 1995.
91. SJ Fang, R Salovey, SD Allan. Polym Eng Sci 29:1241, 1989.
92. G Huadong, GA Voth. J Appl Phys 71:1415, 1992.
93. SI Anisimov, VA Khokhlov. In Instabilities in Laser-Matter Interaction. Tokyo: CRC Press, 1995, chap 1.
94. SR Cain, FC Burns, CE Otis. J Appl Phys 71:4107, 1992.
95. RJ Anderson. J Appl Phys 64:6639, 1988.
96. NP Ffurzikov. SPIE Proc 1503:231, 1991.
97. AV Ravi Kumar, G Padmaja, P Radhakrishnan, VPN Nampoori, CPG Vallabhan. J Physics 37:345, 1991.
98. PE Dyer, GA Oldershaw, D Schudel. Appl Phys B51:314, 1990; PE Dyer, ST Lau. J Mater Res 7:1152, 1992.
99. S Kuper, J Brannon, K Brannon. Appl Phys A56:43, 1993.
100. H Mashuhara, H Fukumura. Polym News 17:5, 1991.
101. JRJ Wingfield. Int J Adhesion Adhesives 13(3):151, 1993.

85.

9

Adhesion Enhancement of Metallic Films to Ceramic Substrates Using UV Lasers and Low-Energy Ions

A. J. Pedraza
The University of Tennessee, Knoxville, Tennessee

I. INTRODUCTION

Metals bonded to insulators such as Al_2O_3 and SiO_2 are widely used in microelectronics and hybrid circuit technologies and have important applications as thermal barriers, as coatings, and also in catalysis. In microelectronics, a strong bond between the conductor film and the insulator is vital for the reliability of advanced integrated circuits. Strong metal-insulator bonding is also required for the integration of devices in advanced electronic packages. The roadblock in the usage of ceramics as heat exchangers, heat engines, and wear-resistant parts can be removed by joining metals to ceramics in new composite materials. However, when the composite performance is poor, it is generally due to insufficient strength of the interfacial bonding rather than to the structural integrity of each material. In film deposition, along with the species of metal being deposited, the adhesion strength is related to the surface characteristics of the substrate. Therefore, it can be expected that the state of this surface may be engineered to meet adhesion requirements of deposited films for specific uses. Alternatively, attention could be focused upon improving adhesion by modifying the interfacial region of the metal-insulator interface after it has been produced. In this chapter, the effects of surface modifications of the substrate prior to film deposition and of already produced interfaces on metal-ceramic bonding will be discussed.

In order to evaluate the kind of modification that is effected on a surface after a given procedure, it is necessary to establish the starting condition of that surface. Moreover, it would be desirable to have as a reference point an initially perfect surface. A perfect surface must be free of chemically or physically adsorbed species and have a crystallinity very close to that of the bulk. These surfaces are difficult to obtain and also require an ultrahigh vacuum chamber for their preparation and characterization [1]. This kind of surface preparation is almost never realized in studies of adhesion. Insulator materials such as Al_2O_3 are often used in the as-sintered, polycrystalline form and have a roughness related to the particulate size. Furthermore, insulators may have a chemical composition at the surface that differs from that of the bulk. For example, commercially available AlN substrates normally have a thin Al_2O_3 film on the near-surface region [2]. Impurities also segregate to the surface, changing the chemical composition of the near-surface region [3].

In general, a surface that has been exposed to ambient atmosphere will have a few monolayers of adsorbed carbon and/or oxygen. For instance, in sapphire (single-crystalline Al_2O_3) samples carbon has been found in two nominal forms, namely as free carbon (amorphous) or in hydrocarbons bound to either nitrogen or oxygen [4]. The adsorbed surface layer can be detected by surface analytical techniques such as X-ray photoelectron spectroscopy (XPS), Auger emission spectroscopy (AES), or ultraviolet photoelectron spectroscopy (UPS). The nature and thickness of the adsorbed layer depend not only on the chemical nature of the substrate but also on the physical state of the surface, as well as on its crystallographic and topographical features, and can vary significantly from sample to sample. The adsorption can be physical, in which case there is a weak electrostatic interaction between the adsorbed species and the substrate, or chemical. In this latter instance, there is electron transfer or sharing between the atoms of the substrate and the adsorbed species [5]. Commonly, contaminants can be removed by ion bombardment; for instance, most carbon is removed from a sapphire surface by etching with 7-keV argon ions for 30 s [4].

Clearly, it is important to be able to characterize the initial state of the surface in order to establish the type of surface modification that a treatment has rendered. There are only a few instances in which both the initial and final states have been studied by surface-sensitive techniques. Nonetheless, on a relative basis, the change in adhesion strength after the surface treatment can provide valuable information.

This chapter is not a comprehensive review of any of the subjects discussed. A core group of ideas and techniques is analyzed and carried through the chapter until its final conclusions. In the first section, studies of the bonding of metallic films to insulators are briefly reviewed, starting with more fundamental approaches and continuing with research that is more oriented toward applications. Next, a brief outline is given of three widely used surface analytical techniques: XPS, AES, and UPS. In the third section, techniques for modifying the metal-ceramic interface to enhance adhesion are described. The fourth section deals with surface and near-surface modifications of the substrate that lead to adhesion enhancement, in some cases with additional heat treatment of the metal-ceramic couples. An overall discussion and a recollection of the main conclusions close the chapter.

Two surface/interface modification techniques are analyzed in this chapter: low-energy ion bombardment and pulsed-laser irradiation. Modification of the interface between a metal film and an insulator aimed at promoting adhesion can be performed at any one of the following three significant stages: after the metallic film has been deposited, during deposition, or before deposition. In Fig. 1, a schematic is shown of the three different ways to affect the metal film–substrate interface. In the case illustrated in Fig. 1a, the ions or photons used for the modification must penetrate the film completely in order to reach the interface. This

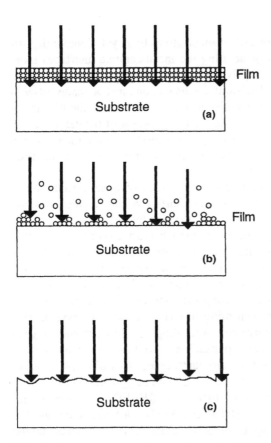

Figure 1 Schematics of surface/interface modification practices to enhance metal-insulator bonding. (a) After film deposition. (b) During film deposition. (c) Before film deposition.

action is not required if the metal-insulator interface is modified by the heat generated during the treatment. As discussed later, this is the case of laser irradiation, except when the photons are used as ionizing radiation to promote reactions at the interface level. For this exception the film thickness must be very small, of the order of 10 nm, because the absorption length of photons in metals is very short [6]. The penetration of ions in films is a function of the ion energy, ion nature, and nature of the film, but in most cases an ion accelerator will be needed. For instance, 500-keV Ar^+ ions are required to penetrate an $\sim$200-nm-thick copper film, as estimated using the TRIM computer code [7].

II. BOND FORMATION BETWEEN METALLIC FILMS AND INSULATORS

The continuous and gradual deposition of metal atoms by physical vapor deposition is the method of choice for generating the metal-ceramic interfaces more amenable to study using surface-sensitive techniques. When analyzing the nature of the bonding between metal and substrate, a distinction must be made between true chemical bonding and other mechanisms that do not involve the transfer or sharing of electrons between atoms of the metal and atoms of the ceramic. The experimental signature of chemical bonding at the metal-ceramic interface is a change in the electronic structure—and hence bond energy—of the atoms relative to the atoms in the bulk materials.

One of the approaches to the experimental study of metal-ceramic interfaces starts with the preparation of a ''perfect'' surface of the ceramic substrate under ultrahigh vacuum (UHV) and its characterization by surface-sensitive techniques. The procedure commonly used to obtain such a surface consists of cutting and polishing the substrate. After this step, contaminants are removed by ion etching in a vacuum chamber. Because ion etching causes structural damage, the crystal perfection is restored by a high-temperature annealing in UHV [1]. Some contaminants could also be removed during the annealing. The combination of ion etching and high-temperature annealing in UHV produces an oxygen deficiency at the surface of oxide substrates. Annealing in a low-pressure oxygen atmosphere helps to restore the crystal perfection as well as to maintain surface stoichiometry and remove adsorbed contaminants, especially carbon [8]. The degree of surface perfection can be assessed using low-energy electron diffraction (LEED), reflection high-energy electron diffraction (RHEED), or scanning tunneling and atomic force microscopies (STM and AFM). Cleavage of a crystal in UHV is still another way of getting a ''perfect'' surface [1]. The changes taking place as metallic monolayers are being deposited on these special surfaces are monitored by various spectroscopic techniques such as XPS, AES, UPS, and low-energy electron loss spectroscopy (LEELS).

The changes occurring at metal-ceramic interfaces and in the metallic films during their deposition on polished and UHV annealed single crystals have also been analyzed using surface analytical techniques. The formation of the Al/α $-$ Al$_2$O$_3$ interface was studied using LEED, high-resolution electron energy loss spectroscopy (HREELS), XPS, and AES [9,10]. It was found that when depositions were conducted at room temperature and at 470°C a strong interaction between the first Al monolayer and the oxygen anions of Al$_2$O$_3$ developed, as revealed by the formation of a suboxide, whereas the subsequent monolayers of Al were metallic. The sticking coefficient of the aluminum atoms to the sapphire surface was 1 at room temperature but decreased with increasing temperature, becoming null at 720°C (60°C above the melting point of solid metallic aluminum). There was no deposition at 720°C; instead, the aluminum atoms etched away the surface imperfections created by ion etching of the sapphire substrate. The first monolayer of Al deposited at room temperature was found to induce a surface reconstruction from the (1 $\times$ 1) LEED pattern to a ($\sqrt{31}$ $\times$ $\sqrt{31}$) R $\pm$ $\tan^{-1}$ ($\sqrt{3}/11$) LEED pattern. It is interesting to point out that in c-oriented Al$_2$O$_3$ a similar surface transformation takes place by thermal annealing under vacuum at 1200°C; this reconstruction has been attributed to loss of surface oxygen [11]. In contrast to the surface effect observed after deposition at room temperature, no reconstruction occurred when the aluminum deposition was performed at 470°C [9]. This difference was ascribed to different preparations of the substrate surfaces [10]. For the experiment performed at room temperature the sample preparation included a 15-min sputtering with an Ar gun at 1 keV followed by a flash annealing at $\sim$1000°C, whereas for the experiment at 470°C only a 1000°C annealing was performed prior to deposition. It was argued that the annealing produced a clean substrate and the ion etching produced a higher density of surface defects that facilitated the ($\sqrt{31}$ $\times$ $\sqrt{31}$) R $\pm$ $\tan^{-1}$ ($\sqrt{3}/11$) structure. The remarkable result of this research is the finding that ion bombardment prior to deposition may determine the structure of the Al$-$Al$_2$O$_3$ interface.

Surface reactions at the substrate and reactions between the metal being deposited and the ceramic substrate are frequently observed and shed light on the nature of interfacial interactions. Thus, formation of metallic aluminum was detected by XPS when Ti was being deposited on α-Al$_2$O$_3$, indicating that the oxide was being reduced [8]. Not only was there an electron transfer from Ti to O but also Al and Ti reacted to form the intermetallic compound Ti$_3$Al. The interfacial reactions taking place when copper is deposited on α-Al$_2$O$_3$ and on AlN were investigated using UPS [12–14]. The shift of the Cu ($3d$) orbital peak relative to that of bulk (metallic) copper was followed as a function of the substrate coverage. In the case of α-Al$_2$O$_3$ substrates the shift was small and quickly tended toward the value of bulk copper. For aluminum nitride (apparently devoid of a surface layer of Al$_2$O$_3$), the copper peak shifted by 2 eV, indicating a strong binding of copper to AlN at submonolayer coverage.

Also, the deposition atmosphere can play a significant role in the type of interface generated. The presence of oxygen can change completely the interaction of the metal and the oxide, as will be emphasized later. For instance, when a small amount of oxygen was introduced during Ni deposition on α-Al_2O_3, an $NiAl_2O_4$ spinel phase formed which was absent when the deposition was carried out in UHV [15].

A surface science approach to studying adhesion such as the one just referred to in the examples is powerful because it provides basic knowledge to understand the complex interactions between metals and ceramics at the interface level. It also allows a direct comparison between theory and experiments [16–19]. One of the limitations of such studies as they relate to adhesion is the assumption that a strong or a weak interaction between a one- or two-monolayer-thick metallic film and a substrate at the interface level is equivalent to a strong or a weak adhesion of a thicker film to the substrate. There are other factors that may affect the adhesion strength as well, such as internal stresses in the deposited film, relaxation of coherency strains when a film thicker than a few monolayers is grown on the ceramic substrate, and the mechanical properties of an intermediate compound, when it forms [20,21].

Other factors that could be studied using those basic tools are the effects of contaminants or impurities on adhesion. In particular, oxygen as an impurity can strongly enhance the bonding of metals to oxides because it may promote formation of a spinel or of a double oxide at the metal-substrate interface [21]. Also, as we shall discuss later, the state of the surface and the near-surface region can have a strong influence on the adhesion strength.

Johnson and Pepper [22] distinguished between the bonding of an oxidized metal to Al_2O_3 and the direct bonding between metal atoms and oxygen anions at the Al_2O_3 surface. Experiments related to the first type of bonding for copper are described below. To represent the direct metal-Al_2O_3 bonding, they calculated the interaction between metallic atoms (Fe, Ni, Cu, and Ag) and an Al_2O_3 cluster using a molecular orbital model and found that a chemical bond could be established at the metal-sapphire interface. Bonding and antibonding molecular orbitals between the metal (d) states and the oxygen (p) states, as well as a metal-to-oxygen charge transfer, were found as indications of true chemical bonding. They also obtained an increase in the occupancy of the metal-Al_2O_3 antibonding orbitals through the series Fe, Ni, Cu, and Ag. The occupancy of the antibonding orbitals tends to cancel the effect of the bonding orbital occupancy, thus decreasing the metal-sapphire bond strength. These results are consistent with the experimental measurements of a static coefficient of friction for the various metals on Al_2O_3 carried out by Pepper [23]. The bonding between Cu and Al_2O_3 is very weak due to a small residual charge transfer. Experimental results of Ohuchi et al. [12] using UPS are also consistent with a weak interaction between Cu and Al_2O_3.

In the case of oxides, the bond strength can be increased if the metal in contact with the substrate is oxidized. O'Brien and Chaklader [24] found that the work of adhesion between liquid copper and sapphire reached a maximum at an oxygen partial pressure of 7.6×10^{-7} torr (10^{-9} atm). They suggested that the formation of $CuAlO_2$ as the oxygen pressure in the atmosphere was increased caused the increase of copper wettability of sapphire. Reactions in the CuO-Cu_2O-Al_2O_3 system to give $CuAl_2O_4$ and Cu_2AlO_4 were reported by Gadalla and White [25]. Mattox [26] obtained strong adhesion between gold films and fused silica when gold was sputter deposited in the presence of oxygen. Also, Moore and Thornton [27] obtained very strong bonding when gold was melted on a fused silica surface in the presence of oxygen. Similar adhesion enhancement of platinum films deposited on alumina surfaces were obtained by Budhani et al. [28] in the presence of oxygen. The first 20–30 nm of the platinum film were deposited by dc magnetron sputtering using in one case a pure argon plasma and in another case an oxygen-diluted argon plasma. After these initial conditions, growth of both films continued under pure argon plasma. The films grown under oxygen exhibited strong bonding, and Rutherford backscattering analysis showed an oxygen concentration of 30 at% in the initially grown 20 to 30 nm.

The process developed by Burgess et al. [29] for direct copper-alumina bonding in hybrid electronic packages yields strong adhesion. In this process, thick copper films are patterned to circuitry requirements, placed in contact with alumina, and heated in a slightly oxidizing atmosphere to a temperature higher than 1065°C but lower than the melting point of copper (1083°C). The authors attributed the strong adhesion to wetting by a very thin film of molten copper in contact with the alumina. Melting of this layer would occur if dissolved oxygen caused the formation of a eutectic occurring at 0.39 wt% oxygen and 1063°C in the copper oxygen phase diagram. Another possibility is the formation of a mixed aluminum-copper oxide that promotes strong adhesion

In all the processes described above, oxygen was introduced while the metal was at temperatures close to or above the melting point or was incorporated during film deposition. If the oxygen entered during film deposition, an intimate contact of oxygen with the substrate surface was ensured. For the case of high-temperature annealing, a substantial amount of oxygen was supplied to the metal-substrate interface by diffusional flow from the oxygen-rich surrounding atmosphere into the (solid or liquid) metal. The formation of oxide is thus expected to take place at the interface and bonding to be enhanced. The next question is how much oxygen is required to form an oxide at the interface. In the case of depositing metallic aluminum on Al_2O_3 it is sufficient to have the surface of Al_2O_3 terminating in oxygen to promote the formation of a nonstoichiometric oxide one monolayer thick [9,10].

Thermodynamic calculations of the oxygen activity required to form metal-aluminate spinel at the metal-Al_2O_3 interface were performed for Ni, Fe, and Cu

[30]. It was concluded that in all cases the oxygen concentration required to form double oxides was below the solubility limit of oxygen for any of the three metals. However, kinetic factors may inhibit the formation of an oxide at the interface of a relatively thin film. For instance, an 80-nm-thick copper film was ion beam sputter deposited on Al_2O_3 at a base pressure of 2×10^{-6} torr and an Ar pressure of 1×10^{-5} torr [31]. The high-purity argon used in this process was additionally circulated through a purifying system. No oxide was detected by AES at the metal-ceramic interface of the as-deposited system or even after the couple had been annealed at 300°C for 1 h in a pressure as high as 1×10^{-5} torr. When the annealing temperature was increased to 500°C, in some specimens a thin oxide layer was detected that strongly enhanced adhesion.

A clear separation emerges between the experiments performed using surface science techniques and the experiments described later in this section. The first experiments are carefully performed and there are analytical tools to monitor in situ the changes taking place. In the second set of experiments the formation of compounds or reactions at the interface are mostly inferred. The latter experiments, however, show the rich variety of results that can be obtained under different experimental conditions, which are generally lacking in the first type of experiments. UHV experimental conditions naturally limit the kind of experiments that can be performed in situ. It is hence desirable to establish surface-sensitive techniques to study metal-ceramic interfaces of specimens prepared outside the analysis chamber. It is important to emphasize that it is the chemical interaction at the interface that relates to adhesion, and for this reason the most important tools for studying these interactions are XPS, UPS, AES, and related techniques. On the other hand, structural features may be secondary to producing a strong bond. For instance, epitaxial growth will not warrant strong bonding; e.g., Baglin et al. [32] described an experiment in which Cu grown epitaxially on Al_2O_3 exhibited weak bonding.

In summary, the main experimental and theoretical results emphasized in this section are the following:

1. Adhesion of a metal to a ceramic by a reactive process can take place via two different mechanisms: direct interaction of the deposited metal and the ceramic across a flat interface, or formation of an intermediate compound between the metal and the ceramic. However, this difference can become very tenuous when the intermediate compound is just one or two monolayers thick.
2. Oxygen has a strong influence in the bonding of metals to oxides. Thermodynamic calculations indicate that in the case of Cu, Fe, or Ni with Al_2O_3, the formation of double oxides requires a low oxygen activity. Kinetic factors such as oxygen diffusion in the metal may, however,

limit the formation of an interfacial oxide even in the case of relatively thin films.

3. The physical state of the surface, as well as its crystallographic and topographical features, can significantly affect the structure that forms when metallic atoms are deposited on it.

4. The interaction of Cu, Au, and transition metals with Al_2O_3 is generally weak, but it can be strongly enhanced by excess oxygen at the metal-ceramic interface.

III. ANALYTICAL TECHNIQUES AND ADHESION MEASUREMENTS

A. Auger Emission Spectroscopy [33–36]

In Auger emission spectrometers the particles incident on the specimen are 1–10-keV electrons and the emitted particles are electrons that have unique energies for each atomic species and are in the range of 0 to 2 keV. The impinging electrons knock out a core electron from a specific atomic level E_1 of atoms within a depth of ~1 μm from the surface of the specimen. An electron in the atom falls from a level E_2 filling the core hole and the excess energy is transferred by an internal process to an electron at another level E_3. This electron, called an Auger electron, is then ejected with a kinetic energy E_a given by the energy balance

$$E_a = E_1 - E_2 - E_3^*$$

where E_3^* is the electron binding energy at level 3 in the presence of a hole at level 2. The Auger electron energies depend solely on the atomic energy levels, and for this reason they are unique for each element. The most universal nomenclature used for Auger energies is given in terms of the X-ray notation for the electron states with $n = 1, 2, 3, 4$ designated as K, L, M, N, respectively. A detailed description of the different ways of accounting for the Auger transitions is given by Briggs and Rivière [34].

Most of the Auger electrons generated within the 1-μm depth remain inside the material because in their energy range there is intense inelastic scattering. Only Auger electrons from the outmost atomic layer are ejected and can be detected. The attenuation length, which is the distance that an Auger electron can travel in the matrix before being inelastically scattered, is a function of the emitted electron energy for each element [36]. As the energy of the Auger electron increases the signal comes from a deeper layer as well; for instance, for an electron energy of 100 eV the Auger electrons are ejected outside the sample from the first two monolayers, whereas if the electron energy is 1300 eV the attenuation length is six or seven monolayers.

Because the magnitude of the core electron energy for a given atom is a function of the type of chemical bonding with other atoms, chemical changes in the bonding will produce shifts in the Auger transitions. This is one of the most attractive features of AES for chemical analysis. The chemical environment of a given atom can also be detected by variations in the shape and in the fine structure associated with a given atom. In Sec. V an example of how chemical changes at the metal-insulator interface can be detected using these changes in the Auger transitions will be shown. Another relevant feature of AES, due to the fact that the probing particles are electrons, is the high spatial resolution, which can reach 5 nm.

B. X-Ray and Ultraviolet Photoelectron Spectroscopies [33–36]

In XPS the incoming particles are photons generated by an X-ray source that commonly uses Al or Mg as the target material. Photoelectrons are ejected by incident X-ray photons from one of the core levels of the atoms with a kinetic energy E_k given by

$$E_k = h\nu - E_B - \phi$$

where $h\nu$ is the photon energy, E_B is the binding energy of the electron in the solid, and ϕ is the spectrometer workfunction. The binding energy of the ejected photoelectron is the difference in energy between the atom with the electron and the atom without it. The photoelectron from a given atomic level is denoted using spectroscopic notation that is related to the n, l, j quantum numbers for each energy level. As with Auger electrons, the photoelectrons in XPS have a kinetic energy in the range of 50 to 2000 eV These low-energy electrons have a very high probability of undergoing inelastic scattering, and only electrons originating from the surface or a few atomic layers below the surface have any possibility of leaving the specimen. The spatial resolution in XPS is of the order of 5 μm. Chemical effects can also be detected by XPS and UV photoelectron spectroscopy (UPS) because the magnitude of the core electron energy for a given atom is a function of the type of chemical bonding with other atoms.

UPS is a technique widely used to study the band structure of metals and semiconductors. Photons of energy much lower than those used in XPS are required, and thus the photon source is a discharge lamp using an inert gas, e.g., He.

C. Interface Analysis by Surface-Sensitive Techniques

When a metal is deposited on a ceramic, the interface that forms can be as thin as one monolayer or can consume the film entirely if a strong reaction takes place. Thick intermediate compounds of the order of a few hundred nanometers can be analyzed by cross-sectional high-resolution transmission electron microscopy (HRTEM) and also by surface-sensitive techniques. In the case of interfaces a few monolayers thick, cross-sectional HRTEM can be used to study the structure of the interface. This technique involves a difficult sample preparation step and does not give any direct information about the nature of the bonding.

Because of the short range of both the Auger electrons and of the photoelectrons, two techniques have been used to study film-substrate interfaces. In one of them, the interface is built in situ by depositing the film in the analysis chamber and the chemical changes are monitored as a function of film coverage. One problem with this technique is that the nature of the film can change as the film is built up monolayer by monolayer [37]. Therefore, the chemical shift due to an interfacial reaction may combine with charging shifts [31]. Another problem is that the deposited atoms will continue to experience changes in their surroundings as atomic monolayers are being deposited. Atomic relaxations at the interface level could take place. Also, it is sometimes very difficult to reproduce processing conditions in the AES chamber and an additional chamber connected to the main one may be required.

In the other method the deposited film is removed layer by layer by sputter etching until the interface is revealed and the substrate is exposed [38]. The main problem with this destructive method is the ion beam–induced damage of the interface under study. Although the damage problem can be mitigated by reducing the ion beam energy for sputter etching, the destructive method requires careful control of the sputtering rate, which depends on the interface thickness. Therefore, several trial-and-error runs may be needed to detect elements in the interface region. In the case of interfaces that are monolayers thick, the ion bombardment itself could produce very important modifications, even using low-energy ions. One way to overcome this problem is to study films deposited on ceramic substrates without surface treatment and compare them with films deposited under equivalent conditions on treated substrates. At any rate, it would be desirable to have an analytical procedure with which the interface could be analyzed after being fully constructed, avoiding the introduction of misleading artifacts in the procedure used for analysis.

Following this idea, a procedure was developed that substantially reduces ion bombardment damage to the interface [31]. For these analyses, approximately 10-nm-thick metallic films are deposited on the sapphire substrates. Sputter etching of the film is performed in the AES vacuum chamber (approximately 10^{-9} torr) using the ion gun at an angle of 53° with respect to the specimen normal. An area of 200×200 μm^2 is bombarded with a 0.5-keV ion beam until a portion of the substrate in the selected area becomes exposed, forming a crater-like region surrounded by metallic film. The metal-ceramic interface is investigated using a technique that takes advantage of the variable film thickness in the vicinity of the crater. It was found by profilometry that the wedge produced by sputter etching had a thickness that increased almost linearly from less than a monolayer up to 10 nm, the full film thickness. A schematic drawing of the wedge and a way to select the analysis position are shown in Fig. 2a. The edge of the sputter-produced crater can be observed by secondary electron imaging and the analysis position can be approximately determined assuming that the slope of the crater edge is linear. The slope generated by this sputtering gun is very shallow: the film thickness increases by 0.5 nm when the electron beam is displaced 5 μm in the radial direction away from the film-denuded area. As illustrated schematically in Fig. 2b, different analysis positions on the edge of a sputter-produced crater can provide different elemental and/or chemical data because the Auger electron escape depths from different Auger transitions are different. The advantage of this technique is that an interface buried only a few monolayers deep can be analyzed, avoiding significant ion damage. The use of this technique for the specific instance of Au films deposited on ion-bombarded sapphire will be presented in Sec. V.A and B.

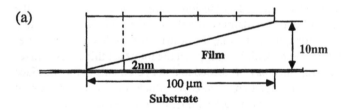

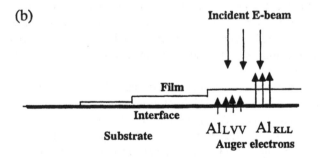

Figure 2 (a) Schematic of the wedge produced in the metallic film by sputter etching in the AES chamber. (b) Sketch illustrating the Auger electron depth close to the wedge center as a function of local film thickness. The wedge is pictured as ledge shaped, as expected from the high wedge length-to-film thickness ratio ($\sim 10^{7}$). (From Ref. 54.)

D. Adhesion Measurements

The adhesion strength results described in this chapter refer to measurements of practical adhesion as defined by Mittal [39], namely the force per unit area required to produce a rupture of the film-substrate specimen. Such a rupture may occur as a partial or total detachment of the film from the substrate, as a failure across the film, or as a failure across the substrate. In the two last cases, the failure should be related to the film-substrate interface and hence is expected to take place in its vicinity. Not all of the techniques for measuring practical adhesion (see Ref. 39 for an extensive listing) are quantitative. Generally, they cannot be precisely correlated, in the sense that different measurements will not differ by a single scaling factor, nor will they necessarily produce the same type of failure. For this reason, it is essential in systemic studies of adhesion to conduct the same test or set of tests for all the systems under investigation. In all our work, we have mainly used the pull test to measure practical adhesion. The technique consists of gluing a metallic pin (e.g., aluminum) to the film or to an intermediate coating covering the film and then pulling the pin at an increasing force per unit area until failure occurs. The adhesion strength measured by this method has an upper limitation given by the failure strength of the epoxy used to glue the pin to the film or coating. As discussed by Mittal [39], practical adhesion measurements cannot by themselves provide information on the specific atomic interactions bonding film to substrate or on the bonding mechanisms that lead to the different magnitudes of adhesion strength for film-substrate couples of given materials. Correlations between the practical adhesion strength and the more fundamental aspects of adhesion have been established via surface science studies using spectroscopy techniques (XPS and AES) and microscopy methods (e.g., TEM, HRTEM, SAD, analytical TEM techniques), as described later.

IV. ADHESION ENHANCEMENT BY MODIFICATION OF THE METAL-INSULATOR INTERFACE

A. Adhesion Enhancement by Ion Bombardment and Implantation After Film Deposition

Two techniques are distinguished when using energetic ions: reactive ion implantation and ion beam stitching. In the first instance, reactive ion species are used, and in the second case inert ions are passed across the interface. Each ion produces first ionization and a high density of electrons, and, as the energy of the ion is decreased, atomic displacements are produced. A collision cascade is generated at the end of the trajectory of the primary ion and the primary displaced atoms. This abrupt release of energy promotes high atomic mobility with atomic rearrangements that occur in $\sim 10^{-9}$s. These rearrangements may promote chemical interactions between the atoms in the film and the substrate when the cascade occurs at the interface and enhance the film-substrate bonding. In some systems a decrease in the Gibbs free energy may favor these interfacial reactions. It is also possible that reactions that would not take place in the bulk may occur at surfaces or interfaces [40]. The formation of metastable complexes at the interface during a collision cascade event is also worth considering because of the very short time available for the reaction. We will come back to the formation of metastable arrangements at the substrate surface during low-energy ion bombardment. Several reviews cover in much detail the effects of ion bombardment after or during film deposition, as well as the mechanisms responsible for adhesion improvement [40, and references therein].

B. Adhesion Enhancement by Pulsed-Laser Irradiation After Film Deposition

Adhesion enhancement of metal-ceramic couples after film deposition has also been achieved by pulsed-laser irradiation of metallic films deposited on insulators, e.g., Al_2O_3 and SiO_2 [41–47]. Irradiations were performed using a 308-nm excimer laser with a 42-ns full width at half-maximum (FWHM) pulse duration. An argon–4% hydrogen gas mixture was used in the chamber during the pulsed-laser treatment (PLT).

Excimer lasers generate an intense pulsed photon beam, tens of nanoseconds in duration, in the ultraviolet range (most commonly at 193, 248, or 308 nm). When a solid metal surface is irradiated with UV light, only 10 to 40% of the pulse energy is reflected. The remainder is absorbed in an ~30-nm-thick near-surface area [48]. Power densities of the order of tens of MW/cm^2 are generated by an excimer laser and deposited in this near-surface region. This energy can be dissipated in photochemical or photothermal processes. In the first instance photons can interact with matter (ionizing radiation), generating molecular excitations and, eventually, producing atomic desorption. However, in the case of metals and semiconductors, very rapidly (in $\sim 10^{-12}$ s) this energy transforms into heat [49], which is removed by conduction into the substrate and, at higher laser energy densities, also by explosive vaporization [45,50,51]. Due to the very short time involved in the irradiation process, surface radiation dissipates a negligible amount of the generated heat.

Figure 3 is an optical micrograph of a 5-μm-thick copper film deposited on a sapphire substrate and laser irradiated at 1.75 J/cm^2 ($4 \times 10^7 W/cm^2$). After the irradiation the film is completely cracked and mostly removed from the substrate. When an 80-nm-thick copper film is irradiated with 1 J/cm^2, the film is completely melted and breaks down into small globules, which, after solidifying,

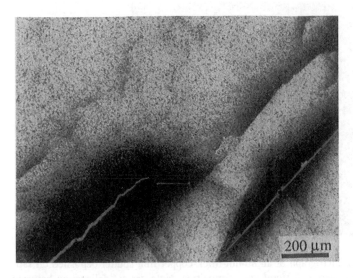

Figure 3 Optical micrograph of a 5-μm-thick copper film sputter deposited on sapphire and laser irradiated at 1.75 J/cm² after deposition. The film is completely cracked and has separated from the substrate. (From Ref. 52.)

are well bonded to the substrate (Fig. 4). The dramatic difference between the two experiments is due to the fact that during irradiation at most 1 μm of the copper film can be melted. For higher laser energy densities intense explosive evaporation takes place, generating a shock wave. In the example of Fig. 3 the metal-sapphire interface was negligibly heated but the shock wave generated during irradiation separated the weakly bonded film and mechanically ruptured it. On the other hand, the 80-nm-thick film was completely melted and reached the boiling temperature [52].

Based on these observations, a process was developed that combines an excimer laser and a sputtering module. A 1-μm-thick copper film was grown on a sapphire substrate by alternating discrete sequential sputter depositions with laser irradiations performed in air. After two sequential depositions, each one followed by laser irradiations, the 30-nm-thick copper film was broken into an unconnected network of copper islands, with the substrate exposed in the neighborhood of these particles. After four sequential depositions followed by irradiation, which added up to a thickness of 180 nm of flat film, the substrate surface was completely covered with copper. The laser-treated films remained bonded to the substrate after being scratched with a 400-g load, demonstrating that this method produces a strong bonding between the film and the substrate. As illus-

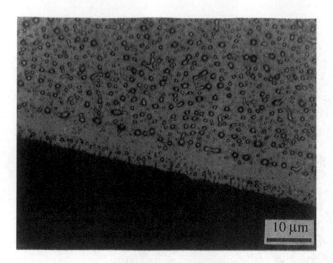

Figure 4 Optical micrograph of an 80-nm-thick copper film sputter deposited on sapphire and laser irradiated at 1 J/cm² after deposition. Note perfect delineation between laser-treated region and the untreated (black) region where the film was stripped by the tape. (From Ref. 46.)

trated in Fig. 5a, an as-sputtered film was totally detached from the groove region and in some places completely removed. The white areas seen in the micrograph of Fig. 5a were produced by transmitted light. By contrast, the sharp groove marked on a laser-treated film (Fig. 5b) shows no indication of detachment [41,42]. Similar results were obtained with nickel films deposited on sapphire [46].

   For the case of metallic films deposited on SiO₂, adhesion enhancement was not the only significant effect achieved by the sequential method just described. Another very remarkable effect, first observed with gold, was encapsulation of the metal taking place when a very thin (few tens of nanometers) sputter-deposited metallic film was laser irradiated. During the irradiation the gold film melted and broke down into small particles. Computer calculations showed that almost any metallic film in this thickness range should melt and thereafter attain a condition of explosive boiling. Instead, after melting, the thin metallic film broke into small particles in order to minimize the interfacial area. Since the gold-silica interface has a very high free energy, the overall surface free energy of the system was greatly reduced. Heat was transferred to the silica by thermal conduction from the liquid particles, in turn melting the silica in the regions adjacent to the gold particles. The particles were propelled into the silica owing

25 µm   25 µm

Figure 5 Scratches produced with a Vickers stylus using a 10-g load on a 300-nm-thick as-sputtered copper film (a) and on a 480-nm-thick copper film built by the sequential deposition and laser treatment process (b). Substrate: sapphire. (From Ref. 46.)

to the pressure developed by the very high rate of evaporation on their free surface. Figure 6 is a cross-sectional transmission electron micrograph of a 20-nm-thick gold film deposited on high-purity fused silica and irradiated at a laser energy density of 0.5 J/cm^2 in air. After the laser processing, a second gold film, 80 nm thick, was deposited to delineate the surface after the irradiation. The 20-nm-thick gold film initially deposited clustered into small spheres that became embedded in the oxide and, as can be seen in Fig. 6, became buried ~25 nm deep in the silica. Copper and iron films deposited on silica undergo the same transition upon irradiation, ending as encapsulated particles. By contrast, a titanium thin film completely reacts with silica upon laser irradiation, forming an amorphous layer [47].

Copper and gold films 75 nm thick were sputter deposited on SiO_2 and the couples were laser irradiated at laser energies from 0.5 to 3 J/cm^2. Following the irradiation, another 75-nm-thick layer of the same metal, namely copper or gold film, was deposited and the system annealed at 300°C. Adhesion of the films was measured by pull testing. In Fig. 7 the pull test results are plotted as a function

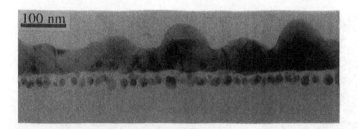

Figure 6 Cross-sectional transmission electron micrograph of a Au-SiO_2 specimen laser irradiated at 0.5 J/cm^2 after deposition showing encapsulation of the film as fine particles. The continuous gold film was deposited a posteriori to delineate the specimen surface. Subsequently, a 1-h annealing at 300°C was performed.

of the laser energy density for the case of copper. It can be seen that a sevenfold increase in adhesion strength took place when the couple was irradiated at a laser energy density of 2 J/cm^2. A similar behavior is seen for gold films, i.e., an adhesion strength increase with increasing laser energy density at low energy densities and a decrease in the high laser energy density range. In this case, a maximal adhesion strength of ~20 MPa was reached in gold-silica couples irradiated at 1 and at 1.5 J/cm^2 [47].

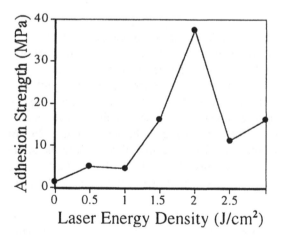

Figure 7 Adhesion strength of copper films prepared as follows: (1) 75-nm-thick films were sputter deposited on SiO_2; (2) the couples were irradiated at laser energy densities from 0.5 to 3 J/cm^2; (3) after irradiation another 75 nm of copper was deposited on the specimens; and (4) the specimens were annealed at 300°C. (From Ref. 47.)

C. Mechanisms of Laser-Enhanced Adhesion After Film Deposition

Transmission electron microscopy was used to analyze the interfacial region between a 100-nm-thick copper film deposited by the sequential procedure described above and an Al_2O_3 substrate. The deposition was performed in three stages, the first stage to a partial thickness of 10 nm, and the three irradiations following each partial deposition were performed in an Ar–4% H_2 atmosphere. The copper-Al_2O_3 couples were ion thinned to transparency from the substrate side. A schematic cross section of the area analyzed by transmission electron microscopy is presented in Fig. 8. Selected area diffraction (SAD) patterns were taken in regions of decreasing film thickness A, B, and C shown in the figure.

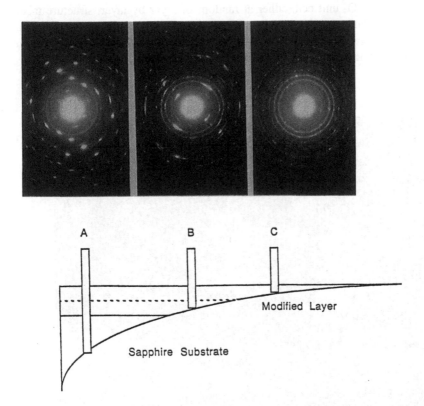

Figure 8 Copper film 100 nm thick deposited in three sequential steps, each followed by laser irradiation at 0.5 J/cm^2 in an Ar–4% H_2 atmosphere. Schematics of a cross section of a specimen thinned for TEM analysis and SAD patterns of each of the three regions of different thickness, A, B, and C. (From Ref. 46.)

In the thicker part of the film (A), a ring pattern can be seen superimposed on the $10\bar{1}0$ spot pattern coming from the substrate. In region B, the ring pattern is more prominent and the spots coming from the substrate show streaking. In C, the thinnest part of the film, the pattern from the substrate has disappeared and a well-developed ring pattern is observed. The analyses of these three SAD patterns are consistent with three regions: the single crystalline substrate, a resolidified Al_2O_3, and an intermediate compound at the interface. Figure 9 is a high-magnification image of the intermediate compound. The clusters shown in the figure are made of very fine particles. The diffraction pattern of the intermediate compound is very similar to the sapphire structure, but careful in situ chemical analysis showed that it contained a significant amount of copper. This explains why the $10\bar{1}3$ and the $20\bar{2}3$ reflections that are forbidden in the sapphire structure are present in the SAD pattern of the compound. If copper substitutes for aluminum in the Al_2O_3 unit cell, either at random or layer by layer, structure factor calculations show that these extra reflections should appear. If the laser irradiations are performed in air, the structure of the intermediate compound is different and can be indexed as a trirutile compound [41,43].

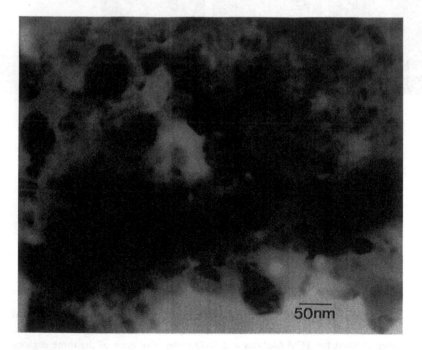

Figure 9 Microstructure of intermediate compound formed by laser irradiation of a 10-nm Cu film deposited on Al_2O_3. (From Ref. 43.)

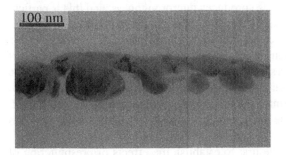

Figure 10 Cross-sectional TEM of an Au-SiO$_2$ specimen laser irradiated at 1.5 J/cm^2 after deposition showing partial encapsulation of the film clustered into large particles. The continuous gold film that was deposited a posteriori to delineate the specimen surface adhered to these particles. Subsequent thermal annealing for 1 h at 300°C strongly enhanced adhesion. (From Ref. 47.)

In conclusion, excimer laser irradiation of copper films on sapphire promotes the formation of an intermediate compound that most likely is responsible for the strong adhesion enhancement. The nature of the intermediate compound is a strong function of the amount of oxygen present during irradiation.

In the case of a metal-silica couple, the first layer of metal deposited is fully encapsulated upon laser irradiation at low energy density, 0.5 J/cm^2. The second layer deposited on top is then in contact with SiO$_2$ and the adhesion is weak as shown in Fig. 7. When the irradiation following the first deposition is done at a higher energy density, e.g., 2 J/cm^2, significant silica ablation takes place together with the clustering and boiling of the metal film, and the metallic particles formed are only partly immersed in the substrate. When the second film is deposited on top, the emerging metallic particles act as anchoring sites, enhancing the bonding. This effect is clearly apparent from Fig. 10. The decrease in adhesion for even higher laser energy density can be understood as due to the intense ablation of SiO$_2$ that results in the total expulsion of the metallic particles [47].

V. ADHESION ENHANCEMENT BY SURFACE MODIFICATION OF THE SUBSTRATE

A. Surface Modifications with Low-Energy Ions

One of the simplest procedures used to improve adhesion consists of bombarding the substrate with low-energy ions prior to film deposition. This procedure is performed to clean the substrate by inducing desorption of surface contaminants.

Low-energy ion bombardment also produces microroughness, and this is particularly noticeable in polymeric substrates. It can be argued that both cleaning and microroughness will promote closer contact of the metallic atoms with the first monolayer of the substrate, thus improving adhesion. However, the formation of interfacial compounds raises a set of questions that will be discussed in the section devoted to the analysis of interfaces. In this section a series of experiments will be described that show the remarkable improvement of the metal-insulator bonding that takes place when the substrates are bombarded with low-energy ions prior to film deposition.

Baglin et al. [32] were the first to establish the effects of presputtering of the substrate on metal-insulator adhesion. In their experiments, Cu was electron beam deposited on sapphire substrates presputtered with 0.5-keV Ar ions. Postdeposition annealing at 450°C produced a six-fold increase in the adhesion strength (measured by peel test) relative to the film as deposited on the presputtered substrate.

Is the annealing step after irradiation required to improve the adhesion? In order to answer this question, tape cast alumina substrates were bombarded in situ with Ar$^+$ ions and neutrals, at energies ranging from 0.5 to 10 keV. A Paulus and Reverchon type of ion gun [53] was used in these experiments instead of the more commonly used Kaufman gun. In the Paulus and Reverchon gun more than 90% of the particles that arrive at the substrate are neutrals. After bombardment, copper films were ion beam sputter deposited using a second Paulus and Reverchon gun pointing toward the target. The Cu-alumina couples were tested for adhesion strength using a customized pull tester. This pull tester has a rigid frame, and the samples were aligned to avoid shear stresses at the interface during pulling. For the adhesion tests an epoxy-coated aluminum pin was bonded to the metallic films using an adhesive having a nominal strength of 70 MPa when cured at 150°C. The pull tests showed that, for ion energies of 1 keV or higher, the resulting bond strength was higher than 70 MPa and that all the tested specimens failed in the glue and not at the metal-ceramic interface. The films deposited without substrate presputtering failed at a tensile strength in the range of 7 to 10 MPa. Ion bombardment at an energy of 500 eV resulted in very poor bonding. In Fig. 11 it is shown that the substrate must be bombarded 3 to 4 min with 1 keV to reach the maximal adhesion strength that can be measured with the pull tester. With 2-keV argon ions, the maximal adhesion is obtained after presputtering for 1 min. These experiments demonstrate that the bonding is a function of both the ion energy and the fluence. The energy spectrum and flux of particles as well as the neutral-to-ion ratio depend on the characteristics of the gun. For this reason, the values of ion energy and ion flux required to achieve maximal adhesion strength could be different for different ion gun types.

As already emphasized, the presence of oxygen can strongly modify the bonding of metals to oxides. Is the presence of oxygen required to enhance adhe-

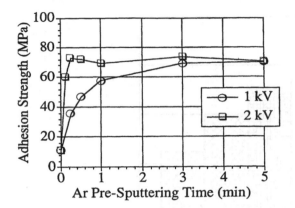

Figure 11 Adhesion strength of copper films deposited on Ar^+-bombarded alumina as a function of presputtering time and accelerating potential. (From Ref. 4.)

sion under ion bombardment as well? Figure 12 shows the effects of several ion bombardment procedures on the adhesion of copper films deposited on alumina. The first bar labeled PS w/Ar shows the adhesion of copper films to alumina substrates that have been presputtered with Ar^+ ions. The second bar labeled PS w/O indicates the adhesion strength of copper films deposited on alumina substrates presputtered with oxygen ions. In the third set of two bars, copper was deposited in two stages: during the first 5 and 15 s, the targets were sputtered with oxygen ions, and in the following ~3 min sputter depositions to produce 80-nm-thick copper films were performed with Ar^+ ions. The last two substrates were *not* sputter cleaned before these depositions. Finally, copper was sputter deposited with Ar^+ ions under a partial pressure of oxygen of ~5 $\times 10^{-4}$ torr. In the last experiment the substrate was not sputter cleaned either. These experiments indicate that the very strong bonding that presputtering with Ar promotes is not related to the presence of oxygen at the interface. All the other four experiments performed with oxygen had a lower adhesion strength than those performed with Ar. A definitive proof that the strong bonding promoted by low-energy ion bombardment is not related to oxygen excess, but in fact the opposite, will be given in the section on the interface analysis.

The effects of ion bombardment with 7-keV Ar^+ ions on the strength of adhesion of gold films to sapphire were studied in another series of experiments. Gold films were deposited on cleaned as-received substrates, on substrates annealed in air at 1350°C, and on substrates first annealed at 1350°C and then etched with 7-keV Ar^+ ions. Table 1 summarizes the pull test results of the Au-Al_2O_3 couples for these three substrate preparations. The adhesion strength of gold films deposited on cleaned as-received substrates is very weak, but it is

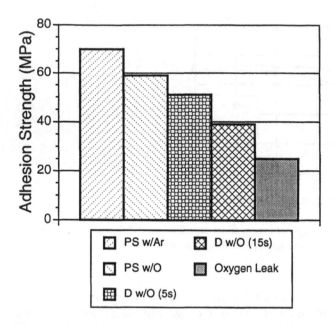

Figure 12 Effects of several ion-neutral bombardment procedures on the adhesion strength of copper films sputter deposited on alumina substrates. PS w/Ar, substrate presputtered with 7-keV Ar$^+$ ions, Cu target sputtered with 7-keV Ar$^+$ ions; PS w/O, substrate presputtered with 7-keV oxygen ions, Cu target sputtered with 7-keV Ar$^+$ ions; D w/O, substrates as received, copper targets initially sputtered with 7-keV oxygen ions for 5 s (left bar) and for 15 s (right bar) followed in both cases by sputtering with 7-keV Ar$^+$ ions for 3 min; oxygen leak, no sputter cleaning, residual oxygen present in deposition chamber at partial pressure of 5×10^{-4} torr.

Table 1 Pull Test Results for Various Conditions of the Gold-Sapphire Couples

Substrate treatment	Adhesion strength (MPa)	Remarks
As received	4.7 ± 4.7	Failure at film-substrate interface
Ion etched	41.9 ± 15.6	Failure at film-substrate interface
Annealed (1350°C for 1 h)	31.7 ± 13.9	Failure at film-substrate interface
Annealed (1350°C for 1 h) and ion etched	70.4 ± 10.5	Failure in the epoxy

Source: Ref. 54.

strongly increased when the substrates have been annealed. The maximal measurable adhesion strength was reached for substrates annealed and ion etched [54].

B. Use of Surface Analytical Techniques to Study the Interface Between a Metallic Film and an Ion-Bombarded Alumina Substrate

1. An XPS Investigation

As discussed previously, the approach taken by surface scientists is to study in situ film deposition and monitor the changes that take place as a function of substrate coverage. This approach was adopted by Baglin et al. [32] when studying the effects of low-energy ion bombardment of Al_2O_3 on the adhesion of copper-Al_2O_3 couples. Sapphires substrates with a c-axis orientation were introduced in a UHV chamber and annealed at 1000°C in a pure oxygen atmosphere. After annealing, the substrates were subjected to ion bombardment with 500-eV or 8-keV Ar^+ ions. The doses for the ion bombardments were chosen to give the maximal adhesion of copper films, as obtained previously by the authors. A very thin copper film, of submonolayer thickness was vapor deposited on all of the bombarded substrates and also on a reference substrate that had been only annealed. XPS analysis was performed using photons from an Mg K_α source. Sample charging was avoided by measuring the peaks as a function of the Auger parameter [55]. The Cu *LMM* peak was found to be considerably broader in presputtered substrates than in the reference nonbombarded substrate. A 500°C annealing sharpened the spectrum in the reference sample and the peak position became very close to the standard value for metallic copper. The same annealing of the sample with a presputtered substrate produced a splitting of the broader peak into two separate peaks: a higher kinetic energy peak that corresponded to metallic copper and a new strong peak shifted by 4 eV toward a lower energy. The shifted peak corresponded to copper located in a different chemical environment, suggesting that an interfacial ternary Cu—Al—O compound had formed. The authors concluded that the strong bonding they measured in annealed specimens of copper films deposited on presputtered substrates was due to the formation of a thin interfacial intermediate compound as detected by XPS.

2. An AES Investigation

The "wedge technique" described in Sec. III.C was used in the AES studies described in this section. Gold films 10 nm thick were deposited on presputtered and nonpresputtered sapphire substrates [54]. A 1.1-eV shift toward lower kinetic energy was detected in the Au NVV Auger peak at the Au-Al_2O_3 interface only in specimens with presputtered substrates. Together with this shift, the O KLL peak was shifted by 0.5 eV toward higher kinetic energy. Figure 13 shows a series of AES differential spectra of Al KLL at various depths from the interface. The first spectrum from the bottom corresponds to metallic aluminum and is used as a reference. The second spectrum from the bottom was taken away from the center of the wedge, where the film thickness is large enough to mostly block the Auger electrons coming from the substrate but not those generated at the gold-sapphire interface (see Fig. 2b). As we move closer to the center of the wedge, the spectra include information on the chemical environment of aluminum deeper inside the substrate. This is simply because as we move to the center of the wedge the gold film thickness decreases. Finally, the spectrum at the top of the figure corresponds to bulk sapphire. The spectrum taken at the interface shows that the ion etching performed on the substrate prior to gold deposition induced decomposition of Al_2O_3, leaving some metallic aluminum on the surface. Practically no metallic aluminum is detected in the fourth spectrum from the bottom taken at 1.5 nm from the interface; instead, the Al peak coming from Al_2O_3 is broader, suggesting that a nonstoichiometric Al_2O_3 is formed deeper, inside the

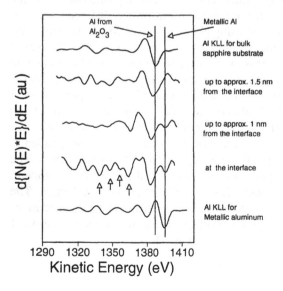

Figure 13 Aluminum KLL AES differential spectra at various depths from the interface of strongly bonded Au-sapphire specimens. Metallic aluminum features with a complex Auger fine structure can be seen. Cu target sputtered with 7-keV Ar^+ ions. (From Ref. 54.)

substrate, by ion bombardment. The fine microstructure of the Al KLL spectrum coming from the interface has several peaks that differ from both the Al KLL spectrum due to metallic aluminum and the Al KLL spectrum from Al_2O_3. These peaks are indicated by arrows in the interface spectrum. This result suggests that aluminum may also be present in a chemical environment different from that of Al_2O_3 or metallic aluminum, consistent with an interfacial reaction involving gold. This interfacial metastable compound seems to be the cause of the strong bonding between gold and sputter-etched Al_2O_3.

3. Discussion

The differences and similarities between the two studies described above reflect the complexities in the analysis of the process of adhesion when surface modifications are introduced. Both studies support the idea originally put forward by Baglin et al. [32] that one of the effects of the low-energy ion bombardment of the substrate was to generate a surface environment conducive to the formation of an interfacial compound. However, annealing of the couples was not required to enhance the bonding of gold films to Al_2O_3 as it was in the experiments performed with copper films deposited on Al_2O_3. A difference between the two experiments is that the gold film was ion beam sputter deposited whereas the copper film was vapor deposited. The atmosphere in the case of copper deposition had a lower residual gas content (UHV). The deposition of gold was performed at a base pressure of 1×10^{-6} torr and the Ar gas pressure was 5×10^{-5} torr during deposition. Although the Ar gas was of high purity and was additionally circulated past a purifier before entering the deposition chamber, the atmosphere present during gold deposition could still contain some oxygen. However, no significant amount of oxygen was present because, as already mentioned, no shift of the Auger lines was detected when the deposition was performed on as-received specimens. Another difference between the two deposition techniques is the energy of the incoming neutrals during substrate sputtering. The energy of the atoms from the ion sputter deposition chamber employed for gold deposition should be of the order of 10 to 15 eV, an energy much larger than the few tenths of electron volts obtained during vapor deposition. Mattox [56] pointed out that more adherent deposited films are obtained by sputter deposition than with vapor deposition. Possibly, the energy of the incoming atoms could help to generate an interfacial compound and hence substantially enhance the bonding to low-energy bombarded substrates.

C. Surface Modifications with UV Lasers

When the very high power that is deposited in the near surface of an insulator is transformed into heat, not only can melting and evaporation occur but also a high concentration of point and linear defects can be produced and, eventually, decomposition of the insulator can take place [57–60]. Point defects in insulators can also be photogenerated [61]. In many insulators such as AlN, Al_2O_3, MgO, and alkaline halides, there are effective mechanisms of absorption of the laser light associated with defects resulting in the generation of heat [62]. The effects on adhesion strength of pulsed-laser irradiation of alumina substrates prior to film deposition, and of thermal annealing following deposition, were studied for gold, copper, and nickel films [63–66]. Alumina substrates were irradiated using laser energy densities from 0.5 to 3 J/cm^2 and three different atmospheres. At a threshold of ∼0.7 J/cm^2 a thin layer of polycrystalline alumina is melted by 308-nm wavelength laser irradiation. At 1 J/cm^2 a 0.2–0.3-μm layer of alumina is melted, and at 2 J/cm^2 ablation of alumina takes place after several "incubation" pulses.

Figure 14a shows the dependence of the adhesion strength of 80-nm-thick gold films deposited on laser-treated alumina as a function of the laser energy density. The three curves correspond to three different irradiation atmospheres: pure oxygen, air, and Ar–4% H_2. When the gold-alumina couples are annealed for 1 h at 300°C there is a significant adhesion improvement (Fig. 14b). The adhesion strength of gold films to as-received alumina substrates and to substrates laser irradiated up to a laser energy density of 0.5 J/cm^2 is extremely weak. There is a strong increase in adhesion in samples irradiated in oxygen, and the adhesion reaches 50 MPa when the couples are annealed at 300°C [63].

In order to study the relation between adhesion strength and thermal stabil-

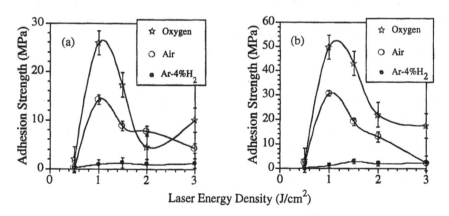

Figure 14   Adhesion strength of gold films to laser-irradiated alumina substrates as a function of laser energy density employed, for different irradiation atmospheres. (a) As-deposited condition; (b) after thermal annealing for 1 h at 300°C in Ar–4% H_2. (From Ref. 63.)

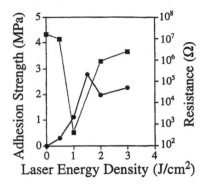

Figure 15 Electrical resistance (■) of 20-nm-thick gold films deposited on Al_2O_3 irradiated in Ar–4% H_2 as a function of laser energy density. The gold-Al_2O_3 couples were annealed at 550°C for 45 min. In this figure, the adhesion strength (●) of 80-nm-thick gold films deposited on substrates laser irradiated in Ar–4% H_2 and annealed at 300°C is also plotted. (From Ref. 67.)

ity of the film, 20-nm-thick gold films were deposited on laser-irradiated alumina. After deposition, the gold-alumina couples were annealed at 550°C for 45 min. Figures 15, 16, and 17 show the electrical resistance as a function of the laser energy density for irradiations performed in each of the three atmospheres. In these figures, the adhesion strength of 80-nm-thick gold films deposited on laser-

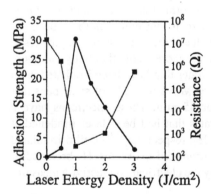

Figure 16 Electrical resistance (■) of 20-nm-thick gold films deposited on Al_2O_3 irradiated in air as a function of laser energy density. The gold-Al_2O_3 couples were annealed at 550°C for 45 min. In this figure, the adhesion strength (●) of 80-nm-thick gold films deposited on substrates laser irradiated in air and annealed at 300°C is also plotted. (From Ref. 67.)

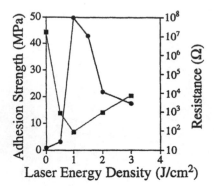

Figure 17 Electrical resistance (■) of 20-nm-thick gold films deposited on Al_2O_3 irradiated in an oxygen atmosphere, as a function of the laser energy density. The gold-Al_2O_3 couples were annealed at 550°C for 45 min. In this figure, the adhesion strength (●) of 80-nm-thick gold films deposited on substrates laser irradiated in oxygen and annealed at 300°C is also plotted. (From Ref. 67.)

irradiated substrates and annealed at 300°C is also plotted. In all the cases shown in the figures, the highest adhesion strength corresponds to the lowest electrical resistance. The electrical resistance varies from 88 Ω for an adhesion strength of 50 MPa to 16 MΩ for an adhesion strength very close to zero (1 MPa or less). The electrical resistance measures the integrity of the film after annealing. At this relatively high annealing temperature, the very thin films tend to break into small clusters. As more and more clusters are formed and are separated by portions of insulating substrate, the electrical circuit that connects them becomes more and more convoluted, increasing the path and therefore increasing the resistance. These results show that the films that are more stable thermally and more strongly bonded are those deposited on substrates that have been irradiated in an oxygen atmosphere. Figures 18 and 19 are SEM micrographs of specimens that were postdeposition annealed at 550°C for 45 min. These specimens have a 20-nm-thick film each deposited on substrates that had been laser irradiated at 1 J/cm^2 in oxygen and 2 J/cm^2 in Ar–4% H_2, respectively. The film irradiated in oxygen is continuous, whereas the film irradiated in Ar–4% H_2 broke completely into small islands, as also revealed by the electrical conductivity measurements (Fig. 17).

   Similar results were obtained for copper films deposited on irradiated alumina substrates. Figure 20 shows the results of the adhesion tests performed on 80-nm-thick copper films deposited on alumina substrates laser irradiated in air. Specimens annealed at 500°C for 1 h had adhesion strengths increased apprecia-

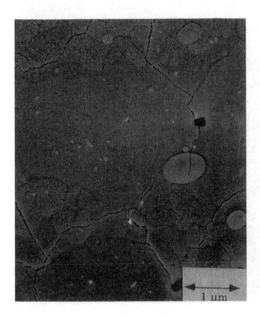

Figure 18 SEM micrograph of a 20-nm-thick gold film sputter deposited on alumina. Substrate was laser irradiated in oxygen at 1 J/cm^2 before deposition. After deposition, the specimen was annealed at 550°C for 45 min. (From Ref. 67.)

bly in comparison with equivalent but unannealed couples. In Fig. 21 similar results are shown but the irradiations were performed in Ar–4% H$_2$ [64]. Pulsed-laser irradiation of alumina substrates prior to nickel deposition also improved substantially the bond strength of nickel films [66,67].

These results single out several factors that contribute to the adhesion of metallic films to alumina. First, adhesion strength is determined by the chemical nature of the metallic film itself as indicated by the very different adhesion strengths of films sputter deposited on unirradiated alumina: 0.1 MPa for gold, 13 MPa for copper, and 32 MPa for nickel. Second, heat evolution during laser irradiation is important for adhesion enhancement; furthermore, maximal adhesion strength is obtained when the near-surface layer of the substrate is melted. Third, the irradiation atmosphere is very important for the kind of surface modification that enhances the bond strength, the maximal strength always being obtained when irradiation is performed in a strongly oxidizing atmosphere.

The adhesion strength of copper and gold films deposited on as-received SiO$_2$ is less than 5 and 0.5 MPa, respectively. No adhesion improvement was

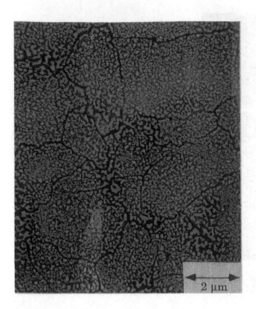

Figure 19 SEM micrograph of a 20-nm-thick gold film sputter deposited on alumina. Substrate laser irradiated in Ar–4% H_2 at 2 J/cm^2 before deposition. After deposition, the specimen was annealed at 550°C for 45 min. (From Ref. 67.)

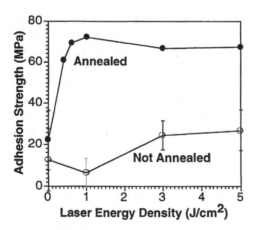

Figure 20 Adhesion strength of copper film sputter deposited on alumina substrate laser irradiated in air, as a function of laser energy density. (From Ref. 64.)

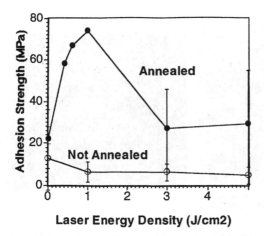

Figure 21 Adhesion strength of copper film sputter deposited on alumina substrate laser irradiated in Ar–4% H_2, as a function of laser energy density. (From Ref. 64.)

detected for gold or copper films deposited on laser-irradiated SiO_2 substrates, even with subsequent annealing. A key difference between Al_2O_3 and SiO_2 is that no significant heat evolution seems to take place during laser irradiation of SiO_2 [68].

1. Morphological Changes in the Near Surface of Al_2O_3 Induced by Laser Irradiation

A cross-sectional transmission electron micrograph of an alumina specimen irradiated at 1.3 J/cm^2 in Ar–4% H_2 is shown in Fig. 22a [69,70]. A uniform, featureless, and amorphous layer about 0.2 μm thick is formed in the near-surface region, as revealed by the convergent beam electron diffraction (CBED) pattern shown in Fig. 22b. A resolidified crystalline columnar structure can also be seen in Fig. 22a between the unmelted substrate and the amorphous layer. As the laser energy is increased, the amorphous layer tends to disappear, leaving the crystalline columnar structure. The energy deposited by the laser increases as the laser energy increases, and it takes more time to dissipate the generated heat by thermal conduction. If the solidification can proceed at a lower rate, there is time to produce a crystalline structure. Perhaps the most interesting effect is that during irradiation in Ar–4% H_2 very fine particles of metallic aluminum are found embedded in the matrix. These particles are not detected during laser irradiation in an oxygen atmosphere.

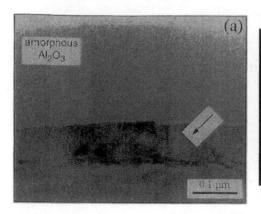 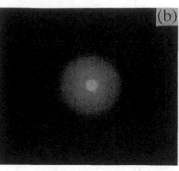

Figure 22 (a) Amorphous layer in the near surface of an alumina substrate formed upon laser irradiation at 1.3 J/cm² in Ar–4% H₂. A resolidified crystalline layer—indicated by the arrow—can be seen between the amorphous alumina and the unmelted substrate. (b) Convergent beam electron diffraction pattern from amorphous layer. (From Ref. 69.)

2. Chemical Changes in the Near Surface of Al₂O₃ Promoted by Laser Irradiation

Auger emission spectroscopy of an Al_2O_3 substrate irradiated in air at 2 J/cm² shows that a thin stoichiometric Al_2O_3 layer is present at the surface. As the surface layers are removed by 500-eV Ar^+ ions, detection of a different small peak reveals the presence of a small amount of metallic aluminum or substoichiometric alumina. This peak is not detected in the surface of the substrate because re-oxidation of the substoichiometric alumina or metallic aluminum immediately takes place once the irradiated substrate is exposed to air. When the irradiation is performed in an Ar–4% H₂ atmosphere a larger peak is detected in the subsurface region at 66 eV. The Al LVV peak positions from this sample at two different locations, the surface (i) and subsurface (ii), are shown in Fig. 23. Because these are differential spectra (first derivative of the Auger spectrum), the peak positions appear as depressions or valleys. On the surface only the 54-eV peak is observed that corresponds to an Al LVV peak in an environment of Al_2O_3 (Fig. 23, spectrum i). In the subsurface region a peak is located at 66 eV (Fig. 23, spectrum ii). This peak is close to the Al LVV for metallic aluminum (68 eV), indicating the presence of metallic aluminum or Al_2O_3 highly deficient in oxygen [70].

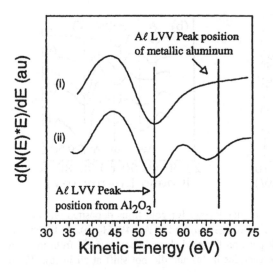

Figure 23 Aluminum LVV Auger differential spectra from the surface (i) and subsurface (ii) of alumina laser irradiated in Ar–4% H_2 at 2 J/cm^2. More metallic aluminum is left after irradiation in a reducing atmosphere than in an oxidizing atmosphere (see text). (From Ref. 70.)

3. AES Analysis of the Interface Between a Metallic Film and a Laser-Irradiated Alumina Substrate

The thicker the intermediate region between a film and its substrate, the more unlikely it is that metal-substrate mixing will occur. In the case of specimens that have been annealed after deposition, the intermediate region is 5 to 20 nm thick and therefore the interface has been reached by conventional ion etching using 500-eV Ar$^+$ ions.

In Figs. 24 and 25 the sequence of numbers 1 through 5 indicates AES spectra taken at increasing depth [38]. The spectra identified by the number (1) indicate a signal coming mostly from the film, giving the reference values of the kinetic energy for the peak position of pure gold. Spectra (5) give the reference values of the kinetic energies of Al and O for bulk Al_2O_3. Figures 24a and b show the differential Au NVV Auger spectra for gold films deposited on alumina substrates laser irradiated at 1 J/cm^2 in oxygen and Ar–4% H_2 atmospheres, respectively. When the film-substrate interface is exposed by ion etching, it can be seen that the Au NVV peak, Fig. 24a, is shifted 1.5 eV toward a lower kinetic energy relative to pure gold. No shift is observed in the Au NVV differential spectrum when the substrate has been irradiated in an Ar–4%

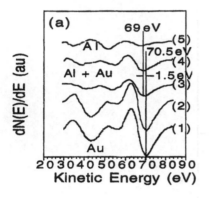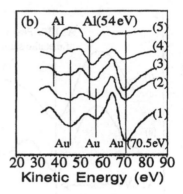

Figure 24 Gold NVV Auger differential spectra for gold films deposited on alumina substrates laser irradiated at 1 J/cm^2 in oxygen (a) and in Ar–4% H$_2$ (b). At the metal-substrate interface, the Auger peak position for gold shifted about 1.5 eV from the film peak position toward lower kinetic energies in (a) and did not shift at all in (b). (From Ref. 38.)

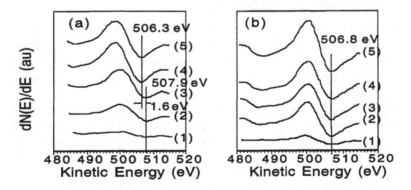

Figure 25 Oxygen KLL Auger differential spectra for gold films deposited on alumina substrates irradiated with a laser energy density of 1 J/cm^2 in oxygen (a) and in Ar–4% H$_2$ atmosphere (b). At the interface, the oxygen peak position shifted 1.6 eV from the peak of bulk alumina toward lower kinetic energies for laser irradiation in oxygen but no shift was detected for the irradiation in Ar–4% H$_2$. (From Ref. 38.)

H_2 atmosphere. Correspondingly, the oxygen KLL peak also shifts 1.6 eV from the peak of bulk alumina toward lower kinetic energies for laser irradiation in oxygen (Fig. 25a) but no shift is detected for the irradiation in Ar–4% H_2 (Fig. 25b).

The reduction in the kinetic energy of the Au NVV peak indicates a tendency to oxidation of the gold. The oxygen peak also indicates a chemical reaction at the interface. This reaction is consistent with the adhesion strength enhancement observed in these couples.

Figure 26 shows three differential peaks of Cu LMM taken at the surface, in the film, and at the metal-ceramic interface, respectively, of a Cu-alumina couple annealed at 300°C. The alumina substrate had been laser irradiated in air at 1 J/cm^2 prior to deposition. The peak position at the surface is 917.3 eV, whereas inside the film it is 919 eV. The difference in kinetic energy between the copper peak measured at the surface and the peak measured inside (1.7 eV) indicates that the oxide at the surface is Cu_2O. The Cu LMM peak is also shifted 1.7 eV to a position very close to Cu_2O, indicating the formation of an oxide at the interface. The oxygen KLL Auger spectra at different depths starting from the surface and finishing in the bulk substrate are shown in Fig. 27. The oxygen from the surface oxide quickly disappears as the specimen is etched away for analysis because this oxide layer is very thin. Further sputtering reveals two oxygen Auger peaks in the transitional (interface) region. The positions of these two peaks are 507.7 and 511.8 eV. The peak near 512 eV is very close to Cu_2O, but the O KLL from Al_2O_3 should be located at 506 eV. This shift in the peak location

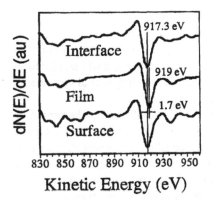

Figure 26 Copper LMM Auger differential spectra from the Cu film deposited on an alumina substrate laser irradiated at 1 J/cm^2 in air. The spectrum at the interface is very similar to that at the surface, revealing the existence of a copper oxide at the interface.

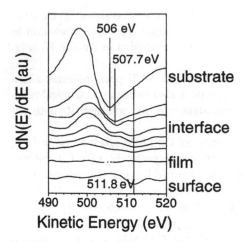

Figure 27 Oxygen KLL Auger differential spectra for the same copper-alumina couple as in Fig. 26. At the interface, the oxygen exists in two different oxidation states, implying that a Cu-Al-O compound was formed at the interface.

suggests that at the interface a Cu-Al-O compound is formed instead of the mixture of two oxides. Studies of Cu-Al_2O_3 with the substrate treated under other irradiation conditions showed that the strong adhesion measured by pull testing correlated with the formation of a Cu-Al-O compound [31].

VI. DISCUSSION AND CONCLUSIONS

The effects of surface modifications of ceramics by low-energy ions and by UV lasers have been analyzed in relation with the adhesion strength of metallic films. In all the cases analyzed here, strong metal-ceramic bonding was always associated with the formation of an intermediate compound.

Two of the most outstanding effects of low-energy ion bombardment are the surface desorption of impurities and the microroughening of the surface. These two effects should certainly help to enhance adhesion; however, associated with them is the generation of surface damage. The two analytical studies reviewed here show the formation of a compound, one or two monolayers thick, involving the metallic film and the ceramic. In these cases, low-energy ion bombardment modified the surface, producing metastable environments that allowed the deposited atoms to be locked in. In one of these studies in which copper was deposited on presputtered sapphire, after annealing at 500°C this structure evolved to a more stable configuration. In the other study using higher energy ions for the bombardment of the sapphire surface, annealing was not required to obtain an interfacial compound when gold was deposited. The copper atoms were vapor deposited on the surface and arrived at the sapphire surface with lower energy than the gold atoms that were ion beam sputter deposited. It is hypothesized that in the gold film case the higher energy of the incoming atoms helps to generate a more stable configuration, without the need for annealing.

Bombardment with low-energy ions seems to be a very general way of improving the adhesion between two dissimilar materials. We have reviewed here the case of metallic films deposited on presputtered Al_2O_3 and SiO_2. It is important to emphasize that annealing the substrate *before* ion bombardment helps to improve the metal-ceramic bonding. This effect is probably related to the removal of contaminants, some of which could become implanted in the near-surface region instead of being desorbed by the bombardment. Although only the case of ceramics was discussed here, it should be noted that ion bombardment of polymeric substrates also enhances the bonding to metallic films [71,72].

The strong influence of oxygen on metal-oxide bonding was analyzed at the beginning of the chapter. The near-surface region of a 7-keV bombarded sapphire shows the presence of metallic aluminum, a clear indication of the reduction of Al_2O_3 during ion bombardment. These results and others discussed in Sec. V.A show that excess oxygen is not required to enhance the bonding between a metallic film and a ceramic, provided the proper modification of the substrate surface is produced.

An intermediate compound was also detected in laser-irradiated substrates when strong bonding took place between the metallic film and alumina. With this surface modification technique, annealing of the metal-ceramic couple at a temperature of 300–500°C is required to produce the strong bonding. The presence of oxygen during irradiation is important for the bonding of gold films to alumina. In the case of copper films deposited on laser-irradiated alumina, a Cu-Al-O compound was detected. The interfacial compound between metal film and the laser-irradiated substrate is much thicker than in the case of ion bombardment. A metastable near-surface region is created by laser irradiation of the substrate. However, in this case a thin surface layer of the substrate is melted and the metastability is generated during the rapid solidification of this layer. The change in color from white to yellow after laser irradiation in both oxygen-rich and oxygen-deficient atmospheres is a clear indication of defect formation during rapid solidification, most probably F centers [57]. The catalytic effect that laser-irradiated alumina has for copper deposition when the substrate is immersed in an electroless solution demonstrates further the profound changes induced by laser irradiation on the physical state of the alumina surface [2,57].

Given the strong dependence of adhesion enhancement on laser energy density for the case of metallic films deposited on alumina, it is clear that if a significant amount of heat does not evolve during irradiation there will be no important surface modification and no adhesion enhancement will take place. This conclusion is supported by the results obtained with SiO_2, where excimer laser irradiation can produce strong ablation of the surface without heat evolution. Thus, laser irradiation *after* deposition of a thin metallic film on SiO_2 significantly improves the film-substrate adhesion strength. Heat evolution takes place because the metallic film strongly absorbs the laser radiation, breaks up into small droplets, and partly dissipates the heat through the substrate, apparently melting it. The droplets become encapsulated in the substrate, enabling the building up of a thicker film. A possibly significant difference between ion bombardment and laser irradiation experiments is that the laser-irradiated substrates have been exposed to air prior to film deposition.

Analytical techniques, especially AES and XPS, have greatly helped in understanding the adhesion mechanisms. Further advances are needed to develop fully a completely satisfactory technique that will allow the study of interfaces between dissimilar materials. The technique that utilizes a wedge in the metallic film greatly reduces damage to the interface during the removal of the metallic film and is one step forward in that direction. A possible improvement of this method is to remove the last portions of the wedge film using a very low energy ion gun (less than 100 eV).

Closing up this chapter, the following conclusions summarize some of the relevant procedures for enhancing the adhesion of metallic films to ceramic substrates and the causes leading to the improvement in specific cases:

1. Ion etching of alumina and silica substrates with Ar^+ ions in the energy range 1 to 7 keV before metallic film deposition can strongly enhance the interfacial bond. Annealing the substrate before irradiation improves the bonding strength even further.
2. Very strong bonding between gold and sapphire is promoted by ion etching, in spite of the reduction of sapphire during ion etching. This result clearly indicates that the presence of oxygen is not required to form an intermediate compound. This metastable compound seems to be favored by the highly metastable reduced alumina generated during presputtering and the relatively high kinetic energy of the incoming gold atoms.
3. Pulsed-laser irradiation of alumina prior to deposition can promote strong bonding with metallic films.
4. Laser-enhanced bond strength demonstrated that (a) the chemical nature of the metal plays a relevant role; (b) the type and extent of laser-

generated surface modifications are of great importance, (e.g., for gold significant bonding enhancement is obtained *only* if pulsed-laser melting of the alumina occurs); and (c) the irradiation atmosphere is a determining factor as well, because the adhesion strength of metals (e.g., Au, Cu, and Ni) to alumina is greatest if pulsed-laser irradiation is performed in an oxygen-rich atmosphere.

5. AES results show that an intermediate interfacial compound is formed when a metallic film is deposited on laser-irradiated or ion-etched substrates.

6. In the case of laser irradiation under an oxygen-rich atmosphere, the substrate surface becomes enriched in oxygen and, as a consequence, either a mixed or a double oxide with the metal is formed.

7. Thin films of gold, copper, and iron deposited on silica are driven into the substrate by a laser pulse of 0.5 J/cm^2 energy density.

8. Pulsed-laser irradiation of thin films deposited on silica can produce strong bonding between freshly deposited films and an irradiated film-substrate couple.

It has been shown that UV laser irradiation of polymeric substrates also enhances the bonding of metallic films [73]. Thus, although the presentation of this chapter has drawn on specific systems and only ceramic substrates, surface modification should be classed as a general methodology for enhancing bonding between dissimilar materials.

ACKNOWLEDGMENTS

I would like to thank Dr. M. J. Godbole and Dr. K. L. Mittal for their detailed review of the manuscript and many useful suggestions.

REFERENCES

1. VE Heinrich, PA Cox. The Surface Science of Metal Oxides. Cambridge: Cambridge University Press, 1994, p 4.
2. AJ Pedraza, JW Park, DH Lowndes, S Cao, WR Allen. Mater Res Soc Symp Proc 397:519, 1996.
3. MP Seah. In: D Briggs, MP Seah, eds. Practical Surface Analysis. Vol I. Wiley, Chichester. UK: 1990, p 311.
4. AJ Pedraza, JW Park. In: TS Srivatsan, JJ Moore, eds. Processing and Applications of Advanced Materials IV. Warrendale, PA: TMS, 1996, p 263.

5. Ref. 1, p 247.
6. IV Mitchell, J Nyberg, RG Elliman. Appl Phys Lett 45:137, 1984.
7. JP Ziegler, JP Biersack, U Littmark. The Stopping and Range of Ions in Solids. Vol 1. New York: Pergamon Press, 1985, p 115.
8. FS Ohuchi, M Kohyama. J Am Ceram Soc 74: 1163, 1991.
9. M Vermeersch, R Sporken, P Lambin, R Caudano. Surf Sci 235:5, 1990.
10. M Vermeersch, F Malengreau, R Sporken, R Caudano. Surf Sci 323:175, 1994.
11. TM French, GA Somorjai. J Phys Chem 74:2489, 1970.
12. FS Ohuchi, TM French, RV Kasowski. J Vac Sci Technol A5:1175, 1987.
13. FS Ohuchi, TM French, RV Kasowski. J Appl Phys 62:2286, 1987.
14. FS Ohuchi. J. Phys Colloque C5 Suppl to No 10 49:783, 1988.
15. Q Zhong, FS Ohuchi. J Vac Sci Technol A8:2107, 1990; Mater Res Soc Symp Proc 153:71, 1987.
16. ED Hondros. Proceedings 2nd International Conference on the Science of Hard Materials, Rhodes, Institute of Physics Conference Series No 75, 1986, chap 2, p 121.
17. HJ Fetch. J Phys Colloque C5 Suppl to No 10 49:171, 1988.
18. R Hoel, HU Habermeier, M Ruhle. J Phys Colloque C4 Suppl to No 4 46:141, 1985.
19. MW Finnis. J Phys Condens Matter 8:5811, 1996.
20. P Benjamin, C Weaver. Proc R Soc A 261:516, 1962.
21. JM Howe. Int Mater Rev 38:233, 1993; Int Mater Rev 38:257, 1993.
22. KH Johnson, SV Pepper. J Appl Phys 53:6634, 1982.
23. SV Pepper. J Appl Phys 47:801, 1976; J Appl Phys 50:8065, 1980.
24. TE O'Brien, CD Chaklader. J Am Ceram Soc 57:329, 1974.
25. AMM Gadalla, J White. Trans Br Ceram Soc 63:39, 1964.
26. DM Mattox. J Appl Phys 37:3613, 1966.
27. DC Moore, HR Thornton. J Res (Natl Bur Stand) 62:2345, 1962.
28. RC Budhani, S Prakash, HJ Doerr, RF Bunshah. J Vac Sci Technol A4:3023, 1986.
29. JF Burgess, CA Neugebauer, G Flanagan, RE Moore. Solid State Technol 18:42, 1975.
30. KP Trumble, M Ruhle. Acta Metall Mater 39:1915, 1991; KP Trumble. Acta Metall Mater 40(Suppl):5105, 1992.
31. JW Park. Adhesion Mechanisms of Metallic Thin Films Deposited on Surface-Modified Ceramics PhD thesis, University of Tennessee at Knoxville, 1995.
32. JEE Baglin, AG Schrott, RD Thompson, KN Tu, A Segmuller. Nucl Instrum Methods Phys Res B19/20:782, 1987.
33. MP Seah, D Briggs. In: D Briggs, and MP Seah, eds. Practical Surface Analysis. Vol I. Chichester, UK: John Wiley & Sons, 1990, p 1.
34. D Briggs, JC Rivière. Ref. 33, p 85.
35. S Hofmann. Ref. 33, p 143.
36. MP Seah. Ref. 33, p 201.
37. RZ Bacharach, SB Hagstrom, SA Flodstrom. Phys Rev B10:2837, 1979.
38. JW Park, AJ Pedraza, WR Allen. Mater Res Soc Symp Proc 357:59, 1995.
39. KL Mittal. In: KL Mittal, ed. Adhesion Measurement of Films and Coatings. Zeist, The Netherlands: VSP, 1995, pp 1–13.
40. JEE Baglin. In: P Mazzoldi, GW Arnold, eds. Ion Beam Modification of Insulators. Amsterdam: Elsevier, 1987, p 585; Nucl Instrum Methods Phys Res B39:764, 1989; IBM J Res Dev 38:413, 1994.

41. AJ Pedraza, MJ Godbole, EA Kenik, DH Lowndes, JR Thompson. J Vac Sci Technol A6:1763, 1988.
42. AJ Pedraza, MJ Godbole, DH Lowndes, JR Thompson. J Mater Sci 24:115, 1989.
43. MJ Godbole, AJ Pedraza, DH Lowndes, EA Kenik. J Mater Res 4:1202, 1989.
44. MJ Godbole, AJ Pedraza, DH Lowndes, JR Thompson. J Mater Res. 7:1004, 1992.
45. AJ Pedraza. In: J Mazumder, KN Mukherjee, eds. Laser Materials Processing III. Warrendale, PA: TMS, 1989, p 183.
46. AJ Pedraza, MJ Godbole. Metall Trans A23:1095, 1992.
47. AJ Pedraza, S Cao, DH Lowndes, LF Allard. Mater Res Soc Symp Proc 397:473, 1996.
48. WL Woolfe. Handbook of Military Infrared Technology. Washington, DC: Government Printing Office, 1965.
49. WL Brown. In: CW White, PS Peercy. eds. Laser and Electron Beam Processing of Materials. New York: Academic Press, 1980, p 20.
50. RF Wood, GE Ellison Jr. In: RF Wood, CW White, RT Young, eds. Pulse Laser Processing of Semiconductors. Orlando, FL: Academic Press, 1984, p 166.
51. R Kelly, A Miotello. Appl Surf Sci 96, 98:215, 1996.
52. MJ Godbole. Adhesion Enhancement of Metal-Ceramic Couples Using Pulsed-Excimer Lasers. PhD thesis, The University of Tennessee at Knoxville, 1989.
53. M Paulus, F Reverchon. J Phys Radium Phys Appl Suppl 6 22:103A, 1961; CG Crocket. Vacuum 23:11, 1972.
54. JW Park, AJ Pedraza, WR Allen. Appl Surf Sci 103:39, 1996.
55. SD Waddington. In: D Briggs, MP Seah, eds. Practical Surface Analysis. Vol I. Chichester, UK: John Wiley & Sons, 1990, p 587.
56. DM Mattox. In: KL Mittal, ed. Adhesion Measurement of Thin Films. Thick Films and Bulk Coatings. ASTM STP 640. Philadelphia: American Society for Testing and Materials, 1978, p 54.
57. GA Shafeev. Adv Mater Opt Elect 2:183, 1993.
58. H Esrom. Mater Res Soc Symp Proc 204:457, 1991.
59. H Esrom, ZY Zhang, AJ Pedraza. Mater Res Soc Symp Proc 236:383, 1992.
60. DH Lowndes, M DeSilva, MJ Godbole, AJ Pedraza, DB Geohegan. Mater Res Soc Symp Proc 285: 191, 1993.
61. CH Chen, MP McCann. Opt Commun 60:296, 1986.
62. RL Webb, C Jensen, SC Langford, JT Dickinson. J Appl Phys 74:2323, 1994; RL Webb, C Jensen, SC Langford, JT Dickinson. J Appl Phys 74:2338, 1994.
63. AJ Pedraza, RA Kumar, DJ Lowndes. Appl Phys Lett 66:1065, 1995.
64. AJ Pedraza, MJ DeSilva, RA Kumar, DH Lowndes. J Appl Phys 77:5176, 1995.
65. MJ DeSilva. Excimer Laser Induced Surface Activation of Ceramics for Electroless Deposition and Enhanced Metal Adhesion. PhD thesis, The University of Tennessee at Knoxville, 1994.
66. RA Kumar. Adhesion Enhancement of Metallic Films Deposited onto Excimer Laser Treated Ceramic Substrates. MS thesis, The University of Tennessee at Knoxville, 1994.
67. H Esrom, AJ Pedraza, RA Kumar, JY Zhang. In: WJ van Ooij, HR Anderson Jr, eds. Adhesion Science and Technology (Festschrift in honor of K.L. Mittal). Zeist, The Netherlands: VSP, 1998.

68. C Buerhop, B Blumenthal, R Weissmann, N Lutz, S Biermann. Appl Surf Sci 46: 430, 1990.
69. AJ Pedraza, S Cao, LF Allard, DH Lowndes. Mater Res Soc Symp Proc 357:53, 1995.
70. AJ Pedraza, JW Park, DH Lowndes, S Cao, WR Allen. J Mater Res 12:3174, 1997.
71. Chin-An Chang, JEE Baglin, AG Schrott, KC Lin. Appl Phys Lett, 51:103, 1987.
72. DL Pappas, JJ Cuomo, KG Sachdev. J Vac Sci and Technol, A9: 2704, 1991.
73. M Murahara, M Okoshi. J Adhesion Sci Technol, 9:1593, 1995; M Murahara, K Toyoda. J Adhesion Sci Technol, 9:1601, 1995.

10
Surface Graft Copolymerization and Grafting of Polymers for Adhesion Improvement

E. T. Kang, Koon Gee Neoh, and Kuang Lee Tan
National University of Singapore, Kent Ridge, Singapore

Der-Jang Liaw
National Taiwan University of Science and Technology, Taipei, Taiwan, Republic of China

I. INTRODUCTION

In most engineering applications, a polymer is selected because of its favorable bulk properties, such as mechanical strength, electrical and/or electronic properties, thermal stability, and processability. Often, however, the selected polymer has surface characteristics that are less than optimal for the intended application. For example, the surfaces of most engineering polymers in use today are all fairly hydrophobic. Therefore, it is difficult to bond the hydrophobic polymer surfaces directly with other materials, such as adhesives, printing inks, and paints, which generally consist of polar groups or components. The problem can be overcome, to a large extent, through controlled modification and functionalization of polymer surfaces. The polymer literature is extremely rich in methods and strategies directed toward modification of polymer surfaces for specific applications. Some common methods employed to improve the physicochemical properties of polymer surfaces, together with the nature of the resultant surfaces, are shown in Table 1 [1].

Controlled chemical modification of polymer surfaces has been made possible by advances in two areas. First, there has been a steady increase in the availability of surface analytical techniques and instruments. Second, the expanded

Table 1 Properties of Polymer Surfaces and Methods for Achieving
Such Properties

Property	Related technique	Surface nature and related characteristics
Adhesion	Polar groups anchoring	Hydrophilicity High surface energy Printability
Wettability	Acidic, alkaline treatment	Hydrophilicity Antifogging Adhesion Electroconductibility
Printability	Corona treatment	Hydrophilicity, adhesion
Antistaticity	Metal coweaving Surfactant mixing	Electroconductibility
Antifogging	Hydrophobic-polymer coating Heating Extremely low surface energy	Hydrophobicity Hydrophobicity
Antifouling	Hydrophobic-polymer coating Microvibration Painting with organic tin compounds Extremely low surface energy	Hydrophilicity Hydrophobicity Mechanical removal Sea pollution
Biocompatibility	Enzyme mixing Protein immobilization Surface grafting	Hydrophilic Hydrophobic Bioinert Antithrombogenicity
Lubricity	Hydrophilic or hydrophobic lubricious coating	Hydrophilicity Hydrophobicity

surface analytical capability has resulted in a steady growth in fundamental understanding of relations between surface structure and various aspects of surface performance [2]. Methods for the modification of polymer surfaces, which have direct relevance to adhesion improvement or enhancement, have been summarized [3,4]. In general, they can be classified as dry and wet chemical methods.

Among the dry modification methods, plasma treatment is by far the most widely practiced [5,6]. The plasma may contain ionized gas molecules or radicals of vapor molecules that react at the substrate surface. When a pure noble gas is used, the gas plasma has an "etching" effect and the surface is activated (oxidized) during the subsequent exposure to the atmosphere. In the presence of organic molecules, such as a vinyl monomer, the gas plasma has a "polymerization" effect, resulting in the deposition or coating of a polymer layer on the substrate surface. The coated layer, however, usually consists of polymers with ill-defined chemical structure. When a high electric field is applied under atmospheric pressure, instead of a reduced pressure as in the case of plasma treatment, the process is termed corona discharge treatment [7–10]. An alternative to gas plasma treatment is controlled exposure to flames [11,12], which are probably the oldest plasma known to mankind. High-energy radiations, such as gamma rays [13], electron and ion beams [14] and radiation from an ultraviolet (UV) source [15,16], have also been widely employed for surface treatment of polymers. In the presence of an added monomer, the radicals formed at the substrate surface after irradiation can react with the monomer to result in surface graft copolymerization. The high-energy irradiation, however, can also result in crosslinking and/or chain scission both at the surface and in the bulk of the film.

The simplest wet chemical methods for rendering a polymer surface hydrophilic are acid and alkaline treatments [17–19]. They do not differ substantially from the dry treatments in terms of introducing oxidized or polar groups onto the polymer surfaces. When the polymer surface to be modified possesses reactive groups (inherent or externally introduced) capable of coupling with other components, surface modification can be readily conducted by chemical reactions. Among the various possible synthetic reactions, surface modification via grafting and graft copolymerization appears to be a very effective means of molecularly redesigning the polymer surfaces for specific functions or purposes. In the present work, the recent literature on surface modification of polymers via grafting and graft copolymerization is first reviewed. The surface microstructures and dynamics of the so-modified polymer surfaces are then discussed. Finally, the adhesion characteristics of the graft-modified polymer surfaces are addressed and illustrated with representative examples.

II. SURFACE GRAFTING AND GRAFT COPOLYMERIZATION

Surface grafting can be achieved by two different approaches: grafting by coupling reaction and surface graft copolymerization. Grafting by coupling reaction is the attachment of previously synthesized polymer chains to the substrate surface, whereas graft copolymerization is the synthesis of polymer chains from monomer by initiating chain growth directly from reactive sites on the substrate surface. The net result in both cases is the covalent attachment of polymer chains to the substrate surface.

A. Grafting by Coupling Reaction

Most of experimental investigations of the direct use of polymer chains for surface modification have involved adsorptive attachment rather than covalent at-

tachment or have involved inorganic surfaces rather than organic polymer surfaces [20–25].

The surface grafting mechanism involves chemical coupling between
the reactive groups present on both the substrate and the polymer to be grafted.
Poly(ethylene oxide) chains were grafted onto a silica surface by direct esterifi-
cation of the silanol groups with the hydroxyl end groups of the polymer at 230°C
under a nitrogen atmosphere according to [26]

$$\text{SiOH} + \text{ROH} \rightarrow \text{SiOR} + \text{H}_2\text{O}$$

Dextran was also grafted onto the surface of ethylene–vinyl alcohol copolymer
through the coupling reaction with or without primary amine groups. The coupling reaction took place via a urethane linkage when unmodified dextran was
used and via a urea linkage when the dextran contained amino groups [27].

Tabata et al. [28] carried out the grafting of collagen molecules on the
surfaces of cellulose and poly(vinyl alcohol) (PVA) films via covalent bonding,
using the cyanogen bromide activation method. Han et al. [29] and Park et al.
[30], on the other hand, reported the grafting of poly(ethylene glycol) (PEG) and
heparin onto the surfaces of polyurethane (PU) and polyurethaneureas by the use
of diisocyanate coupling agents or by isocyanate end capped PEG prepolymer.
A decrease in platelet adhesion to the modified surface was observed. In a later
study, Bergstrom et al. [31] grafted PEG on the surface of polystyrene (PS) lab-
ware for the purpose of reducing nonspecific adsorption of protein. The grafting
process was accomplished via a multistep process that included prior amination
of the PS surface and conversion of some of the PEG hydroxyl groups to epoxide
groups, followed by chemical reaction between the epoxide groups and the amine
groups to attach the PEG chains covalently. In yet another study, PEG chains
were grafted onto poly(ether urethane) films in a two-step process [32]. In the
first step, the film surface was treated with hexamethylene diisocyanate in toluene
and in the presence of triethylamine as a catalyst. In the subsequent step, PEG
chains were allowed to react in toluene with the surface-bound isocyanate groups.

The poly(tetrafluoroethylene) (PTFE) film has also been modified through
bonding of polymeric silicic acid [33]. The porous PTFE film was first exposed
to a silicon tetrahalide atmosphere to allow low-molecular-weight species from
the gaseous phase to penetrate the film. The absorbed silicon tetrahalide was
subsequently hydrolyzed to silicic acid in water. Finally, polyethylene (PE) films
functionalized with diphenylmethyl groups have been deprotonated to form a
nucleophilic lithiated surface. The so-modified surface has, in turn, been used
for the covalent binding of C_{60} to a PE film surface [34].

B. Surface Graft Copolymerization

Surface modification via graft copolymerization has been explored much more
extensively than that by coupling reactions. Unlike the surface coupling reactions,

which require reactive groups on both the substrate and the polymer chain, graft copolymerization of monomers requires the generation of active species at the surface to initiate polymerization. Surface graft copolymerization is based most commonly on free radical reaction of vinyl or acrylic monomers, although cationic and anionic polymerization mechanisms can also be used. For polymerization by free radical mechanism, either free radicals or peroxides have to be generated in the surface region of the substrate to initiate chain growth.

The radical species can be generated on the substrate surface in several ways. One is by means of UV photoinitiators or thermoinitiators. In either case, the initiator decomposes to form free radicals. The free radicals in contact with the substrate surface create unpaired electrons by abstraction of hydrogen from the polymer chains at the surface. The unpaired electrons can also be generated on the polymer surface by bombarding it with high-energy radiation, such as γ-ray irradiation from ^{60}Co. Alternatively, peroxide and hydroxyl peroxide species can be generated on the substrate surface by gas plasma treatment, high-voltage corona discharge, ozone, and UV exposures. In any case, the surface treatment must not result in the overoxidation of the substrate surface and must leave the bulk unperturbed.

Surface graft copolymerization proceeds via the diffusion of the vinyl or acrylic monomer, present either in vapor phase or in solution, to the substrate surface and reaction with the active centers there. Because an active center is regenerated after each monomer addition, the reaction continues and the chains propagate from the substrate surface. In addition to chain growth from the substrate surface, ungrafted homopolymer is produced during the reaction. Since these homopolymer chains are adsorbed on rather than covalently bonded to the substrate, they are usually removed by vigorous rinsing or by Soxhlet extraction. The most widely used monomers for surface graft copolymerization include acrylic acid (AAc), acrylamide (AAm), Na salt of styrenesulfonic acid (NaSS), glycidyl methacrylate (GMA), vinyl pyridine, and their derivatives and analogs. Other functional monomers, such as monomers of some polyelectrolytes, and amphoteric monomers have also been used. The surface graft copolymerizations to be described here all proceed via the radical mechanism and will be classified on the basis of the chemical nature and structure of the substrate.

1. Polyolefin and Acrylate Polymer Substrates

High-density polyethylene (HDPE) and low-density polyethylene (LDPE) are by far the most widely used substrates for surface graft copolymerization. The earlier works involved the graft copolymerization of acrylonitrile on HDPE powders. The substrate was pretreated with a high-frequency discharge from a Tesla coil [35,36]. Pioneering work on the surface modification of polyethylene films and silicone rubber substrates involved surface graft copolymerization with the hydrophilic 2-hydroxyethylmethacrylate (HEMA) and AAm using a simultaneous irradiation technique from a ^{60}Co source [37]. The modified substrate surface was carefully characterized by electron spectroscopy for chemical analysis (ESCA), a surface analytical technique also referred to as the X-ray photoelectron spectroscopy (XPS). Other earlier works involved the UV-photoinitiated graft copolymerization of AAm on PE [38] and the high-energy irradiation–initiated vapor-phase graft copolymerization of methyl acrylate and vinyl acetate on PE [39,40].

The preparation and characterization of surface-modified PE, PP, and ethylene/propylene/diene-ter-polymer (EPDM) of different shapes, such as beads, films, tubes, fibers, and microtome slices, from ^{60}Co gamma ray–initiated graft copolymerization with 15 different monomers had been reported [41]. The surface-modified polymer substrates were tested for tissue and blood compatibility. Other ^{60}Co γ-ray–initiated processes include the graft copolymerization of preirradiated PE in deaerated aqueous solution of AAm monomer and Fe^{2+} ion [42] and the graft copolymerization of AAc onto the PE surface [43]. The AAc graft-copolymerized surface, however, was found to be tightly hydrogen bonded and deformed, resulting in substantial monomer penetration into the film.

The effect of HEMA/ethyl methacrylate monomer composition on the ionizing radiation–induced graft copolymerization on PE films was reported by Hoffman et al. [44]. It was found that the graft reaction took place faster in the beginning if HEMA was present at a high concentration in the monomer mixture. Other novel polymeric biomaterials have also been prepared using the γ-rays and glow discharge method [45,46]. To improve blood compatibility, surface modification via graft copolymerization of γ-ray–preirradiated PE films with AAm in aqueous solution and in the presence or absence of ferrous sulfate had also been carried out [47]. A study had also been made on the post–^{60}Co radiation grafting of AAm onto LDPE film in aqueous medium [48]. It was found that the addition of 0.05 wt% Mohr's salt reduced the efficiency of homopolymerization of AAm. Some properties of the graft copolymer, such as the swelling behavior, electrical conductivity, and desalination by reverse osmosis, were also investigated. In a more recent study of the graft copolymerization of AAc and methacrylic acid onto the ^{60}Co preirradiated LDPE powders, it was found that certain transition metal compounds were also able to suppress the formation of homopolymers while still allowing significant levels of grafting to take place [49]. Other monomers that are susceptible to graft copolymerization with the γ-ray–pretreated PE films include the vinylimidazoles [50] and diethylene glycoldimethacrylate [51].

A number of studies involved the use of thermal or photoinitiators for the surface graft copolymerization. In fact, the earliest work on surface modification in the 1950s involved photografting on substrates precoated with benzophenone [52]. Ogiwara et al. [53] later reported on the photoinduced graft copolymerization of AAc and methacrylic acid on benzophenone-coated LDPE film in aqueous medium. However, it was not easy to graft copolymerize other hydrophilic monomers, such as AAm, to a high extent on these initiator-coated surfaces, and a two-step initiator coating process was required for the photoinduced grafting [54]. The same authors also studied the UV-induced vapor-phase graft copolymerization of functional monomers on photosensitized polyolefin films [53]. The photoinitiators used included benzophenone, anthraquinone, and benzoyl peroxide. A number of subsequent studies were also devoted to the surface modification of LDPE, PP, and PS by photoinduced graft copolymerization with AAc and AAm in the presence of a photoinitiator, either in liquid or in vapor phase [55–59]. For the latter process, both the monomer and the initiator were in vapor phase [55]. Grafting was affected by the combination of solvent and carrier used and the type of polymer substrate. The UV-photoinitiated process was also extended to the vapor-phase graft copolymerization of glycidyl acrylate (GA) and glycidyl methacrylate (GMA) on PE, PP, and PS [60,61]. Other monomers that are susceptible to photoinitiated graft copolymerization with PE or PP include 4-vinyl pyridine [62] and HEMA [63]. More recently, radical living graft copolymerization on PE films with AAc and methacrylic acid in the presence of a photoinitiator, such as benzophenone, xanthone, and 9-fluorenone, has been reported [64]. Finally, a process system for the continuous photoinduced graft copolymerization of AAm and AAc on HDPE films has also been developed [65,66].

The work of Suzuki et al. [67] marked a new approach to the process of surface modification via near-UV light–induced graft copolymerization. The LDPE film was pretreated with radio-frequency gas plasma, followed by exposure to air prior to the copolymerization process. The peroxide formed after the plasma treatment was quantified and utilized in the subsequent near-UV light–initiated surface graft copolymerization. In a subsequent study, Iwata et al. [68] oxidized a PE film surface with corona discharge, followed by graft copolymerization with AAm. Several types of peroxides were found on the corona-pretreated PE film, and the most labile one was mainly responsible for the initiation of graft copolymerization. Photoinduced graft copolymerization has also been achieved on UV-preirradiated PE, PP, and ethylene–vinyl acetate (EVA) copolymer surfaces in the absence of photosensitizers and photoinitiators [69]. Graft copolymerization took place in high yield with increasing UV preirradiation time, and the density of the peroxides generated on the film surface was proportional to the extent of UV exposure. In another study, corona discharge–pretreated PE films were graft copolymerized with different water-soluble ionic and nonionic monomers [70]. The Na salt of styrenesulfonic acid (NaSS) and N,N-dimethylaminopropyl acrylamide (DMAPAA) graft copolymerized surfaces exhibited good cell adhesion and protein adsorption, whereas cells did not adhere to the 2-acrylamide-2-methylpropane sulfonic acid graft-copolymerized surface.

To improve the wettability and adhesion, thermal or near-UV light–induced graft copolymerization of AAm and GMA on Ar plasma–pretreated ultrahigh modulus polyethylene (UHMPE) fibers was carried out [71]. Reaction of propyl-amine with the GMA graft-copolymerized surface resulted in the appearance of a nitrogen core-level signal in the XPS spectrum, suggesting the presence of surface-grafted GMA chains. In a subsequent study, near-UV light–induced graft copolymerization of GMA on corona-pretreated HDPE sheets was carried out [72]. Finally, a number of other olefinic [73–77] and acrylate [78,79] polymer substrates have also been modified via surface graft copolymerization with hydro-philic monomers.

2. Halogen-Containing Polymers

One of the earlier studies on the surface modification of halogen-containing sub-strates appears to be that of Landler and Lebel [80]. These authors carried out a detailed study of the thermal graft copolymerization of various vinyl monomers onto ozone-pretreated poly(vinyl chloride) (PVC). More recently, low-frictional catheter materials have also been developed through photoinduced surface graft copolymerization of UV-preirradiated tubes and films of PVC and EVA copoly-mer with N,N-dimethylacrylamide (DMAA) [81].

In the case of fluorine-containing polymers, the influence of grafting pa-rameters, such as preirradiation dose, monomer concentration, and grafting tem-perature, on the graft yield of hydrophilic N-vinyl-2-pyrrolidone polymer on γ-ray–preirradiated PTFE and poly(tetrafluoroethylene-hexafluoropropylene) films has been studied in detail [82]. The γ-ray–irradiated fluoropolymers suitable for preparation of membranes were also susceptible to thermally induced graft copolymerization in vacuum with vinylimidazole [83]. The grafting process started very rapidly, but the limiting graft yield of 13.2% was reached after 100 h at 60°C.

Surface modification of PTFE by Ar plasma pretreatment, followed by near-UV light–induced graft copolymerization with hydrophilic monomers, such as AAm, AAc, NaSS, DMAA, and (N, N-dimethylamino)ethyl methacrylate (DMAEMA), has also been carried out [84]. The surface compositions and micro-structures were analyzed by angle-resolved XPS. The surface-modified PTFE films were further functionalized via coating of an adherent layer of a conductive polymer base or via covalent immobilization of a biologically active species, such as an enzyme [85].

The synthesis and applications of ozonized poly(vinylidene fluoride) (PVDF) with surface graft-copolymerized styrene, AAc, GMA, and methyl meth-acrylate have also been studied [86]. The products were used as emulsifiers in polymer blends and as promoters for the adhesion of PVDF to composites made of epoxide matrices filled with glass fibers. A somewhat different approach in-volved the graft copolymerization of a fluorinated monomer, 2,2,3,3,3-penta-fluoropropyl methacrylate, with corona-pretreated hydrophilic polymer substrates, such as PVA and cellulose films. The graft copolymerization process was accelerated by the addition of alcohol in the reaction mixture [87].

3. Condensation Polymers: Polyesters and Polyamides

One of the earlier studies reported on the surface modifications of ozone-treated films, fibers, and fabrics of polyesters and polyamides via graft copolymerization with various vinyl monomers, such as styrene, methyl methacrylate, methyl-acrylate, acrylonitrile, methacrylic acid, AAc, and 2-methyl-5-vinylpyridine [88]. Other studies included high-energy irradiation–initiated graft copolymerization of 4-vinyl pyridine on nylon [89], methacrylic acid on a poly(ethylene terephthalate) (PET)/cotton blend [90], and AAc on PET [91].

Osipenko and Martinovicz [92] graft copolymerized AAc onto PET films and fibers using benzoyl peroxide. It was found that the preswelling of PET in dichloroethane led to changes in its sorption-diffusion properties and favored an increase in the degree of grafting. The addition of Fe(II), Ni(II), and Cu(II) salts to the monomer solutions decreased homopolymer yield. In a similar study, meth-acrylic acid was graft copolymerized onto PET fibers [93]. The reaction was initiated by radicals formed from thermal decomposition of benzoyl peroxide. The graft yield increased with temperature up to 85°C and then decreased with further increase in temperature.

A large number of studies involved the use of a photoinduced graft copoly-merization technique in either the presence or absence of an added initiator. UV-photoinitiated graft copolymerizations of 4-vinyl pyridine, AAc, and AAm on PET [58,94] in the presence of a photoinitiator have been reported. The peroxide initiators responsible for the graft copolymerization of AAm onto nylon 6 films can also be generated by near-UV preirradiation of the film surface [69,95].

The UV-induced graft copolymerization of AAm on PET films can be carried out in the absence of a photosensitizer and without prior oxygen removal when the monomer solution also contains an appropriate quantity of periodate $(NaIO_4)$ [96], or a small amount of riboflavin [97]. XPS results clearly indicated that graft copolymerization had actually occurred not merely on the outermost surface but also within the thin surface region of the PET films. UV-induced graft copolymerization of PET films with methoxypoly(ethylene glycol) meth-acrylates and hydroxypoly(ethylene glycol) methacrylates has also been carried out in the absence of any photosensitizer and degassing process [98]. The grafted PET films were extremely hydrophilic and exhibited zeta potential values of nearly zero over a wide pH range, indicating that the water-soluble chains were tethered to the outermost surface of the film.

UV-induced graft copolymerization of AAm and GMA onto the Ar plasma–pretreated surface of Kevlar 49 fibers has also been carried out in the presence of riboflavin [99]. The reaction of propylamine with the GMA graft-copolymerized surface suggested the presence of epoxide functional groups on the fiber surface after grafting.

4. Conjugated Polymers

Surface modification via graft copolymerization has also been carried out on a number of conjugated polymer substrates. Because of the reactive nature of the conjugated polymer surfaces, surface graft copolymerization will proceed to some extent even in the absence of surface pretreatment or an added initiator. In fact, surface pretreatment by Ar plasma [100] and ozone [101] may be accompanied by other side reactions, in addition to surface oxidation.

Graft copolymerization of AAc onto the plasma-pretreated poly[1-(trimethylsilyl)-1-propyne] has been carried out to generate a hydrophilic gas-permeable membrane for dissolved oxygen [102]. Surface graft copolymerization effectively occurred above a threshold temperature of 80°C. When a film exposed to UV irradiation was heated in an aqueous solution of AAc, graft copolymerization also occurred in the bulk of the film. Surface modification of other substituted polyacetylenes, such as poly[(o-(trimethylsilyl)phenyl)acetylene] and poly(1-chloro-2-phenylacetylene), via near-UV light–induced graft copolymerization with AAc, AAm, NaSS, and 3-dimethyl(methacryloyloxyethyl)ammonium propanesulfonate (DMAPS) has also been carried out [103]. Near-UV light–induced graft copolymerization in all cases also resulted in surface photodegradation of the substrate and some bulk polymerization.

Both pristine and Ar plasma–pretreated emeraldine (EM) base films of polyaniline (PAN) are susceptible to near-UV light–induced graft copolymerization with AAc, AAc, and NaSS [104,105]. Extended plasma pretreatment can actually result in a reduction of graft yield. Graft copolymerization with AAc and NaSS readily gives rise to a self-doped (self-protonated), and thus semiconductive, EM surface structure. The AAc graft-copolymerized surface can be further functionalized through the covalent immobilization of proteins and enzymes [105]. Similar surface modification and functionalization can also be performed on other N-containing conjugated polymers, such as polypyrrole films [106].

Finally, surface modification of alkyl-substituted polythiophene, such as solution cast films of poly(3-hexylthiophene), poly(3-octylthiophene), and poly(3-dodecylthiophene), via near-UV light–induced graft copolymerization with hydrophilic monomers such as AAc, AAm, and NaSS has also been carried out [107]. The morphology of the graft-copolymerized surface was determined, to a large extent, by the length of the alkyl substituents.

5. Other Substrates

One of the pioneering works on surface graft copolymerization involved the graft copolymerization of HEMA and AAm on silicone rubber substrates [37]. More recent work indicated that the corona discharge–pretreated surface of silicone could be rendered hydrophilic without altering its bulk properties by thermally induced graft copolymerization with AAm [108]. The grafted AAm chains were also converted to AAc polymer chains, via alkaline hydrolysis, for the subsequent surface immobilization of proteins.

Jansen [109] reported on γ-ray–induced graft copolymerization on PU surfaces preswelled with a monomer, such as HEMA or AAm. In a subsequent study, graft copolymerization of hydrophilic or reactive monomers, such as HEMA, 2,3-epoxypropyl methacrylate, 2,3-dihydroxypropyl methacrylate, and AAm with urethane polymer surfaces, was achieved via γ-irradiation [110]. It was found that the water diffusion and uptake of the grafted and subsequently chemically modified PU film increased with graft yield. Ozone-induced graft copolymerization was also carried out on PU films [111]. The peroxide concentration at the surface was controlled by the ozone concentration and exposure time.

Radical graft copolymerization of AAm has also been carried out on ultrafine silica particles by the use of redox system consisting of ceric ion and reducing groups on the surface [112]. The percentage of grafting onto the silica reached about 25%. The silica obtained from this redox graft copolymerization gave rise to a stable colloidal dispersion in water. In a subsequent study, graft copolymerization of other vinyl monomers, such as methyl methacrylate, styrene, and N-vinylcarbazole, was also carried out on silica and other inorganic ultrafine particles, such as titanium oxide and ferrite [113,114]. The graft copolymerization was initiated by the azo groups introduced onto the particle surface. Surface modification of magnetite particles via graft copolymerization of AAc has also been carried out in a redox system consisting of ceric ion and mercapto groups introduced onto the particles [115]. The AAc polymer–modified particle surface was used for the subsequent covalent immobilization of glucose oxidase. The immobilized enzyme retained 95% of its activity in water over a period of 9 months.

III. MICROSTRUCTURES AND PROPERTIES OF GRAFT-MODIFIED SURFACES

The surfaces of polymeric materials are distinctively different from those of the more rigid crystalline materials, such as metals. Polymer molecules at the surface are known to have much greater freedom for rearrangement according to the changes in the surrounding environment. For example, the surfaces of some polymer hydrogels, which contain as much as 98% water, exhibit surprisingly pronounced hydrophobic characteristics and a significant hysteresis between the advancing and receding water contact angles [116,117]. An attempt was made to correlate the mobility of polymer chains and polar groups with the apparent contact angle of water and its hysteresis and with the decay of the hydrophilicity of the surface-modified hydrophobic polymers [118].

XPS was used to study the surface composition and microstructure, both in the dry state and in the hydrated state (frozen at 160 K), of PE and silicone rubber substrates after radiation-induced graft copolymerization with HEMA and AAm [37]. The HEMA and AAm polymers were readily detected in both the hydrated and dehydrated states when grafted to PE substrates. For silicone rubber substrates, both grafts were observed on the hydrated surface, but the surface concentrations were significantly decreased upon dehydration. The unexpected phenomenon observed on the silicone substrate was explained by graft chain penetration into the silicone rubber or clustering together on the surface. Schematic representations of the two models are shown in Fig. 1 [37].

The works of Ratner et al. [37] and Yasuda et al. [118] thus suggest that the nature of the polymer substrate, or more specifically the relative mobility of the substrate chains, plays an important role in the microstructure of the modified polymer surface. For hydrophilic graft on a hydrophobic substrate, the presence of ''surface restructuring,'' when allowed by the mobility of the substrate chains, suggests that the modified polymer surface essentially strives to re-establish its original low-energy state. Accordingly, it should be interesting to compare the surface microstructures of thermoplastics, partial thermosets, and thermoset polymers after surface modification via graft copolymerization.

The surface microstructures of thermoplastic substrates, such as the polyolefin, polyester, and some fluoropolymer films, after modification by graft co-

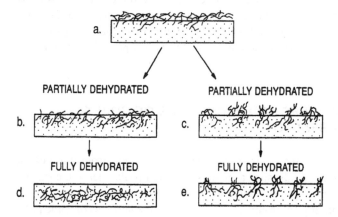

Figure 1 Schematic representations of models to explain the surface depletion of graft upon dehydration for hydrogels grafted onto silicone rubber. Pathway a-b-d, hydrogel graft withdraws below the silicone rubber surface. Pathway a-c-e, hydrogel graft clusters on the surface exposing the silicone rubber substrate.

polymerization with hydrophilic monomers, have been studied by angle-resolved XPS and by contact angle measurements [84,85,119–122]. For a hydrophobic substrate with a substantial amount of hydrophilic surface graft, a stratified surface microstructure with a significantly higher substrate-to-graft chain ratio in the top surface layer than in the subsurface layer is always obtained. Figure 2 shows the angle-dependent C1s core-level and wide-scan spectra of a 5-s Ar plasma–pretreated PTFE film after graft copolymerization in 10% AAc solution [85]. A comparison of the $\underline{C}OOH$-to-$\underline{C}F_2$ component intensity ratios in the C1s core-level spectra, or the C1s-to-F1s peak intensity ratios in the wide-scan spectra, at the two photoelectron takeoff angles (α's) readily reveals the presence of a higher proportion of the PTFE chains in the more surface glancing angle of 20°. The stratified surface microstructure is always observed for PTFE and other thermoplastic substrates with a substantial extent of surface grafted AAc, AAm, NaSS, or GMA polymer. Table 2 summarizes the densities of surface graft, measured at $\alpha = 20°$ and 75°, and the corresponding water contact angles for LDPE, HDPE, PP, PS, polycarbonate (PC), and PET films after graft copolymerization

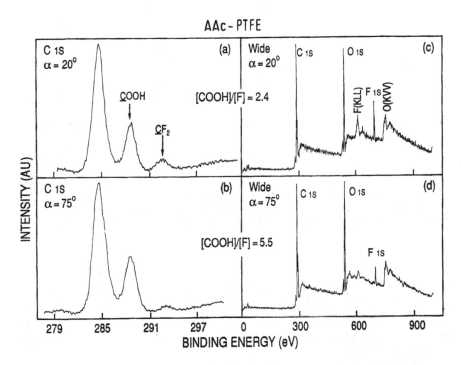

Figure 2 Angle-dependent C1s core-level and wide-scan spectra of a 5-s Ar plasma–pretreated PTFE film after graft copolymerization in 10% AAc solution.

Table 2 Densities of Surface Graft, Measured at α = 20° and 75°, and the Corresponding Water Contact Angles on LDPE, HDPF, PP, PS, PC, and PET Films After Graft Copolymerization with AAm, AAc, and NaSS

Polymer	Density of AAm graft and water contact angle	Density of AAc graft and water contact angle	Density of NaSS graft and water contact angle
Pristine LDPE	0.05(0.09)[a] [30]	0.11(0.15) [86]	0.21(0.33) [15]
Ozone-pretreated LDPE	1.69(3.80) [22]	0.18(0.39) [75]	1.21(6.50) [15]
Pristine HDPE	0.14(0.15) [92]	0.25(0.20) [80]	0.56(0.86) [25]
Ozone-pretreated HDPF	1.21(2.84) [27]	0.65(0.52) [51]	1.08(2.01) [0]
Pristine PP	0.06(0.10) [84]	0.12(0.12) [82]	0.56(0.57) [18]
Ozone-pretreated PP	2.67(7.52) [22]	0.48(0.63) [65]	2.38(8.57) [10]
Pristine PC	0.24(0.11) [82]	4.47(6.91) [72]	1.11(0.82) [42]
Ozone-pretreated PC	2.60(3.02) [60]	11.50(9.73) [25]	9.27(10.36) [0]
Pristine PS	0.36(0.36) [44]	0.46(0.22) [72]	1.74(1.04) [15]
Ozone-pretreated PS	12.22(17.94) [22]	16.68(19.11) [30]	6.84(15.95) [7]
Pristine PET	0.23(0.37?) [58]	0.93(0.32) [58]	1.97(1.16) [5]
Ozone-pretreated PET	0.53(0.37) [55]	2.12(2.17) [25]	2.90(2.04) [5]

[a] Density of graft measured at α = 20°, and the value in parentheses is that measured at α = 75°. Water contact angle in degree is given in square brackets. The static water contact angles on the pristine polymer films are as follows: LDPE, 95°; HDPE, 97°; PP, 90°; PC, 85°; PS, 84°; PET, 75°.

with AAm, AAc, and NaSS [120]. The tendency for the polymer to maintain a hydrophobic surface was also borne out by the fact that for a hydrophilic polymer substrate, such as PVA or cellulose, with a hydrophobic surface graft, such as the 2,2,3,3,3-pentafluoropropyl methacrylate polymer, no surface restructuring was observed [87].

The absence of surface restructuring as a result of limited chain mobility in the substrate matrix is readily observed in cross-linked polymer films after graft copolymerization with hydrophilic monomers. Thus, for the lightly cross-linked EM base film of PAN cast from *N*-methylpyrrolidinone at elevated temperature [123], angle-dependent XPS results suggest that the grafted AAc, NaSS, and AAm polymers are retained predominantly at the top surface [104], as also shown in Fig. 3a–c, respectively [124]. The graft densities in Fig. 3 are expressed as the corresponding [COOH]/[N], [S]/[N], and [H$_2$NC=O]/[N] ratios at the two α's. Similar surface microstructures are observed for EM base film graft copolymerized with other monomers, such as the amphoteric DMAPS. The acquired surface hydrophilicity after graft copolymerization is indicated by the substantial decrease in both the advancing and receding water contact angles as well as in their hysteresis upon increasing the concentration of the surface graft (Fig. 3).

Because of their inherent chemical structures and high glass transition temperatures, polyimides can be regarded as a family of polymers with partial ther-

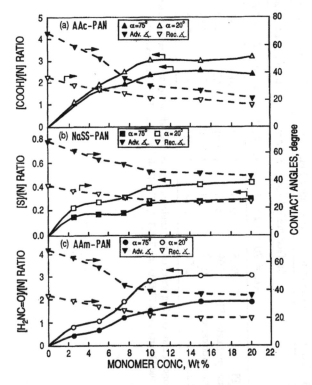

Figure 3 Angle-dependent graft concentrations and water contact angles as a function of monomer concentration for graft copolymerization carried out on cross-linked EM base films.

Table 3 Surface Characterization of Polyimide Films After Graft Copolymerization with AAm, AAc, and NaSS

Sample	Surface graft density of AAm			Surface graft density of AAc			Surface graft density of NaSS		
	($\alpha = 20°$)	($\alpha = 75°$)	$[\theta_{Adv}/\theta_{Rec}]$[a]	($\alpha = 20°$)	($\alpha = 75°$)	$[\theta_{Adv}/\theta_{Rec}]$[a]	($\alpha = 20°$)	($\alpha = 75°$)	$[\theta_{Adv}/\theta_{Rec}]$[a]
Pristine PI	0.38	0.29	[42/15]	1.00	0.80	[41/17]	0.87	0.58	[42/16]
Ar plasma–pretreated PI	5.99	5.88	[19/9]	5.03	4.67	[18/7]	2.39	1.85	[10/6]
O$_2$ plasma–pretreated PI	0.57	0.29	[36/10]	4.88	4.53	[18/6]	2.82	1.20	[15/6]
O$_3$-pretreated PI	2.50	2.12	[21/11]	8.20	7.46	[15/11]	1.49	0.88	[36/16]

[a] The advancing and receding water contact angles in degrees. The advancing and receding water contact angles on pristine (ungrafted) PI, Ar plasma, O$_2$ plasma, and O$_3$-pretreated PI films are 71°/37°, 67°/14°, 48°/8°, and 51°/20°, respectively.

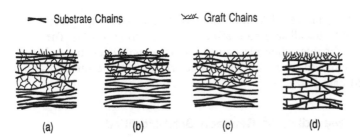

(a)   (b)   (c)   (d)

Figure 4   Schematic models of the graft-copolymerized surfaces: (a) complete penetration model (thermoplastic substrate); (b) partial penetration model (sterically hindered migration of graft chains); (c) intermixing model (partial thermoset substrate); (d) surface graft model (cross-linked substrate).

moset character. Table 3 summarizes the XPS-derived surface compositions of some plasma- and ozone-pretreated poly[(*N*,*N*-oxydiphenylene)pyromellitimide] (Kapton H, DuPont Chemical Co.) films after graft copolymerization with AAm, AAc, and NaSS [125]. Thus, the modified PI surface consists of a uniform mixture of the graft and the substrate chains, or a structure intermediate between that of a similarly modified thermoplastic and that of a thermoset. The lack of penetration of the NaSS polymer graft must have resulted from the sterically hindered migration of the graft chains with bulky substituents. The various surface microstructures that resulted from surface modifications of thermoplastics, thermosets, and partial thermosets via graft copolymerization have been compared and can be summarized schematically as in Fig. 4 [126].

IV. ADHESION CHARACTERISTICS OF THE GRAFT-MODIFIED SURFACES

A. Adhesion of Conductive Polymer Coating

The surfaces of conventional polymer substrates can be modified to improve their adhesion with the coated electroactive polymer layer. The surfaces of PE films have been modified via sulfonation. The covalently bonded sulfonic groups can act as counterions for polypyrrole grown at the surface through oxidative chemical polymerization [127]. The electroactive PAN in its EM base state does not adhere to the pristine HDPE, LDPE, PP, and PTFE surfaces and adheres only poorly on a pristine PET surface, when cast from dilute NMP solution. The adhesion can be substantially enhanced by the charge transfer interaction between the electroactive polymer and the surface-functionalized substrates. The latter substrates were obtained by direct sulfonation (for LDPE, HDPE, PP, and PET), hydrolysis (for PET), and near-UV light–induced surface graft copolymerization with AAc and NaSS (for all substrates after preactivation) [128,129]. The interfacial charge transfer interaction also gives rise to a semiconductive substrate surface. Table 4 summarizes the surface characteristics of these substrates after coating with EM base [128].

B. Adhesive-Free Adhesion Between Graft-Modified Surfaces

Earlier studies have demonstrated the ability·of corona discharge–induced self-adhesion of PE and PET films when joined under conditions of heat and pressure [15]. The self-adhesion property arises from the hydrogen bonding between carbonyl groups and enolic hydrogens (for PE films) and between carboxyl carbonyl groups and phenolic hydrogens (for PET films). PET films can also be made self-adherent by UV irradiation. Thus, it is theoretically possible to achieve adhesion between two polymer surfaces in the absence of any applied adhesives, provided their surfaces possess appropriate functional groups in a favorable spatial distribution.

The concept was first applied by Chen et al. [130,131] to surface-modified PET films and PE films from Ar plasma pretreatment, followed by photoinduced graft copolymerization with water-soluble, nonionic and ionic monomers. In each case, the two surfaces were brought into intimate contact in the presence of water and under applied load. The development of lap shear adhesion strengths between an AAm graft-copolymerized PET film and other polymer surfaces as a function of the junction drying time (adhesion time) is shown in Fig. 5 [130]. In all cases, the adhesion strength increases monotonically with adhesion time, with the highest lap shear adhesion strength being observed between two identical AAm graft-copolymerized surfaces.

When an AAc graft-copolymerized PET film was brought into contact with a DMAPAA graft-copolymerized PET film in the presence of water and under applied load, the films adhered to each other almost instantaneously, due to the presence of ionic interaction at the interface, as shown in Fig. 6 [130]. The lower initial strength observed for the lapped junction with a higher graft density (filled circles) of the AAc polymer is due to the increase in water retention time by the thicker graft layer. As the ionic interaction between the charged groups in aqueous media is greatly influenced by the presence of other salt ions, the adhesion strength between the AAc and DMAPS graft-copolymerized films decreased sharply when 0.2 M NaCl solution was used as the adhesion medium. A more recent study demonstrated the adhesion phenomenon in aqueous media between the AAc and DMAEMA graft-copolymerized polyester films [132]. The shear strength in this case depended on the duration of applied load between the two surfaces in water. The Coulombic attractive force that gave rise to the observed

Table 4 XPS Results for Various Surface-Functionalized Substrates with EM Base Coating

Substrates	Surface modification[a]	Surface composition after modification[a]	Surface composition after EM coating[b]	[N+]/[N]	Surface resistivity, Ω/□
LDPE	Graft copolym. w/AAc	[COOH]/[subst. mono][c] = 0.39	[COO]/[N] = 1.03	0.19	10^7
LDPE	Graft copolym. w/NaSS	[SO$_3$]/[subst. mono] = 6.50	[SO$_3$]/[N] = 0.60	0.46	10^5
LDPE	Sulfonation (1 h)	[SO$_3$]/[subst. mono] = 0.16	[SO$_3$]/[N] = 0.72	0.49	10^4
HPDE	Graft copolym. w/AAc	[COOH]/[subst. mono] = 0.52	[COO]/[N] = 0.66	0.24	10^7
HDPE	Graft copolym. w/NaSS	[SO$_3$]/[subst. mono] = 2.01	[SO$_3$]/[N] = 0.48	0.28	10^6
HDPE	Sulfonation (8 h)	[SO$_3$]/[subst. mono] = 0.16	[SO$_3$]/[N] = 0.80	0.50	10^4
PP	Graft copolym. w/AAc	[COOH]/[subst. mono] = 0.63	[COO]/[N] = 1.64	0.30	10^7
PP	Graft copolym. w/NaSS	[SO$_3$]/[subst. mono] = 8.57	[SO$_3$]/[N] = 0.61	0.40	10^5
PP	Sulfonation (0.5 h)	[SO$_3$]/[subst. mono] = 0.15	[SO$_3$]/[N] = 0.49	0.42	10^4
PET	Graft copolym. w/AAc	[COOH]/[subst. mono] = 2.17	[COO]/[N] = 1.83	0.21	10^7
PET	Graft copolym. w/NaSS	[SO$_3$]/[subst. mono] = 2.04	[SO$_3$]/[N] = 0.41	0.34	10^6
PET	Sulfonation (3 h)	[SO$_3$]/[subst. mono] = 0.05	[SO$_3$]/[N] = 0.11	0.28	10^7
PTFE	Graft copolym. w/AAc	[COOH]/[subst. mono] = 4.64	[COO]/[N] = 2.20	0.29	10^7
PTFE	Graft copolym. w/NaSS	[SO$_3$]/[subst. mono] = 6.08	[SO$_3$]/[N] = 0.52	0.37	10^6

[a] Graft copolymerizations were carried out on O_3-pretreated LDPE, HDPE, PP, and PET and Ar plasma–pretreated PTFE.
[b] Evaluated at the XPS takeoff angle of 75°.
[c] Number of grafted functional groups per repeat unit of the substrate chain.

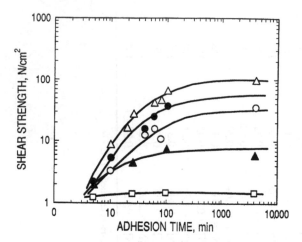

Figure 5 Adhesion time (drying time) dependence of the shear strength between an AAm graft-copolymerized PET film (90 μg/cm²) and another film brought into intimate contact in the presence of water. (○) Virgin PET film; (●) PET film treated with Ar plasma for 10 s only; (△) AAm graft-copolymerized PET film (90 μg/cm²); (▲) LDPE; (□) PTFE.

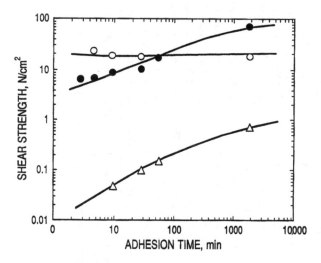

Figure 6 Adhesion time dependence of shear strength between an AAc graft-copoly-merized film and a DMAPAA graft-copolymerized film wetted with distilled water or 0.2 M NaCl solution: (○,●) distilled water; (△) 0.2 M NaCl. Graft density (μg/cm², AAc/ DMAPAA): (○,△) 4/8; (●) 190/8.

adhesion strength was drastically reduced when a salt was added to the water at high concentration.

The results of T-peel measurements for a graft-copolymerized PE surface with other surfaces are summarized in Table 5, together with the result obtained using an adhesive tape (3M Scotch tape) as a counterpart in the dry state [131]. No significant difference in strength was observed between the adhesive tape and another substrate, such as an identical adhesive tape, a pristine PE film, or a surface-modified PE film. In contrast, high strength was not obtained for the wet junction between two graft-modified PE films. High peel strengths developed eventually after 1200 min when the junction became dry. Figure 7 summarizes schematically the various modes of interaction between two graft-modified polymer surfaces in close contact that give rise to the observed phenomenon of adhesive-free adhesion [131].

The adhesive-free adhesion phenomenon has also been demonstrated in chemically inert polymer substrates, such as the PTFE films, whose surfaces had been premodified by graft copolymerization [133]. The lap shear force per unit area at break between two identical AAc graft-copolymerized PTFE films wetted by water is plotted against the adhesion time (drying time) in Fig. 8 for two different surface graft concentrations (in terms of plasma pretreatment times) [134]. In both cases, the adhesion strength increases monotonically with adhesion time and exceeds the tensile yield strength (dashed line in Fig. 8) of the pristine

Table 5 T-Peel Strengths Between Two Similar and Dissimilar Surfaces

Surface pairs	Graft density (μg/cm^2)	Peel strength, mN/cm, at the following contact times	
		5 min	1200 min
Tape-tape		500	
Tape-virgin PE	0	530	
Tape/plasma-treated PE[a]	0	510	
Tape/AAm-grafted PE	90	520	
Virgin PE-virgin PE	0–0	0	
AAm-grafted PE/AAm-grafted PE	10–10	20	70
	30–30		40
	40–40		390
	90–90	12	720
AAc-grafted PE/AAc-grafted PE	90–90	10	630
AAc-grafted PE/DMAPAA-grafted PE	6–6	6	16
	90–20	40	740

[a] Treated with Ar plasma for 15 s.

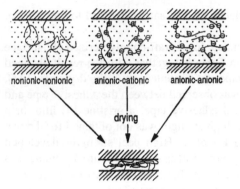

Figure 7   Schematic representation of interactions between two polymer surfaces with graft chains.

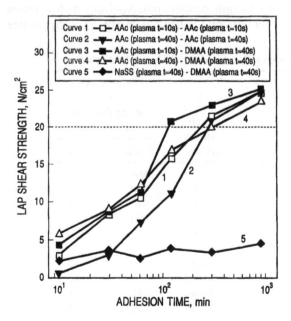

Figure 8   Lap shear adhesion strength at break between surface graft-copolymerized PTFE films wetted by water as a function of adhesion (drying) time and plasma pretreatment time of PTFE.

PTFE film (0.01 cm in thickness) after 300 min. At this point, the PTFE film undergoes a substantial elongation before failure occurs at the lapped junction. Again, diffusion and entanglement of the grafted chains and water retention at high graft concentration (long plasma pretreatment time) play important roles in the observed interfacial adhesion.

When an AAc polymer grafted PTFE film is brought into direct contact with a DMAA graft-copolymerized film, electrostatic interactions, in addition to AAc polymer–dominated chain diffusion and entanglement (as AAc polymer is at a much higher concentration than the DMAA polymer), result in more rapid development of the adhesion strength (curve 3, Fig. 8). The presence of electrostatic interaction is further suggested by the fact that increasing the concentration of the grafted AAc polymer alone does not result in any further improvement in the adhesion strength of this "heterointerface" (curve 4, Fig. 8), as the extent of electrostatic interaction is limited by the low concentration of the grafted DMAA polymer. Finally, curve 5 in Fig. 8 suggests that in the absence of any appreciable chain entanglement due to low graft concentration, the lap shear strength at the heterointerface between the styrenesulfonic acid and DMAA polymer grafted PTFE films is negligible, even though stronger electrostatic interaction may be expected at this interface.

Adhesive-free adhesion has also been observed in surface-modified electroactive polymer films. The lap shear adhesion strength between two identically modified EM base films after graft copolymerization in 10 wt% AAc, NaSS, AAm, and DMAPS solutions is plotted against adhesion time in Fig. 9 [135]. In the case of the DMAPS graft-copolymerized EM films with 300 min of adhesion time, the adhesion strength reaches about 340 N/cm^2. This lap shear adhesion strength is substantially higher than the tensile yield strength of 120 N/cm^2 for the lightly cross-linked EM base film used. The high adhesive-free adhesion strength observed can be attributed to the amphoteric nature of DMAPS units, as well as the high efficiency of graft copolymerization involving the DMAPS monomer [135]. The fact that the final adhesion strength of the AAc-AAm heterojunction is not substantially higher than that of the corresponding AAc-AAc or AAm-AAm homojunction suggests that the electrostatic interactions must have been partially offset by the reduced miscibility between the grafted AAc and AAm polymer chains.

Adhesive-free adhesion can also be achieved between a conductive (electroactive) polymer film and a conventional insulating polymer film. Figure 10 illustrates the phenomenon of adhesive-free adhesion between an AAc graft-copolymerized PTFE film and the AAc, AAm, and DMAPS graft-copolymerized EM base films [135]. Again, electrostatic interactions give rise to a substantial initial adhesion strength. A significantly higher adhesion strength between the DMAPS graft-copolymerized EM base film and the AAc graft-copolymerized PTFE film, or the DMAPS (EM)–AAc (PTFE) pair, than those of the AAc (EM)–

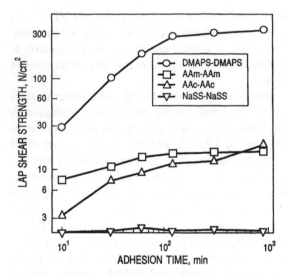

Figure 9 Comparison of the lap shear adhesion strength at break between two identical AAc, AAm, NaSS, and DMAPS graft-copolymerized EM films wetted by water as a function of adhesion (drying) time. Graft copolymerizations were carried out in 10 wt% of the respective monomer solution.

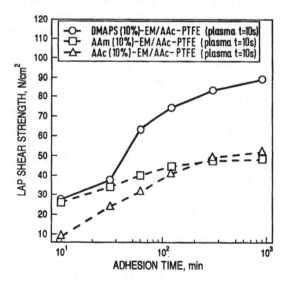

Figure 10 Lap shear adhesion strength at break between AAc (EM) and AAc (PTFE), between AAm (EM) and AAc (PTFE), and between DMAPS (EM) and AAc (PTFE) graft-copolymerized films wetted by water, as a function of adhesion time.

AAc (PTFE) or the AAm (EM)–AAc (PTFE) pairs is also observed at long adhesion time. The phenomenon has been attributed to the high densities of the grafted DMAPS and AAc polymer chains on EM and PTFE surfaces, respectively, as well as the amphoteric nature of the DMAPS chains. Similar adhesion strength has also been observed between the DMAPS (EM)–DMAA (PTFE) film pair. The strong interactions between the graft chains have also resulted in adhesion failure being observed predominantly at the graft-substrate interface of the PTFE film [135].

Finally, it is possible to achieve direct lamination between two polymer surfaces or in composite materials through photografting. In this process, a photopolymerizable composition consisting of monomers, photoinitiator, and additives is sandwiched between polymer substrates. The reactive liquid layer is then cured by UV irradiation, resulting in the simultaneous photografting on both substrate surfaces and lamination [135]. The cured layer has a supermolecular architecture, consisting of either a hyperbranched or a cross-linked network, connected by covalent bonds to the substrates. Direct lamination by photografting can also be achieved in the absence of a photoinitiator, provided that the substrate surfaces have been preactivated to introduce the peroxide species.

C. Adhesive-Promoted Adhesion

Surface modification via graft copolymerization with epoxide-containing monomers, such as GMA, has been carried out on Ar plasma–pretreated UHMPE [71] and Kevlar 49 [99] fibers to improve the interfacial adhesion through covalent bonding in the fabrication of composites. It is conceivable that good interfacial properties of the fibers can be obtained if the fiber surface carries grafted chains that can be molecularly incorporated in the matrix polymer of the composite.

When the surfaces of corona-pretreated HDPE sheets are further modified by graft copolymerization with GMA, the modified surfaces exhibit a substantially enhanced interfacial adhesion when cured in the presence of an epoxy resin [72]. Figure 11 shows the resulting shear adhesion strengths of the various PE surfaces with an epoxy resin [72]. It is apparent that the GMA graft-copolymerized PE surface exhibits the highest adhesion strength with the epoxy resin, almost twice that of the ethylenediamine-treated or HCl-hydrolyzed surfaces. The observed adhesion strength is also about twice that of the PE surface treated only by corona discharge, which is the most widely used industrial surface treatment for improving the adhesion of polyolefins. The enhanced adhesion strength for the GMA graft-copolymerized surface can be attributed to the formation of covalent bonds between the epoxide groups of the grafted chains and the amine groups of the resin hardener. Another important factor is the increase in surface wettability of the modified PE surface for the epoxy resin fluid before curing. The

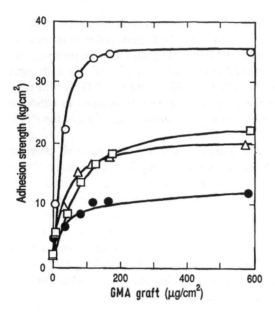

Figure 11   Adhesion strength between surface-treated PE film and epoxy resin: (○) GMA graft-copolymerized PE, (●) GMA graft-copolymerized PE after treatment with 2.0 M butylamine, (△) GMA graft-copolymerized PE after treatment with 2.0 M ethylene-diamine, (□) GMA graft-copolymerized PE after treatment with 2.0 N HCl.

observed sessile contact angle decreased from about 45° for the pristine film to only about 12° for the GMA graft-copolymerized surface.

The effect of surface photografting on the adhesion properties of poly-olefins and polyesters in various forms has also been reported [136]. Molded LDPE plates were surface photografted with AAc, AAm, 4-vinylpyridine, and glycidyl acrylate polymers in a batchwise process in which photoinitiators and monomers were transferred to the substrate surface through the vapor phase. The adhesion of Scotch tape to the LDPE plates increased by five to eight times after graft copolymerization with the four monomers, as measured by the 90° peel test. Strips of PP film and filaments and yarns of PE, PP, and polyester were also photografted with the four polymers in a continuous process in which the photoinitiator and monomer in solution were transferred to the substrate via liquid phase in a presoaking step. The adhesion of the modified substrates to epoxy resin, as measured by pull-off tests from the cured resin, increased by three times for the AAc graft-copolymerized surface and six to seven times for the AAm graft-copolymerized surface.

The adhesion characteristics of even the most inert polymer surfaces, such as those of the fluoropolymers, can be substantially enhanced when these surfaces are modified by graft copolymerization. The lap shear adhesion forces at break between two identical GMA graft-copolymerized PTFE films in the presence of either an epoxy adhesive or an amine curing agent are plotted against the XPS-derived surface [Epoxide]/[F] ratios in Fig. 12 [122]. The adhesion strength increases rapidly with the amount of grafted epoxide groups. The adhesion strength is enhanced to the extent that, with an [Epoxide]/[F] mole ratio greater than 2, the shear strength between two PTFE films has reached about 300 and 80 N/cm^2, respectively, for the epoxy adhesive– and curing agent–promoted adhesion. Two pristine PTFE films do not adhere to one another in the presence of the epoxy adhesive or the curing agent. The rapid increase in adhesion strength associated with the increase in epoxide content at the surface suggests that the adhesion strength is controlled predominantly by chemical reaction between the epoxide groups of the grafted GMA polymer chains and the amine moieties, either

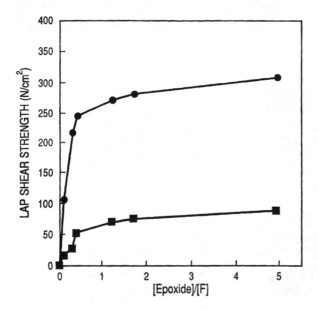

Figure 12 Lap shear adhesion strength at break between two identical GMA graft-copolymerized PTFE films as a function of the surface concentration of the grafted epoxide functional groups. The PTFE films were adhered together in the presence of either an epoxy adhesive (epoxy resin plus curing agent, filled circles) or an amine curing agent alone (filled squares).

in the hardener of the epoxy adhesive or in the curing agent. The reaction readily gives rise to a cross-linked polymer network at the interface. The mechanism is illustrated schematically in Fig. 13.

In the case of the epoxy adhesive–promoted adhesion between the two GMA graft-copolymerized PTFE films, the XPS wide-scan spectra, obtained at $\alpha = 20°$, for the two delaminated films are similar to that of the pristine PTFE film [122]. Thus, the XPS results clearly indicate that the failure mode of the epoxy adhesive–promoted adhesion is cohesive in nature and takes place in the bulk of either film. This result supports the proposition that good interfacial adhesion can be achieved if the grafted chains can be molecularly incorporated via covalent bonding into the adhesive during the bonding process. The shift in failure site to below the graft-substrate interface also helps to account for the fact that the adhesion strength levels off at an [Epoxide]/[F] ratio above 2.

Finally, the surface modification via graft copolymerization technique has been applied to improve the adhesion between the epoxy molding compound and ball-grid-array (BGA) substrate in the packaging of microelectronics [137]. In this case, the surface of the epoxy-based BGA substrate was activated first by ozone treatment, followed by near-UV light–induced graft copolymerization with GMA. The adhesion enhancement has been attributed to the formation of covalent

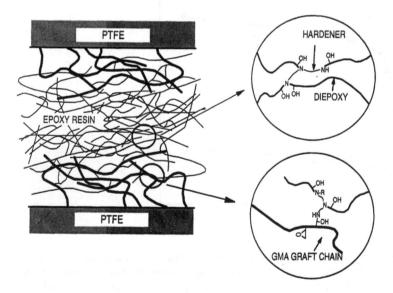

Figure 13 Schematic representation of the interfacial cross-linking reaction between two GMA graft-copolymerized PTFE films in the presence of an epoxy adhesive.

bonds between the epoxide groups of the grafted GMA chains and the curing agent in the epoxy overmold.

V. SUMMARY

The present chapter started with a comprehensive survey of the recent literature on surface modification of polymers via grafting and graft copolymerization. It was subsequently demonstrated that the surface modification method was an effective and viable adhesion improvement technique for polymer surfaces. The modified polymer surfaces exhibit substantially improved adhesion to other polymers (1) during a solution coating process, (2) in the absence of an adhesive (adhesive-free adhesion), and (3) in the presence of an adhesive (adhesive-promoted adhesion).

REFERENCES

1. Y Ikada, Y Uyama. Lubricating Polymer Surfaces. Lancaster, PA: Technomic, 1993, p 76.
2. LS Penn, H Wang. Polym Adv Technol 5:809, 1994.
3. KL Mittal, ed. Polymer Surface Modification: Relevance to Adhesion. Zeist, The Netherlands: VSP, 1995.
4. S Wu. Polymer Interface and Adhesion. New York: Marcel Dekker, 1982.
5. M Strobel, CS Lyons, KL Mittal, eds. Plasma Surface Modification of Polymers: Relevance to Adhesion. Zeist, The Netherlands: VSP, 1994.
6. CM Chan, TM Ko, H Horaoka. Surf Sci Rep 24:1, 1996.
7. H Steinhauser, G Ellinghorst. Angew Makromol Chem 120:177, 1944.
8. B Catoire, P Bouriot, O Bemuth, A Baszkin, M Chevrier. Polymer 26:766, 1984.
9. H Iwata, A Kishida, M Suzuki, Y Hata, Y Ikada. J Appl Polym Sci 36:3309, 1988.
10. D Briggs, CR Kendall. Int J Adhesion Adhesives 2:13, 1982.
11. I Sutherland, DM Brewis, RJ Heath, E Sheng. Surf Interface Anal 17:507, 1991.
12. D Briggs, DM Brewis, MB Kanieczko. J Mater Sci 14:1344, 1979.
13. VT Stannett. Int J Radiat Appl Intrum Part C 35:82, 1990.
14. PG Klein, DW Woods, IM Ward. J Polym Sci Polym Phys Ed 25:1359, 1987.
15. DK Owens. J Appl Polym Sci 19:3315, 1975.
16. CO Kunz, PC Long, AN Wright. Polym Eng Sci 12:209, 1972.
17. SH Zeronian, HZ Wang, KW Alger. J Appl Polym Sci 41:527, 1990.
18. SR Holmes-Farley, RH Reamey, TJ McCarthy, J Deutch, GM Whitesides. Langmuir 1:266, 1985.
19. MA Mohammed, V Rossback. J Appl Polym Sci 50:929, 1993.
20. DK Bridger, D Fairhurst, B Vincent. J Colloid Interface Sci 68:190, 1979.
21. J Clarke, B Vincent. J Colloid Interface Sci 82:208, 1981.

22. T Nakatsuka. J Appl Polym Sci 34:2126, 1987.
23. N Tsubokawa, K Maruyama, Y Sone, M Shimomura. Colloid Polym Sci 267:511, 1989.
24. N Tsubokawa, A Kogure. J Polym Sci Polym Chem Ed 29:697, 1991.
25. P Auroy, L Murray, L Leger. Phys Rev Lett 66:719, 1991.
26. H Hommel, AP LeGrand, P Tougne. Macromolecules 17:1578, 1984.
27. M Taniguchi, RK Samal, M Suzuki, H Iwata, Y Ikada. Am Chem Soc Symp Ser 187:217, 1982.
28. Y Tabata, SV Lanikar, F Horii, Y Ikada. Biomaterials 7:234, 1986.
29. DK Han, KD Park, KD Ahn, SY Jeong, YH Kim. J Biomed Mater Res 23:87, 1989.
30. KD Park, T Okano, C Nojiri, SW Kim. J Biomed Mater Res 22:977, 1988.
31. K Bergstrom, K Holmberg, A Safranj, AS Hoffman, MJ Edgell, A Kozlowski, BA Hovanes, JM Harris. J Biomed Mater Res 26:779, 1992.
32. C Freij-Larsson, B Wesslen. J Appl Polym Sci 50:345, 1993.
33. MA Mohammed, V Rossbach. J Appl Polym Sci 50:929, 1993.
34. DE Bergbreiter, HN Gray. J Chem Soc Chem Commun 645, 1993.
35. CH Bamford, JC Ward. Polymer 2:277, 1961.
36. CH Bamford, AD Jenkins, JC Ward. Nature 186:712, 1960.
37. BD Ratner, PK Weathersby, AS Hoffman. J Appl Polym Sci 22:643, 1978.
38. S Tazuke, H Kimura. J Polym Sci Polym Lett Ed 16:497, 1978.
39. F Yamamoto, S Yamakawa, Y Kato. J Polym Sci Polym Chem Ed 16:1883, 1978.
40. F Yamamoto, S Yamakawa, Y Kato. J Polym Sci Polym Chem Ed 16:1897, 1978.
41. JP Fischer, JR Becker, SP Von Halasz, K-F Miick, H Piishner, S Rossinger, A Schmidt, HH Suhr. J Polym Sci Polym Symp 66:443, 1979.
42. VA Postnikov, NJ Lukin, BV Maslov, NA Plate. Polym Bull 3:75, 1980.
43. JP Lawler, A Charlesby. Radiat Phys Chem 15:595, 1980.
44. AS Hoffman, D Cohn, SR Hanson, LA Harker, TA Horbett, BD Ratner, LO Reynolds. Radiat Phys Chem 22:267, 1983.
45. AS Hoffman. Adv Polym Sci 57:141, 1984.
46. D Cohn, AS Hoffman, BD Ratner. J Appl Polym Sci 29:2645, 1984.
47. M Suzuki, Y Tamada, H Iwata, Y Ikada. In: KL Mittal, ed. Physicochemical Aspects of Polymer Surfaces. Vol. 2. New York: Plenum, 1983, p 923.
48. EA Hegazy, MM El-Dessouky, SA El-Sharabasy. Radiat Phys Chem 27:323, 1986.
49. K Gorgan, E Kronfli, KV Lovell. Radiat Phys Chem 36:757, 1990.
50. Z Xu, A Chapiro, N Schmitt. Eur Polym J 29:1435, 1993.
51. J-L Gineste, J-L Garound, G Pourcelly. J Appl Polym Sci Polym Chem Ed 34:2969, 1993.
52. G Oster, O Shibata. J Polym Sci 26:233, 1957.
53. Y Ogiwara, M Kanda, M Takumi, H Kubota. J Polym Sci Polym Lett Ed 19:457, 1981.
54. Y Ogiwara, M Takumi, H Kubota. J Appl Polym Sci 28:3743, 1982.
55. K Allmer, A Hult, B Ranby. J Polym Sci Polym Chem Ed 26:2099, 1988.
56. PY Zhang, B Ranby. J Appl Polym Sci 40:1647, 1990.
57. PY Zhang, B Ranby. J Appl Polym Sci 41:1469, 1990.

58. Z Feng, B Ranby. Angew Makromol Chem 195:17, 1992.
59. LM Hamilton, A Green, S Edge, JPS Badyal, WJ Feast, WF Pancynko. J Appl Polym Sci 52:413, 1994.
60. K Allmer, A Hult, B Ranby. J Polym Sci Polym Chem Ed 27:1641, 1989.
61. K Allmer, A Hult, B Ranby. J Polym Sci Polym Chem Ed 27:3405, 1989.
62. CH Ang, JL Garnett, R Levot, MA Long. J Macromol Sci Chem A17:87, 1982.
63. S Edge, S Walker, WJ Feast, WF Pancynko. J Appl Polym Sci 47:1075, 1993.
64. W Yang, B Ranby. Macromolecules 29:3308, 1995.
65. ZP Yao, R Ranby. J Appl Polym Sci 40:1647, 1990.
66. B Ranby, ZM Gao, A Hult, PY Zhang. Am Chem Soc Polym Prep 27:38, 1986.
67. M Suzuki, A Kishida, H Iwata, Y Ikada. Macromolecules 19:1804, 1986.
68. H Iwata, A Kishida, M Suzuki, Y Hata, Y Ikada. J Polym Sci Polym Chem Ed 26:3309, 1988.
69. Y Uyama, Y Ikada. J Appl Polym Sci 36:1087, 1988.
70. A Kishida, H Iwata, Y Tamada, Y Ikada. Biomaterials 12:786, 1991.
71. M Mori, Y Uyama, Y Ikada. J Polym Sci Polym Chem Ed 32:1683, 1994.
72. J Zhang, K Kato, Y Uyama, Y Ikada. J Polym Sci Polym Chem Ed 33:2629, 1995.
73. IR Bellobona, F Tolusso, E Selli, S Calgari, A Berlin. J Appl Polym Sci 26:619, 1981.
74. PY Zhang, B Ranby. J Appl Polym Sci 43:621, 1991.
75. K Kaji. J Appl Polym Sci 32:4405, 1986.
76. K Kildal, K Olafsen, A Stori. J Appl Polym Sci 44:1893, 1992.
77. S Tazuke, H Kimura. Am Chem Soc Symp Ser 121:217, 1980.
78. H Ichijima, T Okada, Y Uyama, Y Ikada. Angew Makromol Chem 192:1213, 1991.
79. MC Daries, RAP Lynn, SS Davis, J Hearn, JF Watts, JC Vickerman, D Johnson. J Colloid Interface Sci 156:229, 1993.
80. Y Landler, P Lebel. J Polym Sci 28:477, 1960.
81. Y Uyama, H Tadokora, Y Ikada. Biomaterials 12:71, 1991.
82. E-SA Hegazy. J Polym Sci Polym Chem Ed 22:493, 1984.
83. Z Xu, A Chapiro, N Schmitt. Eur Polym J 29:301, 1993.
84. KL Tan, LL Woon, HK Wong, ET Kang, KG Neoh. Macromolecules 26:2832, 1993.
85. ET Kang, KL Tan, K Kato, Y Uyama, Y Ikada. Macromolecules 29:6872, 1996.
86. B Boutevin, JJ Robin, A Serrani. Eur Polym J 28:1507, 1992.
87. A Hoebergen, Y Uyama, T Okada, Y Ikada. J Appl Polym Sci 48:1825, 1993.
88. VV Korshak, KK Mozgova, MA Shkolina. J Polym Sci Part C 32:75, 1964.
89. IA Sykes, JK Thomas. J Polym Sci 35:721, 1961.
90. T Mares, JC Arthur, JA Harris. Textile Res J 46:563, 1976.
91. PD Kale, HT Lokhande, KN Rao, MH Rao. J Appl Polym Sci 19:461, 1975.
92. IF Osipenko, VI Martinovicz. J Appl Polym Sci 39:935, 1990.
93. M Sacak, F Serthaya, M Talu. J Appl Polym Sci 44:1737, 1992.
94. PY Zhang, B Ranby. J Appl Polym Sci 41:1459, 1990.
95. Y Uyama, H Tadokaro, Y Ikada. J Appl Polym Sci 39:489, 1990.
96. E Uchida, Y Uyama, Y Ikada. J Appl Polym Sci 41:677, 1990.
97. E Uchida, Y Uyama, H Iwata, Y Ikada. J Polym Sci Polym Chem Ed 28:2837, 1990.

98.   E Uchida, Y Uyama, Y Ikada. Langmuir 10:481, 1994.
99.   M Mori, Y Uyama, Y Ikada. Polymer 35:5336, 1994.
100. ET Kang, K Kato, Y Uyama, Y Ikada. J Mater Res 11:1570, 1996.
101. ET Kang, KG Neoh, X Zhang, KL Tan, DJ Liaw. Surf Interface Anal 24:51, 1996.
102. T Masuda, M Kotoura, K Tsuchihara, T Higashimura. J Appl Polym Sci 43:423,
      1991.
103. ET Kang, KG Neoh, KL Tan, DJ Liaw. Polym Degrad Stab 40:45, 1993.
104. ET Kang, KG Neoh, KL Tan, Y Uyama, N Morikawa, Y Ikada. Macromolecules
      25:1959, 1992.
105. FC Loh, KL Tan, ET Kang, K Kato, Y Uyama, Y Ikada. Surf Interface Anal 24:
      597, 1996.
106. X Zhang, ET Kang, KG Neoh, KL Tan, DY Kim, CY Kim. J Appl Polym Sci 60:
      625, 1996.
107. ET Kang, KG Neoh, KL Tan. Macromolecules 25:6842, 1992.
108. T Okada, Y Ikada. Angew Makromol Chem 192:1705, 1991.
109. B Jansen. Polym Sci Technol 23:287, 1983.
110. B Jansen, G Ellinghorst. J Biomed Mater Res 18:655, 1984.
111. K Fujimoto, Y Takebayashi, H Inoue, Y Ikada. J Polym Sci Polym Chem Ed 31:
      1035, 1993.
112. N Tsubokawa, K Maruyama, Y Sone, M Shimomura. Polym J 21:475, 1989.
113. N Tsubokawa, A Kogure, K Maruyama, Y Sone, M Shimomura. Polym J 22:827,
      1990.
114. N Tsubokawa, Y Shirai, H Tsuchida, S Handa. J Polym Sci Polym Chem Ed 32:
      2327, 1994.
115. M Shimomura, H Kikuchi, T Yamaguchi, S Miyauchi. J Macromol Sci-Chem A33:
      1687, 1996.
116. AS Hoffman, C Harris. ACS Polym Prepr 13:740, 1972.
117. FJ Holly, MF Refojo. J Biomed Mater Res 9:315, 1975.
118. H Yasuda, AK Sharma, T Yasuda. J Polym Sci Polym Phys Ed 19:1285, 1981.
119. ET Kang, KG Neoh, KL Tan, FC Loh, DJ Liaw. Polym Adv Technol 5:837, 1994.
120. FC Loh, KL Tan, ET Kang, KG Neoh, MY Pun. J Vac Sci Technol A 12:2705,
      1994.
121. FC Loh, KL Tan, ET Kang, Y Uyama, Y Ikada. Polymer 36:21, 1995.
122. Tie Wang, ET Kang, KG Neoh, KL Tan, CQ Cui, TB Lim. J Adhesion Sci Technol
      11:679, 1997.
123. K Tzou, RV Gregory. Synth Met 55–57:983, 1993.
124. ZF Li, ET Kang, KG Neoh, KL Tan, CC Huang, DJ Liaw. Macromolecules 30:
      3354, 1997.
125. FC Loh, CB Lau, KL Tan, ET Kang. J Appl Polym Sci 56:1707, 1995.
126. FC Loh, KL Tan, ET Kang, SFY Li. J Vac Sci Technol B 14:1611, 1996.
127. C Arribas, DR Rueda, JL Fierro. Langmuir 7:2682, 1991.
128. ET Kang, KG Neoh, MY Pun, KL Tan, FC Loh. Synth Met 69:105, 1995.
129. MY Pun, KG Neoh, ET Kang, FC Loh, KL Tan. J Appl Polym Sci 56:355, 1995.
130. KS Chen, Y Uyama, Y Ikada. J Adhesion Sci Technol 6:1023, 1992.
131. KS Chen, Y Uyama, Y Ikada. Langmuir 10:1319, 1994.
132. JF Zhang, E Uchida, Y Uyama, Y Ikada. Langmuir 11:1688, 1995.

133. ET Kang, KG Neoh, W Chen, KL Tan, CC Huang, DJ Liaw. J Adhesion Sci Technol 10:725, 1996.
134. ET Kang, KG Neoh, KL Tan, BC Senn, PJ Pigram, J Liesegang. Polym Adv Technol 8:683, 1997.
135. WT Yang. Lamination by Photografting, PhD thesis, Department of Polymer Technology, Royal Institute of Technology, Stockholm, 1996.
136. R Ranby. J Adhesion Sci Technol 9:599, 1995.
137. T Wang, ET Kang, KG Neoh, KL Tan, CQ Cui, KK Chakravorty, TB Lim. Mater Res Bull 31:1361, 1996.

11
Microbial Treatment of Polymer Surfaces to Improve Adhesion

Elena V. Pisanova
Academy of Sciences of Belarus, Gomel, Republic of Belarus

I. INTRODUCTION

By now, the effect of microorganisms on polymer materials has been well studied. However, most of the investigations were aimed at polymer protection against biocorrosion or, on the contrary, biodegradation of polymer wastes. Using microbial treatment for polymer adhesion improvement was initiated only in the past decade. Nevertheless, such treatment, being a variant of chemical surface modification, has a number of advantages in comparison with other known treatment techniques:

It needs no expensive chemicals and solvents.
It is conducted at moderate temperatures and needs no energy expenditure.
It is ecologically clean.
Because of the great variety of existing microorganisms, it can offer the desired degree of treatment for different polymer materials.

Practically, there is no polymer that can resist attack by microorganisms (bacteria and/or microscopic fungi). This is due to the ability of microorganisms to adapt to new (for them) nutrition sources—synthetic substances. For a particular application of microorganisms to polymer surface modification, one should take into account characteristics of metabolism of different microorganisms as well as chemical structure of the polymer. The treatment time, composition of the nutrient medium, and structure of the polymer are also important factors influencing the degree of modification.

323

II. ACTIVITY OF DIFFERENT MICROORGANISMS WITH RESPECT TO POLYMERS

Active agents in microbial treatment are the products of the microorganisms' metabolism. For instance, in some cases organic acids given off by microorganisms erode polymer surfaces [1,2]. The pH variation resulting from the microbial life process can give rise to hydrolysis of polycondensed polymers (polyesters, polyamides), and the variation in the redox potential induces oxidation of macromolecules on the surface [1,3]. However, the main destructors of polymer chains are ferments (enzymes) given off by the microbes. Enzymes characteristic of a given microorganism are called constitutive enzymes; these are produced irrespective of the medium composition. The existence of constitutive enzymes accounts for the high specificity of the microorganisms' effect on various polymers. At the same time, when placed in a new medium, microbes begin to synthesize other enzymes called adaptive enzymes. In this way, it is possible to make a microorganism adapt to a nutrition source that is new for it, e.g., a synthetic polymer.

Knowledge of constitutive enzymes characteristic of different microorganisms makes possible target-oriented selection of a strain capable of affecting chemical bonds present in macromolecules of a given polymer. For instance, destruction of polyamides requires proteolytic enzymes produced by many *Bacillus* bacteria (e.g., *B. subtilis* and *B. mesentericus*) [4,5] as well as by some other microorganisms. Decomposition of polyester surface macromolecules can be achieved under the action of enzymes called lipases, given off by many microscopic fungi (e.g., *Aspergillus niger*) and micromyceta [1,5,6]. More rarely such constitutive enzymes occur as dehalogenases, which take away halogen atoms from macromolecules; oxireductases, capable of destroying aromatic rings; transferases, which can catalyze the transfer of various groups from one substrate to another; and so forth.

However, for practical modification of polymer surfaces there is no need to select for every material its "own" specific strain of microorganisms. An easier way seems to be to use adaptive enzymes, which the microorganisms begin to produce gradually under scarcity of nutrition and/or energy to adapt themselves to a substrate that is new for them. High adaptive ability is inherent in many bacteria, abundant in nature, such as *Bacillus* and, to an extreme degree, *Pseudomonas* [4,5]. These bacteria are capable of using polyamides as the only source of carbon and nitrogen; they can destroy polyacrylate, raw and vulcanized rubber, and many other polymeric materials. No worse is the adaptive variability of some microscopic fungi, among which emphasis should be placed on the *Aspergillus* family, which shows the highest enzymatic potential [5].

"Adaptation" of pure strains of microorganisms to a new substrate is time consuming (it takes, as a rule, several weeks). However, when microorganisms are taken from a polymer surface already damaged by them, these adaptive strains are much more active. They settle faster on a new surface and cause deeper decomposition [4,5]. The strongest decomposer of synthetic substances is the microflora of active silt from sewage. This is due to the presence in silts of a great variety of microorganisms distinguished by high adaptability to the products of chemical synthesis [5]. It should be noted that the adaptive variability is often reversible, i.e., it is not fixed genetically. With a change in the medium, a microorganism may lose its ability to biosynthesize an enzyme.

III. EFFECT OF THE NATURE OF POLYMER ON THE EFFICIENCY OF THE MICROBIAL MODIFICATION

The resistance of different polymers to the action of microorganisms depends, first of all, on the chemical structure of polymer macromolecules. Natural polymers are subject to microbial attack to the highest degree, because their macromolecules are usual nutrition and energy sources for a large number of bacteria and microscopic fungi.

As to synthetic materials, there are great variations in their behavior on exposure to microorganisms. As a rule, microbes adapt more easily to macromolecules that contain fragments similar in structure to those of natural polymers. For instance, the analogy between the molecular structure of proteins and polyamides enables one to use microorganisms that produce proteolytic enzymes for polyamide decomposition [1,5]. In both cases, the degradation proceeds according to the following scheme:

$$—NH—C— + H_2O \rightarrow —NH_2 + HO—C—$$
$$\quad\quad \overset{\|}{O} \quad\quad\quad\quad\quad\quad\quad\quad\quad \overset{\|}{O}$$

Indeed, as demonstrated by numerous experiments [1,4–10], polyamides are susceptible to microbiological corrosion to the utmost extent. In a similar way, polymers containing ester bonds, e.g., poly(ethylene terephthalate) (PET) and poly(methyl methacrylate) (PMMA), may degrade under the action of lipases produced by many microorganisms [1,5]. Hydrolysis proceeds in a similar way as in natural fats:

$$—\overset{|}{C}—O—C— + H_2O \rightarrow —\overset{|}{C}—OH + HO—C—$$
$$\quad\quad \overset{|}{} \quad \overset{\|}{O} \quad\quad\quad\quad\quad\quad\quad \overset{|}{} \quad\quad\quad\quad \overset{\|}{O}$$

The amines, alcohols, and acids produced in the hydrolysis of polyamides and polyesters are consumed by the microorganisms. Polymer materials such as polyethylene, polypropylene, and fluoroplastics that do not contain active atomic groups that are habitual for microbes are more resistant to biochemical action [3–6]. The polymer resistance to microbial attack is especially high when the material contains —NO$_2$, —Cl, —Br, —I, or —SO$_3$H groups or phosphorus atoms. In such cases, even treatment with microorganisms for several months does not result in appreciable changes [4,5].

Nevertheless, there is no synthetic polymer that can resist every biochemical attack. Even glass fabrics are known to decay in free air under the action of microscopic fungi [1]. The large variety of microorganisms and their ability to adapt to synthetic substances make possible, in principle, modification of all types of polymers.

IV.   FACTORS INFLUENCING THE CONTROL OF MICROBIOLOGICAL ACTION

The polymer treatment with microorganisms does not require any special equipment or complicated techniques. A typical procedure is to place a polymer sheet, film, or fibers into an aqueous suspension of microorganisms and keep it at 25–30°C for a certain period—from several days to several weeks [2–13]. To speed up the process, aeration is desirable, as well as addition of nutrient substances and trace elements needed for the microorganisms' metabolism. Of course, pathogenic microbes should not be used. After the treatment, the specimen is sterilized in boiling water or in an autoclave and then dried.

The first stage of the microbiological action is the sticking (adhesion) of microbes to the polymer surface. Causes and mechanisms of this adhesional interaction are not yet understood in detail; nevertheless, some factors that have an effect on it have been revealed. It was demonstrated that gram-positive bacteria interacted with polymer surfaces more actively, and the adsorption of gram-negative bacteria was less. Besides, a relationship between the adhesion and the value and sign of the electrical charge of the surface has been shown. Consequently, positively charged polymer fibers are characterized by enhanced adhesion to many microorganisms, whereas negatively charged ones practically do not interact with them [4]. The presence of sizing agents on the polymer surface is also important. Sized fibers adsorb several times more microorganisms than unsized fibers do [4].

The intensity of decomposition can be increased by increasing the treatment time and/or employing adaptive strains. The structure of polymers also plays an important role. Surface modification takes place in most cases, but further destruction is possible only when pores and capillaries of sufficiently large size for products of metabolism to penetrate internal layers of the polymer are present at the surface. It should be noted that the size of microorganisms is typically 1–

10 μm [5] and that of enzyme molecules about 10^{-2} μm [14]. The number and size of defects on the polymer surface depend on production conditions as well as on the polymer structure. Any procedure leading to structure ordering, e.g., orientational drawing or thermal treatment accompanied by an increase in the degree of crystallinity, enhances polymer resistance to microbial treatment. Thus, thermally treated poly(vinyl alcohol) (PVA) fibers are much more resistant than freshly made ones [4,15], because the latter show an irregular structure as well as a high degree of pores and hollows. Polishing a polymer surface protects the internal layers against the penetration of the products of metabolism. Fibers that have a smooth, regularly structured surface, e.g., polyester fibers, are also susceptible to surface corrosion only [4,8].

Molecular inhomogeneity of polymers also influences the degree of their biochemical destruction. For instance, an increased content of low-molecular-weight additives and oligomers suppresses the microbiological resistance of polymer materials. A polymer's ability to swell can also be of great importance. The destruction of cellulose fibers begins only after interfibrillar swelling causes diffusion of large enzyme molecules into the fiber's internal layers [16]. Intensive biochemical decomposition of polycaproamide (PCA) is also facilitated by its good swelling ability [4,6].

Of course, an important approach to controlling biochemical modification is the application of various microorganisms. In Refs. 17 and 18, the effects of a wide variety of bacteria and microscopic fungi on PCA fibers were investigated. It was demonstrated that the mechanism of the microbial action was identical in all cases, namely enzymatic hydrolysis of amide bonds. However, electron paramagnetic resonance (EPR) and infrared (IR) study of modified fiber surfaces revealed that not only microtopography and the degree of degradation but also the chemical nature of the surfaces was substantially different depending on the microorganism species. In some cases, active amino groups were present predominantly at the fiber surface, in other cases, carboxyl ones. As a result, fibers differed in their wetting and adhesional ability [18].

V. CHANGES IN MICROTOPOGRAPHY AND CHEMICAL STRUCTURE OF POLYMER SURFACES INDUCED BY MICROORGANISMS

Scanning electron microscopy (SEM) study of polymer surfaces exposed to microbial action revealed the general avenues and the main stages of this process, despite the fact that its particular manifestations can differ depending on the nature of the polymer and the microorganism. The action of a microorganism begins at the site of its attachment to the surface. For instance, a hollow or pore is formed under the mycelium of a microscopic fungus as the effect of enzymes or other products of metabolism [1]. With time, more and more pores and cracks are developed on the surface, turning it rough and nonuniform [2,4]. The initial attack is aimed at amorphous regions and intercrystalline layers. As the fibers degrade, their internal structure (often microfibrillar) is exposed. For example, an 8-day exposure of Kevlar fibers to *Aspergillus flavus* fungi gives rise to development of short cracks parallel to the fiber axis at its surface, and in 30 days fibrils having a very smooth surface are exposed [7]. Fibrillation is also characteristic of some other fibers, e.g., poly(acrylonitrile) (PAN), PVA, PCA [4], and cellulose [15]. It should be noted that the surface microtopography can be different even for fibers having a similar chemical nature when different microorganisms have been employed (Fig. 1a–c). A long-duration exposure (several months) results in splitting of the fibrils and in the appearance of deep cracks (Fig. 1d).

Pisanova et al. [10,11] investigated the surfaces of various aramid fibers by means of secondary electron emission. It was shown that after treatment with *Pseudomonas* and *Bacillus* bacteria, the real surface area of the fibers decreased and the mean inclination of asperities with respect to the fiber surface decreased. At the same time, the number of asperities per unit area increased (by three to six times); i.e., irregularities became smoother but their surface density increased. Thus, large hills disintegrate under the action of microorganisms, and hollows (pores, microcracks) are, in all probability, filled with the products of metabolism, so the characteristic size of asperities on the whole decreases.

By means of selective analysis of the fiber microtopography, the increases in the relative surface area and surface asperity densities corresponding to their different sizes have been determined. An important characteristic of the topography is the surface fraction, δ, occupied by irregularities of a certain diameter, d. It is defined as the product of the characteristic area of the asperity and the number of such asperities, N per unit area (i.e., surface density corresponding to the asperity diameter d):

$$\delta = \frac{\pi d^2}{4} N \cdot 100\%$$

The δ-d dependence is presented in Fig. 2. As one can see, poly-*p*-amidobenzimidazole (PABI) and polyimide (PI) fiber microtopography is altered in different ways upon microbial treatment, although there are common effects, too. In both cases, the total area of large (~ 1–$2 \ \mu m$) asperities increases. These asperities are well discernible in the microphotographs (Fig. 1); they are more than likely products of the microorganisms' metabolism strongly bound to the surface. Besides, for both types of fibers the number of small (0.1–0.3 μm) structural elements increases, although the spectra of irregularities are different for PABI and

PI fibers. In the case of polyimide, a uniform growth of δ in the range 0.07–0.56 μm is observed, whereas PABI fibers are distinguished by a predominant increase of the fraction of the smallest (~0.1 μm) asperities. In these fibers, asperities about 0.3–0.6 μm in size are practically not affected by the microbial treatment; and in PI fibers, the fraction of irregularities of size between 0.6 and 1.2 μm even decreases.

Thus, a microbiological treatment of organic fibers results in smoothing of small asperities accompanied by a simultaneous increase in their number. Besides, new structural elements of size 1–2 μm appear at the surface due to possible nonuniformity of the microbial treatment.

Additional information on structural changes in surface layers of the fibers has been obtained with the EPR technique. The correlation time of nitroxyl radical at the surface of untreated polyimide fibers was 9.1×10^{-10} s. The fiber treatment with the bacteria *Bacillus cereus* reduced the correlation time to 7.5×10^{-10} s, and with *Pseudomonas putida* to 2.4×10^{-10} s. The decrease in the radical correlation time evidences improved mobility of the segments of polymer chains, resulted from loosening of the fiber surface structure in the range of the smallest asperities. This result is consistent with the increase in the fraction of asperities of size 0.07–0.14 μm (Fig. 2). Besides, the fraction of even smaller structural elements probably increases, too.

However, the real surface area of treated polymers not only decreases but can also increase depending on the species of microorganism used and the treatment conditions. For instance, an increase in the real surface area has been observed [19] in the case of exposure of aramid fibers to *Bacillus* bacteria (Fig. 3). Under these conditions, fiber adhesion to both polyarylate and polycarbonate was also improved.

As a result of microbial treatment, not only the microtopography of polymer surfaces but also their chemical structure undergo changes. It has been demonstrated by IR spectroscopy that biomodified polymers (of various structures) reveal, as a rule, increased intensity of absorption bands in the region of 1650–1710 cm^{-1}, corresponding to C$=$O and —COOH group oscillations, as well as in the region 3300–3500 cm^{-1}, attributed to —OH group vibration [4,9,11,13]. For PCA, the appearance of absorption at 3000 cm^{-1} has been reported, which provides evidence for amino group formation [4,9]. In the spectra of rubber treated with micromyceta, the intensity of absorption at 1200, 1400, and 1460 cm^{-1} corresponding to —CH— and CH$_2$— group vibrations decreases significantly, but bands corresponding to OH— group valent oscillations appear [1]. A change in the intensity of the band at 445 cm^{-1} due to breaking of C—C links in the main polymer chain was also noted [4]. Thus, the IR data provide evidence for predominant destruction and oxidation of macromolecules accompanied by the appearance of active groups, such as —NH$_2$, —OH, C$=$O, —COOH, at

(a)

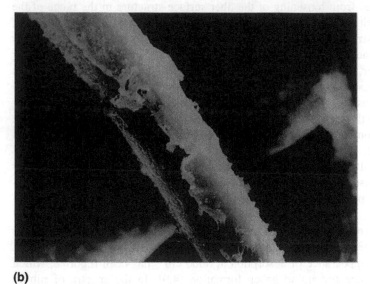

(b)

Figure 1 Surfaces of poly-*m*-phenyleneisophthalamide (a and b), poly-*p*-phenyleneter-ephthalamide (c) and polyimide (d) fibers modified with bacteria *Pseudomonas putida* (a, c, d) and fungi *Aspergillus flavus* (b). Exposure time: 2 weeks (a–c); 8 weeks (d).

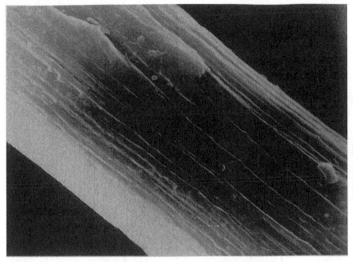

(c)

(d)

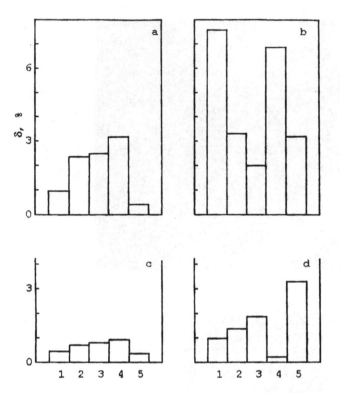

Figure 2 Fractions of PABI (a and b) and PI (c and d) surface area occupied by asperities of diameter 0.07–0.14 μm (1); 0.14–0.28 μm (2); 0.28–0.56 μm (3); 0.56–1.12 μm (4); 1.12–2.25 μm (5). (a and b) Original fibers; (c and d) fibers modified with bacteria *Pseudomonas putida*. (From Ref. 10.)

the polymer surface. The presence of carboxyl groups was also demonstrated by means of fluorescence analysis [13].

The amino groups produced are, as a rule, consumed by the microorganisms, but the carboxyl groups remain on the polymer surface [9]. Depending on the composition of the medium, they may react with different substances present in the solution. For instance, when ions of metals (Ca, Na, Fe, Mg, Al) were added to the nutrient liquid, they deposited on the polyamide surface. Metal ions were then detected in the polymer with a laser microanalyzer [9,11,13].

X-ray diffraction analysis demonstrates that the degree of order of the polymer structure decreases with biochemical treatment [2,4,9]. With longer expo-

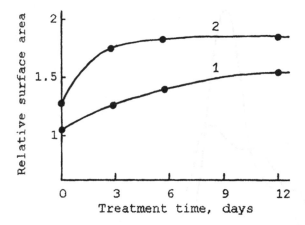

Figure 3 Relative surface area for PPIA (1) and PABI (2) fibers versus the time of treatment with *Bacillus cereus* bacteria. (From Ref. 19.)

sures, gradual disorientation of crystallites is observed. By means of X-ray analysis it is possible to distinguish between surface and bulk modifications [9]. Differential thermal analysis also provides information on changes in the polymer structure. For instance, upon exposure to *B. subtilis* bacteria the melting temperature of PCA fibers is lowered by 10°C and the decomposition temperature by 30°C (Fig. 4a).

Biochemical modification is not restricted to the destruction of polymer macromolecules. Adsorption of products of the microorganisms' metabolism at the surface is also possible. Moreover, it is known that immobilized enzymes are able to synthesize various complex substances, e.g., esters [20,21]. Some microorganisms produce thermoplastic polymers, which then accumulate in the substrate pores [21]. In these cases, with time the polymer surface becomes covered with a thin layer of a substance having a chemical nature different from that of the source polymer. Such layers can be created artificially by adding to the nutrient solution chemical compounds capable of grafting onto active groups exposed on the surface. An example of target-oriented fiber modification is the treatment of aramid fibers with microorganisms in the presence of poly(vinyl alcohol) [9–13]. It was shown that the first stage of this process was enzymatic hydrolysis of amide groups:

$$—NH—\underset{\underset{O}{\|}}{C}— + H_2O \rightarrow —NH_2 + HO—\underset{\underset{O}{\|}}{C}——$$

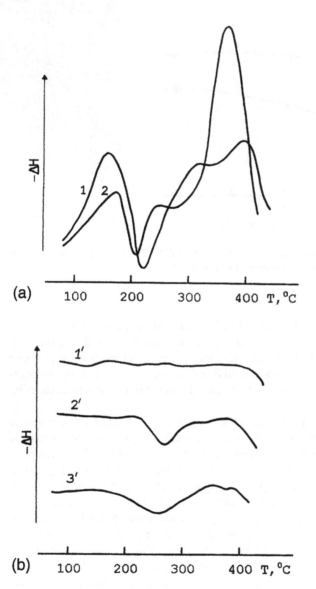

Figure 4 Thermograms of PCA (a) [4] and PPIA (b) [13] fibers: untreated (1, 1') and modified with bacteria *Bacillus subtilis* (2, 2') and fungi *Aspergillus niger* (3'). Treatment time: 6 months (a); 3 weeks (b). (From Refs. 4 and 13.)

Then PVA macromolecules are grafted onto carboxyl groups produced at the fiber surface; i.e., the etherification reaction proceeds:

$$\underset{\overset{\|}{O}}{-C}-OH + HO-\overset{|}{\underset{|}{C}}- \rightarrow -\underset{\overset{\|}{O}}{C}-O-\overset{|}{\underset{|}{C}}- + H_2O$$

As a result, properties of the fiber surface undergo considerable alteration, but the strength and internal structure of the fibers remain unchanged [9–12]. New peaks at 250–270°C are present in thermograms for modified fibers, demonstrating that new organic compounds have been synthesized at the surface (Fig. 4b). In the IR spectra of modified fibers, bands characteristic of PVA appear as well as absorption bands of ester groups (1028, 1244, and 1080 cm^{-1}). At the same time, the intensity of the bands attributed to amide group vibration is decreased [9,11,13].

Thus, there are two substantially different methods for biochemical modification of polymers: enzymatic destruction of macromolecules, and coating the surface with a thin layer of grafted polymer. These processes have different effects on the strength of polymeric materials.

VI. EFFECT OF MICROORGANISM ACTION ON POLYMER STRENGTH

For the purpose of surface modification, it is important to determine the optimal conditions of biochemical treatment. The degree of destruction depends on the nature of the polymer and its structure as well as on the microorganism species and nutrient medium composition. The treatment time also plays an important role. In the first stages of the treatment only surface effects take place; but with time changes deeper in the polymer structure are observed, accompanied by loss of strength. So, damage to the surface layers, swelling, filamentation, and finally total decomposition are sequentially observed for a fiber [4].

The degradation of internal structure causes loss of strength of fibers as well as of bulk specimens [1,2,4]; it also results in faster swelling and better solubility of the treated polymers [9,11]. For example, exposure of PCA to *A. niger* fungi for 21 days gives rise to considerable loss of its strength and elastic modulus in static bending as well as to a decrease in surface hardness by 40%. Under identical conditions, polyethylene (PE), polypropylene (PP), and polystyrene (PS) practically do not undergo changes in their strength properties. The degree of swelling of PE, PP, and PS is below 1%, whereas that of PCA about 7% [6]. On exposure to *Aspergillus awamori* fungi, PMMA loses about 35% of its strength [2]. It has been shown that one of the results of the fungal action is an increase in moisture adsorption at the polymer surface; after this, the rate of diffusion of the microorganisms' metabolic products into the polymer material increases and biocorrosion speeds up.

In spite of similar general rules of biodecomposition, the time needed to switch from surface to bulk corrosion differs considerably for different polymer-fiber pairs and may be from several days to several months. For bulk polymers, because of their low-ordered structure, surface modification is often accompanied by internal degradation and strength loss. On the contrary, the structure of many types of organic fibers is distinguished by high order; therefore, fibers can easily be surface modified without deterioration of their strength [9–11]. For instance, the strength of PCA fibers decreases only by 20% after an 8-week treatment with *B. cereus* bacteria [11], and PAN and PET fibers lose practically no strength after a 4-week exposure to a set of specially selected active microorganisms (the test set) (Fig. 5). Even when adaptive strains of the species *B. mesentericus* and *Bacterium herbicola* are employed, the fiber tensile strength and rupture strain do not undergo considerable changes within the first month of the treatment (Fig. 6). On the contrary, in cases in which adsorption of the metabolic products by the fiber surface is possible, an increase in the fiber strength is often observed (Table 1), provided that the treatment time is no longer than 2–3 weeks [9–12,18,22]. This phenomenon can be explained by "healing" of the cracks on the fiber surface. The failure of the fibers is mostly initiated by surface flaws, so their healing during the microbial treatment increases the mean fiber strength and, at the same time, reduces the scatter in the strength values. As a result, polymer composites reinforced with modified fibers are characterized by improved strength [11,19,22].

VII. IMPROVEMENT OF ADHESIONAL ABILITY OF POLYMER SURFACES SUBJECTED TO BIOCHEMICAL TREATMENT

Hills, hollows, and microcracks developing at the polymer surface as a result of biochemical treatment, as well as alterations of its chemical structure, in particular, emergence of various active groups, improve the adhesional ability of modified polymers. So far, most of the investigations in the field of microbiological modification have been concerned with organic (polyamide, polyester, etc.) fibers used as reinforcing elements in composite materials. This is for two main reasons. First, there is an urgent need for surface modification of many types of organic reinforcing fibers, especially when solvent-free technologies for their combination with polymer matrices are to be employed, because of their poor wetting by

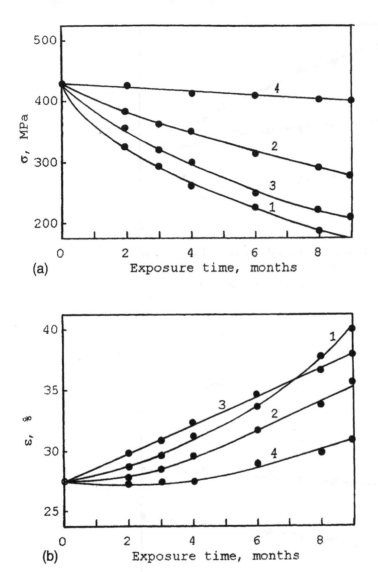

Figure 5   Tensile strength (a) and elongation to break (b) for poly(vinyl alcohol) fibers versus the time of treatment with *Bacillus mesentericus* (1), test set of bacteria (2), and *Bacterium herbicola* (3); control fibers (4). (From Ref. 4.)

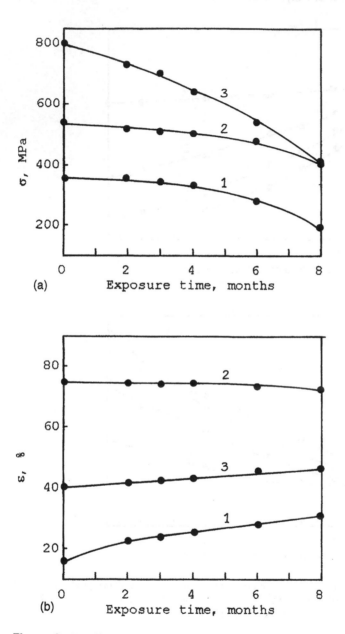

Figure 6 Tensile strength (a) and elongation to break (b) for PAN (1), PET (2), and PCA (3) fibers versus the time of exposure to the test set of bacteria. (From Ref. 4.)

Table 1 Characteristics of Untreated and Modified Polycaproamide (PCA) Fibers

Microorganism[a]	Tensile strength (GPa)	Elongation to break (%)	Contact angle (degree) of molten HDPE[b]	Bond strength in the PCA/ HDPE system[b] (MPa)
—	1.52	16.0	35	13.1
B. subtilis	1.61	18.5	32	15.8
P. putida	1.67	17.9	31	15.0
B. vulgaris	1.74	17.8	31	21.0
B. megaterium	1.74	18.2	32	17.6
B. cereus	1.79	20.2	29	20.0
B. cereus[c]	1.23	15.5	25	19.9

[a] B., Bacillus; P., Pseudomonas. Treated for 2 weeks.
[b] Thermal treatment conditions: 180°C for 15 min.
[c] The treatment time was 8 weeks.

polymer melts and low adhesional ability with respect to matrix polymers. Second, thin organic fibers are a good model material for microbial treatment because of their specific structure. Because of the high degree of order of their internal layers, biochemical modification is restricted to the surface, and the strength of the fibers remains the same or is even increased to some extent [9–12]. In Table 1, properties of PCA fibers exposed to microbiological treatment are given. For not very long exposure times (2 weeks), most of the *Bacillus* bacteria have modified the fiber surface so that its wetting by the binder polymer melt has improved and the bond strength in the composites (measured by the fragmentation technique [11,18]) has increased. The most pronounced effect was achieved when *Bacillus vulgaris* or *B. cereus* was used. In all cases, the fiber strength was improved to some extent; however, increasing the treatment time to 8 weeks resulted in a loss of strength (Table 1).

Similar results were obtained for other fiber-matrix systems as well. For instance, the action of microorganisms on poly-*p*-amidobenzimidazole (PABI) and poly-*m*-phenylisophthalamide fibers gave rise to an increase in their adhesion to polyarylate (by 21 and 36%, respectively) [23].

As a rule, the action of microorganisms results in improved wetting of the fiber surface by polymer melts (Fig. 7) [10,12,22].

Table 2 presents the results of fiber-polymer adhesional bond strength measurements by the pull-out method (pulling a fiber out of a polymer matrix droplet). The adhesional bond strength increases considerably with microbial treatment of the fibers. This effect is especially pronounced when the treatment is conducted in a medium containing poly(vinyl alcohol). In this case, a thin layer of grafted

(a)

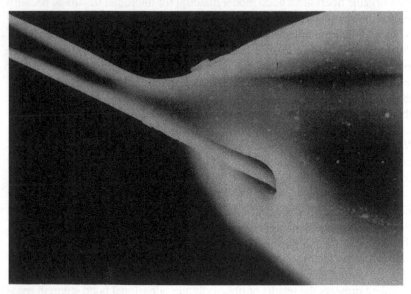

(b)

Figure 7   Wetting of PABI fibers untreated (a) and modified with *Bacillus mesentericus* (b) with polysulfone melt. (From Ref. 12.)

Table 2 Properties of Untreated and Modified PI and PABI Fibers

Fiber	Microorganism	Diameter (µm)	Mean strength (GPa)	Bond strength in the fiber/PC[a] system (MPa)[b]
PI[a]	—	15.7	0.78	39.3
PI[a]	*Caulobacter bacteroides*	15.7	0.72	51.4
PABI[a]	—	13.5	3.76	45.7
PABI[a]	*Pseudomonas putida*	13.5	3.67	57.0
PABI[a]	*Bacillus cereus*[c]	13.5	4.06	86.1

[a] PI, polyimide; PABI, poly-*p*-amidobenzimidazole; PC, polycarbonate
[b] Thermal treatment conditions: 275°C for 15 min
[c] In a nutrient medium containing poly(vinyl alcohol).
Source: Ref. 11.

polymer is formed on the fiber surface [9,11,22], increasing not only the adhesion but also the fiber tensile strength.

The improved fiber adhesion can be attributed to both increased real fiber-matrix contact area due to a fiber surface area increase and chemical affinity of polyesters for active groups formed at the surface as a result of the microbial treatment [9–13]. The SEM study of the fiber and matrix surfaces after failure of fiber-matrix contacts has demonstrated that there were no traces of the matrix polymer on the surface of source fibers; i.e., the failure was always interfacial. At the same time, composites containing modified fibers often showed cohesive failure in the fiber (separation of outer fibrils, exfoliation of fiber skin) or in the matrix (Fig. 8) [12,22]. A major proportion of fiber-matrix composites showed a mixed failure mode evidencing good interfacial bonding.

These findings are further corroborated by the results of mechanical testing of plastics containing organic fibrous reinforcement. The fiber surface treatment with microorganisms intensifies fiber-matrix interactions at the interface and thus improves the strength of the composites. As seen in Table 3, thermoplastic polymers reinforced with modified fibers are superior to composites containing nontreated fibers in basic mechanical properties [23,24]. In particular, fibrous composites based on microbially treated reinforcement exhibit higher strength in quasistatic and dynamic loading, impact strength, and durability to wear, including wear at high-speed sliding [23–25]. The mode of the composite failure also undergoes changes. Nontreated fibers are typically pulled out of the matrix with their surface clean; but on modification a major portion of the load is carried by the interlayer and the fibers, which fail without being pulled out [19,24,26]. Figure 9 shows typical failure surfaces of a polymer composite containing original (a) and modified (b) fibers. Data obtained by means of optical and electron micro-

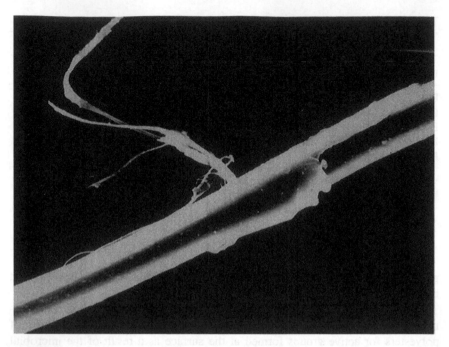

Figure 8 Surface of PABI fibers modified with *Bacillus mesentericus* bacteria after failure of their joint with polycarbonate. (From Ref. 12.)

Table 3 Some Properties of Fiber-Reinforced Plastics

	Composite					
	PPIA/polyarylate		PABI/ polycarbonate		PABI/polyarylate	
Characteristic	Original fibers	Treated fibers	Original fibers	Treated fibers	Original fibers	Treated fibers
Density, g/cm³	1.30	1.34	1.41	1.46	1.31	1.33
Tensile strength, MPa	115	170	95	150	130	185
Compressive strength, MPa	195	300	135	240	220	310
Ultimate compressive deformation, %	12	25	18	29	19	28
Impact strength, kJ/m²	80	135	110	215	90	150

Source: Refs. 23 and 29.

(a)

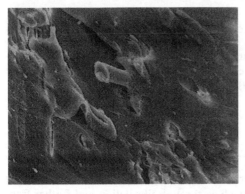

(b)

Figure 9 Typical failure surfaces of PPIA fiber-polyarylate matrix composites: (a) untreated fibers; (b) fibers modified with fungi *Aspergillus oryzae*. (From Ref. 26.)

scopies are consistent with those from the acoustic emission investigation of the composite failure. Composites containing modified aramid fibers generate 30 to 40 times more acoustic pulses than untreated fibers, and total emission energy increases by 10–15 times. This provides evidence for more fibers being involved in stress transfer in the composites, thus improving their mechanical properties [24,26,27].

In Fig. 10, the compressive stress and the acoustic emission (AE) rate are plotted against the deformation of the composites based on original and modified poly-*m*-phenyleneisophthalamide (PPIA) fibers and polyarylate matrix. The considerable increase in the composite compressive strength upon microbial treatment of fibers is due to the change in the mode of composite failure, which can

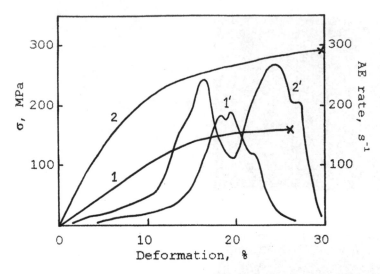

Figure 10 Compressive stress (1, 2) and acoustic emission rate (1′, 2′) versus compressive deformation for PPIA/polyarylate composites with original fibers (1, 1′) and fibers modified with *Bacillus subtilis* (2, 2′). (From Ref. 24.)

be deduced from the AE data. The appearance of the second peak in curve 2′ and the increase in the total number of AE events provide evidence, according to Ref. 24, for failure of fibers themselves through fibrillation and further fragmentation of microfibrils. The increase in the number of total AE events is also attributed to considerably improved bond strength at the fiber-matrix interface, which results in complex composite failure under compressive loading, including simultaneous cohesive failure of the matrix and the reinforcement as well as failure at the interface. Investigation of the microtopography of failure zones by means of secondary electron emission corroborates that there are different failure mechanisms for composites containing original and biochemically treated fibers [19,26,27].

Microbiological modification facilitates combining the polymer binders with reinforcing fibers (continuous as well as short-cut ones). Because of the improved wetting and impregnation of modified fibers and fabrics by polymer melts, it is possible to carry out the combining process at lower temperatures and pressures as well as to reduce processing time and to increase the fiber content in the composites obtained [25,28]. The increase in the adsorptional ability of organic fibers after biochemical treatment gives rise to better adhesion at the fiber-matrix interface [28].

High-strength plastics based on thermoplastic matrices (polycarbonate, polyarylate, polysulfone) and biochemically treated fibers can be produced by extrusion, molding, and compression molding. A comparative study [28] has shown that the strongest plastics are those obtained by compression molding with a fiber content of 37–40% and fiber lengths in the range of 10–12 mm.

VIII. CONCLUSION

Biochemical (microbiological) treatment is a simple and efficient way to alter the surface properties of polymers. At present, the major field of application of microbial modification is the surface treatment of aramid reinforcing fibers to facilite their combining with thermoplastic matrices and improve the properties of the composites obtained. Because this process is very sensitive to the microorganism strain, the chemical nature of the fiber, and the composition of the nutrient medium, a wide-ranging, target-oriented variation of the fiber surface properties, without degradation of the strength of the fibers, can be obtained. In many cases, the treatment with microorganisms results in an improvement of the adhesional bonding between fibers and thermoplastic matrices, thus providing a way to obtain reinforced plastics with improved characteristics. The main application of these plastics is in machine building (load-carrying parts, bearings in vehicles, electrical insulation, etc. [25,28]).

REFERENCES

1. AY Lugauskas, LI Levinskajte, DK Lukshajte, DN Pechulite. Plast Mater (USSR) No 2:24, 1991.
2. VN Kestelman, VL Yarovenko, EI Melnikova. Int Biodeterion Bull 8:15, 1972.
3. VN Kestelman, YG Naumov, SN Negmatov, G Blohm. Plaste Kautsch 25:545, 1978.
4. IA Ermilova. Theoretical and Practical Basis of the Microbiological Destruction of Chemical Fibres. Moscow: Nauka, 1991.
5. MN Rotmistrov, PI Gvozdyak, SS Stavskaya. Microbiological Destruction of Synthetic Organic Substances. Kiev: Naukova Dumka, 1975.
6. VN Kestelman, VA Belokurova. Microbiol Synth (USSR) No 3:3, 1969.
7. T Watanabe. J Jpn Soc Fibre Sci Technol 43:192, 1987.
8. IA Ermilova, LN Alekseeva. Text Ind (USSR) No 9:55, 1981.
9. AI Sviridenok, TK Sirotina, EV Pisanova. High Mol Weight Compd (USSR) Ser B. 31:571, 1989.
10. EV Pisanova, VS Detsuk, MB Vainshtein, EV Voevoda, VG Grishchenkov, VV Meshkov. Chem Fibres (USSR) No 3:18, 1993.
11. EV Pisanova, SF Zhandarov. J Adhesion Sci Technol 9:1291, 1995.
12. AI Sviridenok, TK Sirotina, EV Pisanova. J Adhesion Sci Technol 5:229 1991.

13. AI Sviridenok, TK Sirotina, VV Meshkov. USSR Acad Rep. 298:666, 1988.
14. CR Cantor, PR Schimmel. Biophysical Chemistry. Vol 1. San Francisco: WH Freeman, 1980, p 12.
15. IA Ermilova, EY Danilova, VP Mazovetskaya, LA Wolf, EM Makarova. Micro-Organisms as Destructors of Materials and Articles. Moscow: Khimia, 1979, p 64.
16. B Philipp. High Mol Weight Compd (USSR) Ser A 23:3, 1981.
17. AI Sviridenok, IM Rinkevich, AI Artsukevich. Belarus Acad Sci Rep 38:52, 1994.
18. AI Sviridenok, VV Meshkov, IM Rinkevich. Belarus Acad Sci Rep 39:113, 1995.
19. AI Sviridenok, TF Kalmykova, VV Meshkov, TK Sirotina. Belarus Acad Sci Rep. 34:327, 1990.
20. ML Lubareva, RS Barshtein. Plast Mater (USSR) No 7:30, 1987.
21. PA Holmes. Gummi Fasern Kunstst 40:377, 1987.
22. AI Sviridenok, TK Sirotina, EV Pisanova, SF Zhandarov. Mech Composite Mater (USSR) No 5:771, 1991.
23. AI Sviridenok, VV Meshkov, TK Sirotina, EV Pisanova. Proceedings of Polymer Composites '90. Leningrad: Science House Publishers, 1990, pp 3–5.
24. VV Meshkov, OV Kholodilov, AI Sviridenok. In: MSJ Hashmi, ed. Proceedings of the International Conference on Advances in Materials and Processing Technologies (AMPT '93). Vol II. Dublin City: Dublin University Press, 1993, pp 935–943.
25. VV Meshkov, AI Sviridenok, EV Pisanova. Proceedings of the European Meeting of the Polymer Processing Society (Stuttgart). Munich: Carl Hanser Verlag, 1995, p 6.7.
26. AI Sviridenok, AY Grigoryev, VV Meshkov, TK Sirotina. Mech Composite Mater (USSR) No 3:444, 1989.
27. AI Sviridenok, VV Meshkov, OV Kholodilov, IS Philatov. Proceedings of the 7th European Conference on Fracture, Budapest, Vol 1, 1988, pp 459–466.
28. AI Sviridenok, VV Meshkov, EV Pisanova. Proceedings of the 24th International Meeting on Chemical Engineering and Biotechnology (ACHEMA '94). Frankfurt am Main: DECHEMA, 1994, pp 108–109.
29. VV Meshkov. Unpublished results, 1996.

12

Silanes on Glass Fibers—Adhesion Promoters for Composite Applications

Leanne Britcher, Scott Kempson, and Janis Matisons
University of South Australia, Mawson Lakes, South Australia, Australia

I. INTRODUCTION

Glass-reinforced composites, based on synthetic resins such as phenolics, ureas, melamines, and unsaturated polyesters, generally became available in the 1940s, when dramatic improvements in strength-to-weight ratios were first observed. Nowadays the crucial importance of composite products in a number of areas, particularly in aircraft and marine applications, is accepted, as today's materials prove superior to their aluminum or steel counterparts. Although the specific dry strength and modulus of recent reinforced composites exceed those of aluminum or steel, upon prolonged exposure to humid environments, a dramatic decrease in these properties can still sometimes occur. Early composite products were very susceptible to moisture, even in ambient humidity. The bonding of polymeric materials to inorganic surfaces such as glass has been intrinsic to many products, but the interfacial bond strength and durability remain susceptible to moisture during processing. Yet it is this same interfacial bond strength that is also linked to the product's acceptance and eventual success in the marketplace [1].

The commercial glass fibers used in reinforced composites are almost always pretreated with a sizing or coupling agent, which is capable of interacting with both the organic polymer resin and the inorganic oxide substrate [1]. Such a coupling agent must not only ensure that the physical properties of the reinforced material remain relatively unaffected by moisture or humidity but also reduce the stress concentrated at the interface during thermal cycling. Trialkoxysilanes (see Fig. 1), which contain organic groups compatible with the polymer resin, are the most commonly used coupling agents or adhesion promoters. Addition of almost any trialkoxysilane coupling agent will improve the water resistance of the reinforced composite. However, it is important to realize that such silanes are usually applied as oligomeric siloxanes (oligomerization taking place in solution several hours prior to application to the substrate). Oligomerization can also occur while transporting silanes over vast distances or during storage. Small amounts of incipient moisture inside sealed silane containers initiate oligomerization. The oligomerization can be controlled if the silane is supplied shortly after

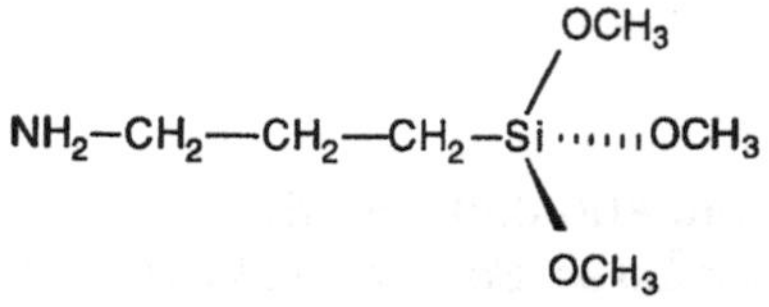

γ-aminopropyl triethoxy silane

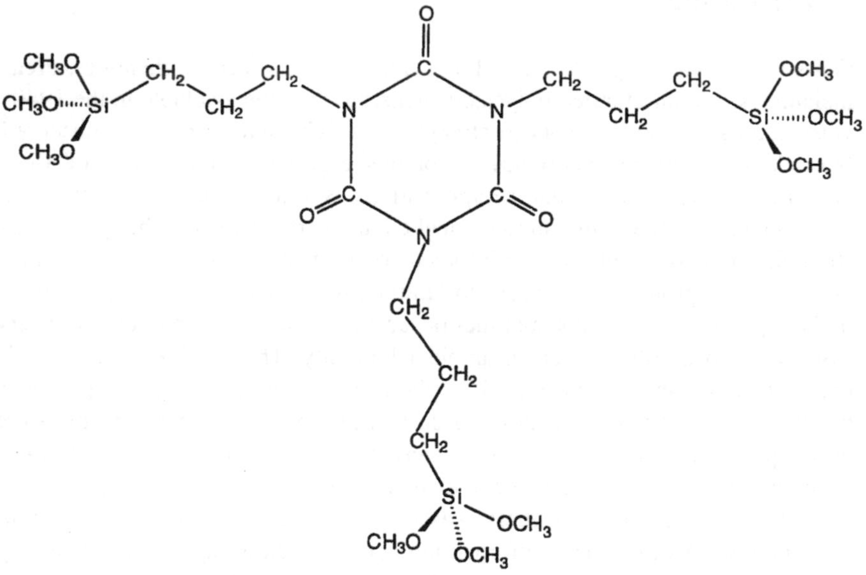

1,3,5-tris[(trimethoxysilyl)propyl]isocyanurate

Figure 1   Two very different silanes.

its manufacture and no opportunity for partial oligomerization and/or cross-linking exists. Silane coupling agents are, however, manufactured in only a few countries. For this reason, when coupling agents are supplied to industry, varying degrees of oligomerization and cross-linking may already have occurred (it is intriguing to monitor by nuclear magnetic resonance (NMR) the vast difference in industrial materials supplied in 200-liter drums against the chemical samples supplied to scientific research laboratories).

Coupling agents have been defined as materials that "promote adhesion between mineral and organic phases." [2]. Plueddemann [1] more specifically defined them as "materials that improve the chemical resistance (especially to water) of the bond across an interface" and further stated that, although any polar groups on the polymer will help to promote adhesion, the use of organofunctional silane coupling agents appears to induce optimal adhesion. Historically, silane coupling agents have developed as the optimal interfacial adhesion promoters, so much so that the generic term "coupling agent," used on its own, usually refers to a silane coupling agent.

Silane coupling agents have hydrolyzable alkoxy groups that are able to interact with surfaces and organofunctional groups able to interact with the polymer matrix or resin, so providing a direct, water-resistant interface between the surface and polymer [2]. The siloxanes that form, as silane coupling agents oligomerize, are also very effective as polymeric coupling agents [3,4]. Figure 1 shows the structure of two silanes; γ-aminopropyltrimethoxy silane, a commercially available silane coupling agent containing three alkoxy groups, and a commercially available trifunctional isocyanurate silane containing nine alkoxy groups [5]. No coupling agent has been so studied as γ-aminopropyltriethoxy silane, yet even now many important aspects of this widely used silane remain controversial or even undetermined.

Many theories have been proposed to explain how silane coupling agents attach to surfaces [6–9]. Adsorption of silanes is a complex phenomenon that depends not only on the silane used but also on the substrate and the treating solution. It is therefore appropriate to take a brief look at two contrasting silanes, such as γ-aminopropyltrialkoxy silane and 1,3,5-tris[(trimethoxysilyl)propyl] isocyanurate (see Fig. 1), and then consider the most common substrate in composite applications, E-glass fibers, before finally considering the treating solution and the effect this has on silane and siloxane adsorption.

II. INTERFACIAL REAGENTS—A TALE OF TWO SILANES

The scientific literature abounds with studies (several thousand, in fact) that have focused on γ-aminopropyltrialkoxy silanes and the influence of pH, silane concentration, temperature, and curing conditions on adsorption. Yet controversy still follows even this much-studied silane, and the application of such silane coupling agents is still viewed as a "black art," to be successfully practiced by those who have acquired sufficient experience to do it properly.

The 1,3,5-tris[(trimethoxysilyl)propyl] isocyanurate (referred to herein as the isocyanurate silane; Fig. 1) has yet to be scientifically studied, although it has been commercially available for several years. It is almost as if researchers are unsure how to unlock the potential inherent in this unusual silane. A systematic investigation of the properties of the isocyanurate silane has now commenced, and indeed our initial studies [10–12] reveal that the silane does show unusual adsorption behavior, a behavior that is highly dependent on the solvent from which the silane is applied. For instance, Fig. 2 reveals scanning electron microscopic (SEM) evidence indicating that 1,3,5-tris[(trimethoxysilyl)propyl] isocyanurate forms different bridging structures between glass fibers, structures that are highly dependent on the solvent from which the trialkoxy silane is applied to the fibers. Applying the isocyanurate silane from aqueous solution results in the typically smooth surface morphology associated with aqueous silane treatments of glass fibers (Fig. 2a) [10]. Unfortunately, aqueous application of isocyanurate silane solutions also results in very low silane adsorption to the glass fibers, and it is the low silane adsorption that has probably limited the use of this particular silane in any composite application. However, applying 1,3,5-tris [(trimethoxysilyl)propyl] isocyanurate from CCl_4 (Fig. 2b) [10], toluene (Fig. 2c) [11], and ethanol (Fig. 2d) [12] results in bridging siloxane structures linking adjoining glass fibers, as well as greatly increased silane adsorption. The bridging structures, however, are different for each solvent from which the silane is applied. Silane adsorption from toluene (Fig. 2c) [11] reveals very fine bridges between adjacent glass fibers. Adsorption from ethanol (Fig. 2d) [12] reveals massive siloxane bridging units between fibers, whereas from CCl_4 (Fig. 2b) [10] the isocyanurate silane forms bridges with dimensions between these two extremes. Such studies have shown that the isocyanurate silane can not only align glass fibers in solution but also break the alignment by simply stirring the aligned fibers in another solvent (such as wet acetone). The ability to align or disperse fibers randomly within an organic matrix has tremendous significance for the mechanical performance and reinforcement of polymeric composites.

The other silane in Fig. 1, γ-aminopropyltrialkoxy silane, adsorbs significantly from aqueous solutions onto various substrates [13]. The molecular structure of the thin films formed by aminosilane deposition onto surfaces (such as iron, aluminum, and copper) from alkaline solution has been monitored and differentiated by Fourier transform infrared (FTIR) spectroscopy. The smooth films initially formed (similar to the scanning electron micrograph shown in Fig. 2a) later adsorbed carbon dioxide and water vapor to form amino bicarbonate salts, characterized by new FTIR bands appearing at 1330, 1470, 1570, and 1640 cm^{-1} [13]. The metallic substrates used in the study significantly influenced the aminosilane adsorption, as the FTIR band intensities varied with the different substrates. The orientation of γ-aminopropyltriethoxy silane was controlled by the solution pH, so the interfacial film formed was stable as long as the pH was between the isoelectric point of the surface (oxide) and the pK_a of the ionizable silane group [14,15]. Iron surfaces are basic, so at pH 8 acidic silanol groups adsorb to the surface, generating moisture-stable siloxane films. However, in alkaline solution, adsorption occurs through the amino groups, generating less stable interfaces. The films formed by γ-aminopropyltriethoxy silane deposited from acidified solutions are different from those deposited at the pH of the silane itself (pH 10.4) [15]. Acidified solutions promote silane polymerization generating cationic amino siloxanes.

To investigate complex phenomena, several instrumental techniques should be used, but this is not always possible. Investigations by X-ray photoelectron spectroscopy (XPS) of γ-aminopropyltriethoxy silane adsorbed onto silica surfaces shows two nitrogen components (N $1s$ signals), attributed to free and protonated amino groups [7,16,17]. Washing with acidic or basic solutions does allow some of the amino groups to be converted from one form to another; however, some amino groups also remain irreversibly protonated, whereas still others remain irreversibly free [16]. Glass surfaces are thought to adsorb aminosilanes (or even aminosiloxanes) through the amino groups rather than the silanol groups, although both the protonated and free amino groups are again clearly seen [3,7]. The potential for aminosilanes to align glass fibers through hydrogen bonding of the free amino groups has not been systematically researched. One could envisage that fiber alignment could be disrupted simply by stirring in acidified solutions, thereby converting some of the free amino groups to protonated amino groups. The protonated amino groups would help disperse fibers randomly in solution, as charge repulsion would keep adjacent fibers apart. In such instances, a different mechanism would operate for aligning or dispersing fibers randomly in a solution (from that for the isocyanurate silane), even though the significance of controlling the fiber orientation for improving mechanical performance and reinforcement may not have diminished.

Each silane is commercially available, yet both have very different technological histories. Both, however, have the potential for improving composite reinforcement through reversible fiber alignment, although different interfacial behavior would exist in each case, i.e., for the isocyanurate silane and the aminosilane. In spite of the thousands of publications on aminosilanes and the controversies that have followed the adsorption behavior, the potential for using aminosilane-treated surfaces to align fibers in polymer resin matrices remains unknown. The isocyanurate silane, with no extant scientific studies, also holds tremendous potential for improving composite reinforcement, yet, in this instance, even the optimal methods for applying the silane to various surfaces have not been established. It seems some basic questions now need to be systematically

(a)

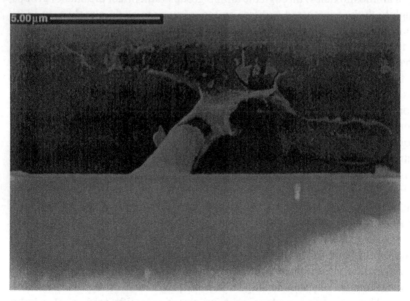

(b)

Figure 2   (a) Isocyanurate silane applied to E-glass fibers from water; (b) isocyanurate silane applied to E-glass fibers from CCl₄; (c) isocyanurate silane applied to E-glass fibers from toluene; (d) isocyanurate silane applied to E-glass fibers from ethanol.

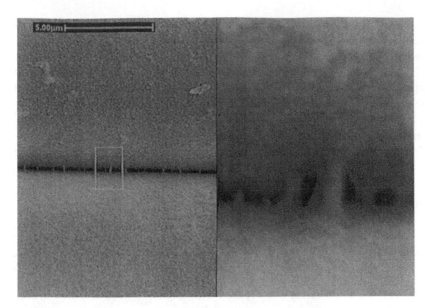

(c)

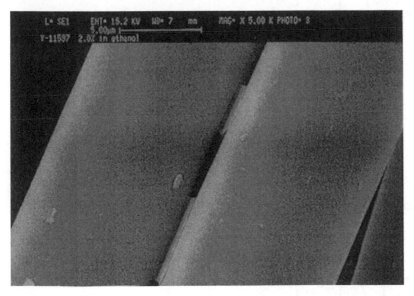

(d)

addressed, if the extensive research into silane coupling agents is to meet real industrial needs or to achieve a molecular understanding of the complex interplay of factors that determine silane adsorption or adhesion to different substrates.

The deleterious effects of water on the mechanical properties of polymeric composites incorporating untreated fibers or fillers are now well documented [1,18–21]. Diffusion and subsequent reaction of water at the filler-polymer interface is responsible for interfacial delamination. Silane coupling agents, however, generate a water-resistant interface between the organic polymer and the inorganic substrate, as they are able to react or interact with both the glass surface

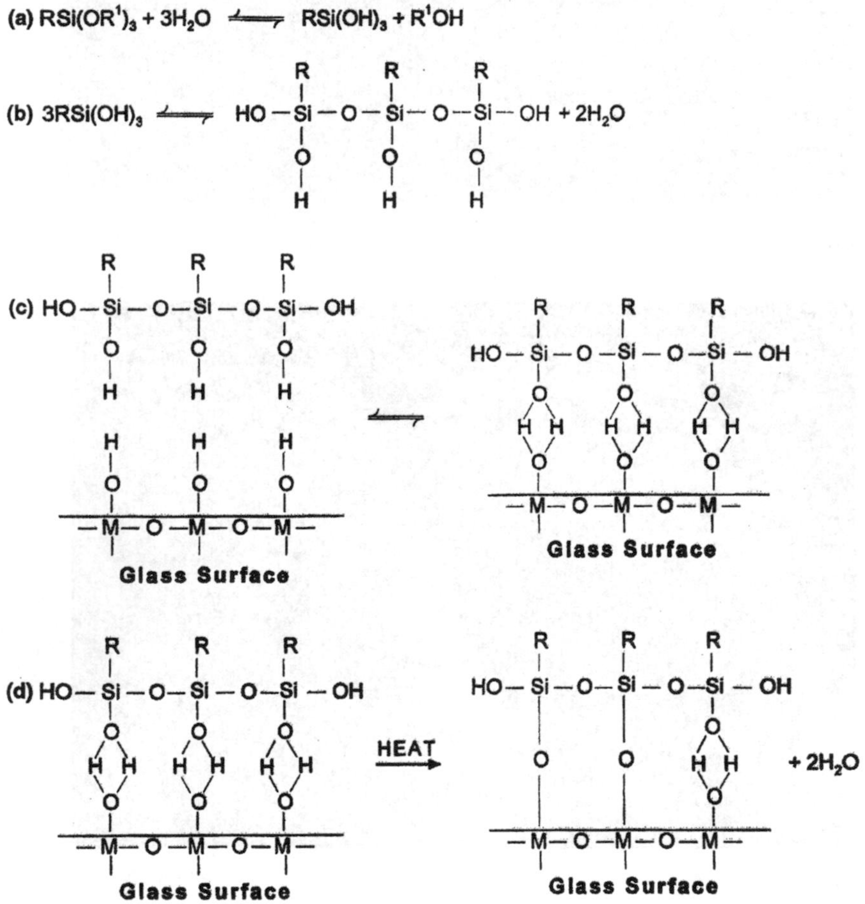

Figure 3 Alkoxysilane coupling agents on glass: (a) hydrolysis of alkoxy groups; (b) condensation; (c) hydrogen bonding with glass silanols; (d) surface bond formation.

and the polymer [22–28]. Even if only a small amount of coupling agent is added, the performance of the resultant composite is improved [13–15]. Almost always, the coupling agents are applied from dilute aqueous solutions, partial hydrolyzates, or organic solvents (generally an alcohol) [1,22–28], and most have undergone initial hydrolysis and oligomerization prior to interacting with the chosen substrate [25,26]. The silanes may interact with the substrates initially through hydrogen bonding to surface hydroxyl groups, with subsequent condensation reactions generating siloxane structures (see Fig. 3). It is also possible, in some systems, that lateral polymerization occurs without the formation of bonds to the surface (see Fig. 3) [29]. The siloxane film formed on the substrate then consists of multiple layers [23,27,28].

It is tempting (and commonly done) in the scientific literature to presume that silane coupling agents are in fact sizing agents for glass fibers. They could be, but generally, on their own, they are not. In fact, they are rarely even the principal component of glass fiber sizing formulations. Glass fiber sizes require more than a simple coupling agent to achieve optimal processing, so studies that investigate silane adsorption or adhesion to various substrates are much too simplistic even to model industrially relevant situations. In a general sense, academic research has continued to investigate silanes on their own, while industrial technology has leaped ahead and the patent literature seldom has sizing formulations that contain only silane coupling agents. Therefore, before proceeding to examine the glass substrate, it is better to consider what glass fiber sizing agents actually are.

III. GLASS FIBER SIZING AGENTS—A TALE WITH MANY PARTS

Industrial situations deal with sizing agents, not just simple silane coupling agents. Silanes do play a key role in sizing agent formulations, but in most instances, they may not even be the major component of the sizing agent formulation. Using molecular approaches alone for understanding sizing behavior is dangerous, because silanes are rarely used alone as the sizing agent. Most commercial sizes are complex proprietary compositions (see Table 1). The general proportions of components in a commercial size, excluding the solvent or carrier used to apply the size, are also listed in Table 1.

A molecular understanding of the complex factors that determine silane adhesion to different substrates becomes even more complex when investigating sizing agent formulations. Systematic experiments are required to make timely progress in developing sizing formulations, where the interactions between the various components may be almost impossible to understand in detail. Problems arise, however, when product performance changes significantly with seemingly

Table 1 Typical Components of a Glass Fiber Size

Component	Percent
Film-forming resin or binder	1–5
Antistatic agent	0.1–0.2
Lubricant	0.1–0.2
Coupling agent (e.g., silane or siloxane)	0.1–0.5

minor adjustments in size composition or with a variable that was not part of initial experiments. It can also be very difficult to achieve incremental improvements in composite performance with a size that has already been optimized for a list of other specifications. Changing product specifications with increasing technological advances drive formulators to make trial-and-error adjustments in existing formulations that eventually become accepted in company research programs as a "black art" practiced by a few experienced professionals. Such events are often the incentive to study the silane alone in university research laboratories, as the silane is usually added to improve composite properties, whereas most of the other components are present to optimize production and processing.

Sizes that are applied to glass fibers are multicomponent systems. The composition of a silane coupling agent–based size is complex, with the coupling agent constituting a relatively small proportion of the material applied to the fibers. The constituents of typical sizing systems (see Table 1) are compatible and *must interact* with each other to be effective. The resulting distribution of the mixed components in the interphase region is not understood, and the formulation of effective sizing systems is an empirical undertaking for the most part. The processing aids present in the size may change the interphase properties in a manner similar or even counter to that of the silane on its own.

The interaction of various silane coupling agents with silica and, to a much lesser extent, glass surfaces, at the chemical or molecular level, has been extensively studied. The structure of physisorbed and chemisorbed silane coupling agents at silica and glass surfaces has been investigated via FTIR [29], solid-state NMR [30–33], and XPS [34–37]. The interaction of silane coupling agents with glass surfaces extends beyond the monolayer range, and the silanes adsorb in multilayers where possible. The presence of covalent bonding has bolstered the chemical bonding concept developed by Plueddemann [1], in which coupling agents act as a bridge between the glass surface and the polymeric matrix.

Plueddemann's chemical bonding theory [1] has not been able to account for the increase in adhesion experienced between unreactive matrices (such as polyolefins) and various inorganic reinforcements, where chemical bonds are not thought to form. Such composites have led to an alternative proposal that a third phase, called the "interphase," forms around the reinforcing filler. This third phase arises by interdiffusion of physisorbed silane and polymer molecules in an interphase region [1] and possibly also by preferential adsorption of both silane and polymer components on the surface. The mutual proximity of the silane and the polymer at the interface has critical implications for both the mechanical properties associated with the formation of a glass-polymer interphase and the way that glass fiber sizes may adsorb during fiber processing. The alteration of interphase mechanical properties near the reinforcement surface, through a synergistic silane-polymer interaction, can explain many of the phenomena associated with glass fiber sizing agents. For instance, the increased hydrolytic stability of treated glass-polymer composites can occur through hydrolyzable bonds, between the silane and the glass surface, that can break and reform easily enough to relieve stress at the interface and preserve the composite integrity [38]. Interaction with a polymeric film-forming agent (see Table 1) then maintains the integrity of the interphase region, preventing water ingress while the stress is being relieved. Such arguments support an interphase model in which the molecular structure of the interphase is often considered to be an interpenetrating network between the silane, the polymer matrix, and the polymeric film formers inherent in the sizing formulation. The current interphase model, unfortunately, does not incorporate polymeric film-forming agents but simplistically considers only polymer matrix-silane interactions.

Plueddemann [1] noted that the diversity of applications for silane coupling agents precluded any single explanation for their efficacy at improving composite properties, mentioning chemical reactivity, interpolymer network formation, and interphase modification as important mechanisms. Investigators have silane-treated glass fiber surfaces, generating submonolayer coverage using both mono- and trialkoxy silanes, to produce a relatively hydrophobic glass surface that still shows improvement in overall composite properties, even in the presence of water [39,40]. The silane deposition does lower the interfacial energy, thereby diminishing the tendency of water to migrate from the polymer matrix to the glass surface [41].

Silane coupling agents can be modified in a number of ways to impart specific characteristics to glass fiber surfaces. First, the number of hydrolyzable alkoxy groups can be varied, to influence the properties of the resultant silane film on the glass surface, through the number of bonds that can be made by each silane molecule to the surface and the degree of cross-linking between the individual silane molecules [42]. The electron-withdrawing ability of the organofunctional group on the silane affects not only the types of polymeric substances the silane can interact with but also the reactivity of the silane alkoxy groups [2,43]. In this way, the organofunctional groups exert a strong influence not only on the reactivity of the silane with various surfaces but also on the ultimate properties of the modified surface.

Silanes can also modify the acid-base characteristics of glass surfaces, enhancing the secondary glass-polymer attractive forces without necessarily promoting direct reaction of silane with the polymer resin. Fowkes et al., [7,44] improved the toughness of cast polymer films by simply modifying glass powder with silanes to enhance interfacial acid-base interactions. Osmont and Schreiber [45] describe the acid-base properties of E-glass fibers treated with various silanes by the ratio of the specific retention volumes of n-butanol and n-butylamine (Lewis acid and base, respectively). Heats of adsorption for acidic and basic probes on bare and silane-treated glass beads can be measured [46,47], and resolved into dispersive and acid-base enthalpy components, from which the Drago C and E coefficients can be estimated (for acidic surface sites). Tsutsumi and Ohsuga [48] also investigated the changes in dispersive and electron donor-acceptor properties of glass fibers after silane treatment.

Many theories have been proposed over the years to explain the molecular mechanisms of silane coupling to glass surfaces; none of these addresses sizing agents. Even the adsorption and adhesion of silane coupling agents to glass surfaces are complex phenomena that depend on the specific silane, the glass, and the treating solution. Most of the fundamental studies have focused on the aminosilanes and the different influences of solution pH, silane concentration, and "curing" conditions. Silica has been used in several studies, but really silica remains another oversimplification (of the more complex, E-glass surface). It seems too difficult for many researchers to investigate and understand the role of specific glass components in the surface coupling mechanisms of complex sizing formulations.

Although the subjects of silane chemistry and its interaction with both the glass surface and the polymer matrix have been extensively studied, little fundamental information of a predictive nature relating sizing application and composite mechanical properties has been established. Sizing application remains therefore a "black art," yet a necessary one, so that acceptable and reproducible mechanical performance results from fiber-reinforced composites. For the past 12 years, the Polymer Science Group at the University of South Australia, as part of several commercial research programs, has investigated the glass fiber-sizing-polymer interphase encountered in commercial fiber "sizing" systems. The industrial research seeks to provide a sound basis for the development of a scientific framework upon which a predictive methodology can be established.

IV. GLASS SURFACES—COMPLEX SUBSTRATES

Glass fibers are still the most common reinforcement used in composite materials. Glass compositions, however, vary with intended application [49–52]. Soda-lime glass (see Table 2), composed primarily of the oxides of silicon, sodium,

Table 2 Constituents of Commercial Glasses

Component	Soda-lime glass (%)	Pyrex (%)	E-glass (%)	S-glass (%)
SiO_2	70–75	80–86	52–56	64–66
CaO	7–10	—	16–25	0–0.3
Al_2O_3	0–1.5	0–2	12–16	24–26
B_2O_3	—	6–18	5–10	—
MgO	0–4	—	0–5	9–11
Na_2O	10–13	2–8	0–2	0–0.3
K_2O	0–1	—	0–2	0–0.3
TiO_2	—	—	0–0.8	—
Li_2O	—	0–1	—	—
SO_3	0–0.5	—	—	—
Fe_2O_3	0–0.2	—	0.05–0.4	0–0.3
F_2	—	—	0–1.0	—

Source: Refs. 49–52.

and calcium, is commonly used for bottles and containers. Pyrex, a borosilicate glass, has a high resistance to thermal shock due to the presence of boron oxide, making it suitable for laboratory and kitchen glassware. E-glass fibers, the most common type of glass employed in textiles or reinforced composites, are also borosilicates. S-glass fibers, however, are aluminosilicates and are used primarily for high-performance materials that require fibers with very high tensile strength. The minor inorganic oxide components are added not only for economic and production considerations but also to control and/or modify some of the glass properties. Calcium and aluminum oxides control or improve the expansion, durability, and chemical resistance of the glass [49–52]. Alkali metal–oxides are added to reduce the melting temperature and viscosity of the glass by disrupting the continuity of the silica network (i.e., breaking some of the Si—O bonds). However, alkali oxides also lower the chemical resistance of the glass. The silica network is retained within the multicomponent silicate glass as it forms, and the resident nonbridging oxygen atoms provide the necessary charge balance for the glass cations. Overall, glass surfaces then reflect the balance of performance expectations that must be met for optimal processing and materials properties, at minimal cost.

The surface concentrations of the various inorganic oxides vary from the bulk glass composition, depending on the thermal history of the glass, the relative humidity during processing, and the surface treatment to which it has been subjected after melting and cooling [51–56]. The strength of a glass fiber is influenced by the nature of its surface, which is defined as the first 0.1 μm^2. Components that lower the surface free energy diffuse toward the surface (surface segregation) of the glass while in the molten state. Hydrolysis and leaching of the alkali and alkali earth metal silicates, together with the volatilization of the alkali oxides (such as Na_2O and B_2O_3) during glass melting and cooling to room temperature, determine the final glass surface composition [53–59].

The chemical heterogeneity that then exists may not only vary from manufacturer to manufacturer but also be determined by the glass history (i.e., how it has been cooled, cleaned, and stored). Thus, in making comparisons between studies of chemisorption on various E-glass fibers, it is important to know the exact surface chemistry of each substrate being compared. First, chemical micro-heterogeneities could exist across the glass fiber surface, which would complicate analysis [60]. Second, there are a number of geometric forms for the same substrate, i.e., plates, cylinders (fibers), and spheres (powder), and each geometric form would have a different surface composition, since each geometric form would have a different thermal history associated with its particular shape.

Silicate glass surfaces contain primarily surface silanol groups ($\equiv$Si—OH), although other glass formers are also present on the surface, such as Al—OH and B—OH [49–52]. Silica surface reactions are usually used to represent glass surface chemistry. However, the differences between glass and silica are significant. Molecular water physisorbed onto silica has O—H vibrations (FTIR) at $\sim$3450 cm^{-1} and 1250 cm^{-1}. Heating silica in vacuo to 150°C for several hours results in both O—H FTIR bands disappearing, with two new FTIR bands appearing at 3660 cm^{-1} and 3747 cm^{-1}. The band at 3660 cm^{-1} arises from silanol groups, close enough to be hydrogen bonded, and the remaining FTIR band, at 3747 cm^{-1}, arises from isolated silanol groups. The latter band at 3747 cm^{-1} is the only band remaining after heating silica to 400°C for several hours. Above 425°C, hydrophilic silica becomes hydrophobic, as all the surface silanols are removed. E-glass fibers, however, have only a very broad O—H band at $\sim$3450 cm^{-1}, ascribed to physisorbed water. Heat treatment reduces the size of the band, but it never completely disappears. The FTIR differences seen between various silanol sites on silica are not seen on glass, as other inorganic oxide O—H sites obscure specific O—H surface interactions. E-glass has an advantage in FTIR analysis, because the first overtone of the B—O vibration occurs adjacent to the C—H stretching vibration region [61].

FTIR studies indicate that water adsorbs on silica through hydrogen-bonded silanol groups, whereas silanes or amines adsorb through isolated silanol groups [62–65]. The density of isolated Si—OH groups on a silica surface is 1.4 groups per nm^2; that of hydrogen-bonded silanol groups is 3.2 groups per nm^2. Precipitated, acid-washed silica has a surface area of 134 m^2/g [66]. Heat-cleaned E-glass cloth (0.12 m^2/g) and E-glass microfibers (0.79 m^2/g) have much lower surface areas [67]. The silica surface area is much greater than that for E-glass fibers. However, the density of surface silanol groups for either heated E-glass cloth (7 SiOH per nm^2) or E-glass microfibers (13 SiOH per nm^2) is much greater than that seen for precipitated silica [67]. Furthermore, the isoelectric point of a glass surface occurs at a neutral pH, allowing both acidic and basic reagents to interact or adsorb, whereas the isoelectric point of silica occurs at an acidic pH generally allowing only basic (or at best neutral) reagents, such as γ-aminopropyltriethoxy silane, to adsorb rapidly. Given this perspective, it is sometimes hard to see how silica surfaces could be considered as models for glass surfaces.

Thin surface films of "low" refractive index form on glass surfaces immediately after manufacture (as detected by optical measurements). The very fine silica film, between 11 and 35 nm thick, arises from the loss of alkali oxides through volatilization and the subsequent hydrolysis and leaching of the alkali and alkali earth metal silicates into solution [68,69]. Such films have chemical and physical properties different from those of the bulk glass and retard further bulk glass deterioration by presenting a barrier through which component ions must diffuse before being lost into solution. The thickness and density of the surface film vary with glass composition, time, temperature, and pH of the sizing solution. A less durable glass has a thicker film than a more durable glass.

What becomes abundantly clear is that water is present in multiple layers on E-glass fiber surfaces, and this water plays a key role in the diffusive transport of ions in the glass. The outgassing of glass in vacuo really represents the diffusion-controlled removal of dissolved water from bulk glass through a surface silicate film. The effects of relative humidity on such glass surfaces can be monitored by contact angle measurements using methylene iodide (a non-hydrogen-bonding organic liquid) [70]. A contact angle of 13° occurs at 1% relative humidity, and a contact angle of 36° occurs at 95% relative humidity (similar to the contact angle of 36° for methylene iodide on water) [70]. However, clean silica surfaces (only surface silanol groups present in ultrahigh vacuum) show a methylene iodide contact angle of just 10° [70]. As water vapor is slowly admitted into the vacuum chamber, the water coverage of these silica surfaces gradually rises, and the contact angles increases from 11 to 20°. If the contact angle rises above 20°, water multilayers have formed on the silica. Subsequent adsorption isotherm studies have confirmed the contact angle conclusions [70].

Glasses at 1–50% relative humidity gradually form a monolayer of adsorbed water on the surface. Above 50% relative humidity, multilayers of water form, until a thick film develops that resembles bulk water at 95% relative humidity. Composite isotherms can be constructed from studies in different pressure regions [70]. Between 10^{-3} and 10^0 torr, corresponding to 0.005–5% relative humidity, such isotherms indicate that only a fraction of a water monolayer remains, but whether this residual moisture is due to molecular water or glass surface silanols is uncertain.

The corrosion of alkali-silicate and alkali-lime-silicate glasses through exchange of the monovalent cations, R^+, with water, has been described by the following reactions [71–73]:

The penetration of a "proton" from water into the glassy network, replacing an alkali ion:

$$\equiv Si—OR + H_2O \leftrightarrow \equiv Si—OH + R^+ + OH^- \tag{1}$$

The hydroxyl ion then disrupts the siloxane bonds in glass:

$$\equiv Si—O—Si \equiv + OH^- \leftrightarrow \equiv Si—OH + \equiv Si—O^- \tag{2}$$

The nonbridging oxygen formed in reaction (2) interacts with a further molecule of water, producing a hydroxyl ion which is free to repeat reaction (2) again:

$$\equiv Si—O^- + H_2O \leftrightarrow \equiv Si—OH + OH^- \tag{3}$$

Reactions (1)–(3) represent the three classes of leaching reactions that can occur: hydration, ion exchange, and hydrolysis. The surface leaching reactions are distinguished by

The extent to which water dissociates
The bonding character of the reactive site (ionic versus covalent)
Whether material is removed from the surface

Modeling such reactions is a complex task. Most glasses contain several potential sites for ion exchange, and the relative abundance of these sites also varies (chemical heterogeneities).

As the hydrogen ion is smaller than other monovalent cations, replacement of an alkali ion in reaction (1) will produce tensile stress at the glass surface, causing enhanced reactivity to further hydration, which will eventually result in cracks forming on the glass surface. Reaction (1) is energetically unfavorable. The hydration energy for H^+ to form H_3O^+, however, is favorable, being large and negative (~ -367 kcal/mol). The SiO groups bonded to the higher valency cations do not generally exchange with hydrogen ions, as the higher valency cations are tightly bound to two or more SiO groups and can play only a minor role in such surface reactions [71,72]. The oxide groups of glass are linked together into "covalent polymer networks" [73]. The metal cations that form these polymers are called network formers, and all tend toward tetrahedral or trigonal oxygen coordination (however, aluminum is octahedral). Metal-oxygen bonds among oxide sites have approximately 50–60% covalent character and link to form a wide range of polymer structures.

The structure of a hypothetical silicate glass is shown in Fig. 4. The glass structure consists of linear silicate chains with a cross-link density of ~ 0.2; i.e., the chains are bonded to each other in approximately one out of every five tetrahe-

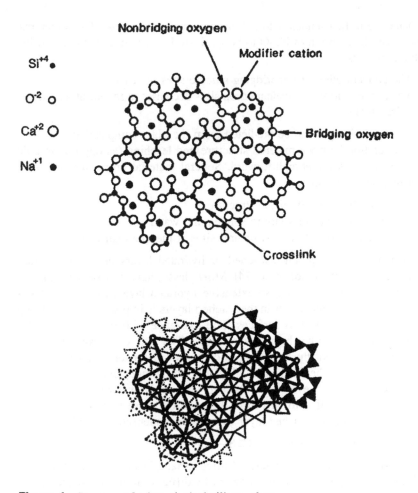

Si^{+4} •

O^{-2} ○

Ca^{+2} ○

Na^{+1} •

Figure 4 Structure of a hypothetical silicate glass.

dra. Calcium ions are coordinated to nonbridging oxygen atoms, which have an unsatisfied negative charge (not immediately apparent in Fig. 4). Metal cations (such as Na^+ or Ca^{2+}) that form ionic rather than covalent bonds to polymeric oxygens are called network modifiers. Network-modifying cations perform two related functions in glass structures: they coordinate to nonbridging oxygen atoms branching from the polymer and they compensate the unbalanced charge stemming from the isostructural substitutions in a polymer. In boro- and aluminosilicates, the unbalanced charge is caused by substitution of trivalent boron and aluminum for tetravalent silicon. Furthermore, network-modifying cations derive their name from their tendency to polymerize or depolymerize glass structures. For instance, the addition of Na_2O to borosilicate E-glass induces the following structural changes:

> The concentration of nonbridging silicate sites increases.
> The concentration of trivalent boron sites decreases in favour of tetrahedral boron sites.

Such structural changes can be predicted from the glass composition.

The relation between material structure and leaching (as opposed to bulk dissolution rates) is not straightforward. Several material properties determine the leaching characteristics:

> The cross-link density of the structure
> The structure transmissivity to solutes and water
> The reactivity of different structural sites at the glass surface

Rates of reactions that produce leached or hydrated layers on silicate glasses vary considerably with temperature [74]. Most glasses have experienced extreme temperatures during diagenesis, so extensive hydrated layers form, and understanding the reactions that occur in the leached layers is important to identifying adsorption processes on glass surfaces. The existence of free water molecules inside the leached layer has been reported [73]. The ratio of entrant ''protons'' to water molecules inside the leached layer changes with the temperature of leaching and the nature of the exchanged alkali ion but is independent of the alkali content of the glass. However, it is not clear whether the water molecules in the leached layer have really diffused from the solution phase as H_2O or H_3O^+ or formed in situ inside the leached layer by an autocondensation reaction (4):

$$\equiv Si—OH + HO—Si \equiv \rightleftharpoons \equiv Si—O—Si \equiv + H_2O \qquad (4)$$

Reaction (4) is well known on a hydrated silica surface. Water may be responsible for relocation of oxygen ions in the network by diffusive transport. A succession of transport steps occur, such that a given ion is alternatively stationary (i.e., as part of the network) or mobile (as part of an interstitial water molecule). The effects of hydration range, therefore, from simple penetration of water into the glass to profound disruption of the glass structure. Factors that control the extent and rate of hydration include

> The structure and size of any pores in the glass
> The degree of covalent connectivity in the polymeric structure
> The hydration energies of cations ionically associated within the structure

Most glasses are not transmissive to water, and so hydration does not extend to a great depth without parallel reactions that relax the glass structure. Materials that are dramatically affected by hydration tend to be held together by ionic forces rather than covalent cross-links. The ionic forces diminish as the ions between the polymers become solvated. In the extreme case, hydration causes dissolution into polymeric fragments. Clearly, glass surfaces are much more dynamic than silica surfaces and vary greatly from the bulk glass composition, depending on the thermal history, subsequent treatment, and even the relative humidity.

Chemisorption of silanes upon a glass substrate will vary substantially between different geometries, such as plates, fibers, and powders. Indeed, chemisorption is affected by the diameter differences in various fibers [75]. The glass surface area not only influences the chemical properties of the surface but also determines the sensitivity required by the surface analytical technique used to study the adsorption processes (the smaller the surface area, the greater the sensitivity required). For all these reasons, studies of the adsorption of small molecules, such as silane coupling agents, have focused on large surface area, homogeneous particles, such as silica. Taking all these factors into account, silica is simply not an adequate model for glass surfaces any longer, and using silica as such will produce serious errors in optimizing glass fiber sizing formulations.

V. SIZING ADSORPTION ON GLASS—BALANCED PERSPECTIVES IN COMPLEX PHENOMENA

Science always needs to maintain a balanced perspective. The plethora of scientific literature on silane coupling agents continues to expand simply because balanced perspectives are not maintained. The time for considering adsorption of silanes as simplified model systems (generally from aqueous solutions) is now past. Industrial applications demanding complex formulations are now commonplace, although the systematic scientific study of such formulations may not yet have commenced. Having therefore considered two silanes, as part of sizing formulations, and then proceeded to consider one particular substrate, E-glass fibers, attention now will focus on the treatment solution and how that affects adsorption processes. It would be useful to consider the kinetics of adsorption onto glass fibers of the different components in a sizing formulation, but to the best of our knowledge this has never been attempted. In the absence of such data, the kinetics and mechanism of silane adsorption to glass surfaces will have to suffice.

Reaction of silane coupling agents with glass surfaces is generally accepted as occurring in two steps: hydrolysis and condensation. Hydrolysis involves the conversion of the alkoxy groups on the silane into silanol groups by reaction with water as shown in Fig. 3. The silanols are very reactive, so in a sense the silane alkoxy groups do in fact serve to protect silane coupling agents from forming siloxane bonds prematurely or oligomerizing during storage [42]. Siloxanes can, however, form before or during silane interaction with a surface, depending on the application solvent and conditions employed. Hydrolysis rates also influence the rate of silane reaction with surfaces, and as such reactions are usually carried out using water or organic solvent–water mixtures, hydrolysis is rapid and subsequent silane film formation is also rapid [76]. Solution studies using high-field ^{13}C and ^{29}Si NMR to monitor the kinetics of silane adsorption have not been done as often as might be anticipated, given the power of the technique [77–83]. Furthermore, submonolayer concentrations on surfaces can now not only be detected but also quantitatively monitored with both solid-state and liquid NMR spectroscopies measured simultaneously [84].

One of the first studies of the homogeneous hydrolysis of a trialkoxy silane coupling agent monitored the progress of a hydrolysis reaction by measuring a change in optical absorbance at 220 nm [85]. Based on this initial work and subsequent data, a kinetic equation, Eq. (5), was developed, describing the rate of total hydrolysis over a wide pH range and monitoring specific acid and base catalysis. At low pH the hydrolysis occurs in three steps, with each succeeding alkoxy group being removed faster than the previous group [58]. At high pH, however, further hydrolysis of the first intermediate is inhibited due to the ionization of the initial Si—OH group.

$$-d[S]/dt = k_{spon}[H_2O]^n[S] + k_H[H^+][H_2O]^m[S] \qquad (5)$$
$$+ k_{HO}[OH^-]^o[H_2O]^p[S] + k_B[B][H_2O]^q[S]$$

The reaction order for spontaneous hydrolysis, n, and for acid-catalyzed hydrolysis, m, depends on both the type of alkoxy group being hydrolyzed and the solvent conditions [86,87]. Where water is used in conjunction with an organic solvent, the solution polarity and the pK_a need to be precisely determined. The solvent affects the order of the reaction. More water is required for spontaneous hydrolysis than for acid-catalyzed hydrolysis [87]. The base-catalyzed hydrolysis order, q, or the hydroxide order, p, has been the subject of some controversy, and the issue is still far from resolved [88–93]. The reaction order in water has been assigned various values between 4.4 and 8. The critical concern remains the effect that the organofunctional silane group has on the alkoxy group hydrolysis rate. In this sort of work it is critical not only that each silane be individually evaluated on its own merits but also that hydrolysis rates be determined for each separate silane-alkoxy group. The large task of doing this for all silane coupling agents today has not even been half accomplished. When considering sizing formulations, the task has not even been started!

The progress of methoxysilane hydrolysis has been monitored using FTIR, generating nonlinear regression curves of [MeOH] versus time, from which hydrolysis rate constants can be determined [42]. The first hydrolysis step occurs far more rapidly than subsequent hydrolysis steps. Only hydrolysis rate constants for the first two hydrolysis steps were determined [42], as the similarity between the second and third hydrolysis steps could not be resolved by the regression analysis. This one study reveals not only the difficulty in assigning accurate rate constants but also the need for more than a single technique to be employed in investigating silane adsorption.

Two steps are involved in the process of condensation (see Fig. 3). The first step is the hydrogen bonding of a surface silanol (or other surface hydroxyl) group with the silanol groups of the hydrolyzed silane. A condensation reaction follows, forming an oxane bond with the surface and releasing one molecule of water. The degree to which such condensation occurs is largely determined by the temperature and duration of the drying conditions employed. Chemisorption often occurs in the first few molecular layers on the surface and is followed by strong physisorption in subsequent layers [94]. Most of the physisorbed molecules can be removed by washing the surface with suitable solvent [95], and for most silanes, wet acetone is extremely effective [1].

Studies conducted by Osterholtz and Pohl [77–79] on silanols established a general kinetic equation for silane condensation, Eq. (6), in terms of the concentrations of silane, acid, and base present.

$$-d[S]/dt = k_H[H^+][S']^2 + k_{HO}[OH^-][S']^2 + k_B[B][S']^2 \qquad (6)$$

Overall, adsorption processes are dependent on the rate of formation of the active (hydrolyzed) species, the ability of the active species to be attracted to the surface, its orientation with respect to the surface, and the type of layer it forms on the surface [39]. Most solid surfaces in aqueous solution develop some electric charge; the magnitude of this charge may vary, but it always exists. Glass surfaces are more hygroscopic than silica surfaces, because of the alkali metal oxides on the surface. Whereas the charge associated with silica surfaces makes them acidic, glass surfaces are predominantly neutral. Therefore, different substrates will induce silane surface condensation to different extents, not only because of the physical and chemical interactions of the silane with the charged surface but also because of the different availability of surface water to promote hydrolysis.

Reactions carried out in systems containing water have a disadvantage in that silane hydrolysis leads to siloxane formation before any interaction with the glass surface. Siloxanes, like most polymers, can strongly physisorb onto metal oxide surfaces, resulting in the formation of multiple physisorbed layers [4]. In order to avoid extensive hydrolysis and oligomerization of silanes, before interaction with glass surfaces, reactions must be performed in dry solvents. Direct condensation between surface silanols and alkoxy groups can still occur through the moisture resident on the glass surface [49], so there is no need for water to be present in the solvent for silanization to proceed. The extent of silane adsorption onto glass surfaces, then, depends on the ability of the chosen solvent to "extract moisture" from the glass surface, leading to eventual silane hydrolysis [96]. The intrinsic ability of various solvents to extract surface moisture can be determined by immersing dry glass fibers in D_2O and subsequently extracting the D_2O before reacting it with phenyl lithium and analyzing the resultant deuterobenzene by gas chromatography–mass spectrometry [96]. Aromatic solvents, such as benzene and toluene, extract the greatest amount of surface water and consequently produce the densest films. Conversely, solvents such as n-pentane extract the least surface water and so form the least substantial films [96].

Water adsorption capacities before and after glass fiber silane treatment have been determined and have shown untreated glass fibers to adsorb around 20 water molecules per nm^2, a significant quantity [97]. Dry solvents allow surface interactions to occur, without the strong hydrogen-bonding interactions that are inherent in aqueous solutions. For example, the orientation of γ-aminopropyltrimethoxy silane on glass varies, depending on whether the reaction is carried out in dry organic solvents or in aqueous solution. For a given silane, then, a dry solvent should induce surface adsorption of the silane in a defined and more readily controlled manner.

Surface condensation using dry solvents occurs, as exemplified by γ-aminopropyltrimethoxy silane in dry toluene, and results in condensation on quartz surfaces in a highly controlled manner [98]. Furthermore, other aminosilanes almost immediately condense from dry toluene to form monolayer films on acid-leached glass surfaces [99], and epoxysilanes will interact with silica from dry CCl_4 without ring opening of the epoxy group (by trace water acting as a nucleophile) [100]. Even from dry solvents it is possible to generate an interpenetrating network of the coupling agent (resident on the fiber) within the polymer matrix [1,23]. There does exist an optimal thickness for each coupling agent, which, if not achieved, results in a substantial decline in the overall performance of any resultant glass fiber-reinforced composite [1,23]. A large flexible polymeric backbone will enable the interphase to adjust to the steric constraints imposed by the glass surface and display a continuous reactive surface to the polymer matrix. It is largely for this reason that film-forming agents constitute the greatest part of a glass sizing agent formulation.

Using a mixture of silanes, the ratio of hydrophobic to hydrophilic groups can be adjusted to optimize the number of polar group interactions (which may also act as sites for water ingress), so that maximal dry strength and durability are achieved [1,23]. For this reason, several recent sizing formulations in the patent literature use at least two silane coupling agents together. It is necessary to control the hydrolysis and oligomerization rates very carefully if consistent and reproducible silane-modified surfaces are to be produced [101]. It is also necessary to control the degree of cross-linking and size of the polymeric silane segments, to ensure their interpenetration into the polymer matrix for optimal composite properties [1,9].

There are many more factors that may influence the adsorption and the resultant structure of the silane coupling agent interphase. The pH of an aqueous solution is important, because basic or acidic conditions can dramatically increase the rate of hydrolysis and condensation of the silane, thereby increasing the amount of silane adsorbed (see earlier). The surface potential of the glass surface also varies with the pH of the applied solution, affecting the orientation of the adsorbed silane layers [60]. This effect of pH on surface potential is more complex on mixed-oxide substrates, such as glass, as surface microheterogeneities may exist, with the resultant surface potential no longer a simple average of the component oxide potentials [60]. The drying conditions used for the silane-treated substrate also influence the final structure of the adsorbed interphase. The temperature and duration of the drying procedure will influence the number of siloxane bonds formed between adjacent silanes (siloxane formation), as well as with the surface. Such "siloxane coatings" with multiple surface bonds generally give improved composite performance [1,60].

Siloxanes, like silanes, are also capable of adhering to a variety of surfaces [4,102–104]. Siloxanes are strongly water-resistant polymers [105] and should, in principle, also be able to produce water-resistant interfaces between glass fibers and organic resins in composite materials. They display considerable backbone flexibility [106–109], allowing them to adjust to the sparse availability of reactive sites on glass surfaces. Siloxanes can be synthesized with various pendant functional groups, and the molecular weight distribution may also be far easier to control than the degree of oligomerization for analogous silanes in aqueous media [60,110–112]. Thus the investigation of siloxanes bearing appropriate functional groups may lead to a whole new class of coupling agents, with all the advantages of silanes, but with greater control over reproducibility in surface modification.

VI. INSTRUMENTAL TECHNIQUES—DIFFERENT APPROACHES PROVIDE CLEARER PERSPECTIVES

Chemical surface design is distinct from surface modification in that it aims at using chemically controllable surface systems through the modification of incipient surface species in well-defined chemical steps. The development of glass sizing formulations has come a long way, but today, with the availability of many surface-sensitive analytical techniques, the treatment of glass fiber surfaces can reach the stage of chemical surface design.

In the 1970s, the physical tools for direct characterization of solid surfaces became available and were widely used to exploit characteristic surface properties. Surface science principally involved the use of clean single-crystal substrates to investigate the behavior of adsorbates. Such experiments probed surfaces that were both static (unreactive) and dynamic (reactive). In the 1980s, surface science directed attention toward the study of the behavior of atoms or molecules on chemically modified surfaces, in preference to the clarification of static phenomena on clean, single-crystal surfaces. Modified surfaces were developed so that changes could be seen using new surface analytical techniques. Future research must now proceed to the next stage, in which surface modification and reaction mechanisms can be discussed on the basis of structures, compositions, and electronic states of active sites or assemblies on surfaces. Finding the changes in active sites and being able to control atoms, particularly on glass surfaces, will lead to the development of precise reaction spheres, essential to molecular level development of not only such industrially important areas such as composite materials but also better catalysts, etc. Improved surface activity and selectivity can then be achieved by applying the results of surface composition analyses to develop structure-activity relationships.

Earlier studies in our laboratories have examined the attachment of a number of functionalized siloxanes to E-glass fibers and compared them with commercial silane coupling agents [3,4,102]. X-ray photoelectron spectroscopy (XPS) and diffuse reflectance FTIR (DRIFT) were used to establish the presence of each functionalized siloxane on the glass surface and to compare the treated glass surfaces semiquantitatively. Interestingly, not only did siloxane-bearing trialkoxy functional groups adsorb onto glass fiber surfaces as effectively as a common silane coupling agent, vinyl-tris(2-methoxyethoxy) silane, but also other functional siloxanes (containing pendant amino, aminohydroxy, hydrido, and methacryl groups) strongly adsorbed [3,4,102].

Analysis of polymer adsorption on E-glass fibers is somewhat difficult, because of the small surface area ($\sim$0.12 m^2/g) and mixture of oxides inherent in glass. XPS is a sensitive technique capable of providing chemical information on the surface to a depth of only a few nanometers that has been applied to a wide range of modified surfaces [113]. Although XPS allows quantification of elements at the glass surface, interpretation of such data in terms of the "degree of coverage" or coating "thickness" is still the subject of much discussion in the literature [114–118]. The two most frequently used XPS methods make use of the fact that the number of photoelectrons detected from the glass atoms is exponentially reduced by an increase in the thickness of the coating [113]. The "overlayer" method compares the atomic concentrations detected at the surface for a particular glass atom present in the clean uncoated glass ($I_{Ca}*$) with that measured with the coating on the glass (I_{Ca}). The thickness of the coating may then be calculated from Eq. (7) [113]:

$$t = -\lambda_{Ca} \ln[1 - I_{Ca}/I_{Ca}^*] \, 2/\pi \tag{7}$$

where λ_{Ca} is the mean free path of the calcium atom's photoelectron. Equation (7) may be modified to accommodate various sample geometries.

Angle-resolved XPS varies the angle of the sample with respect to the detector and so varies the distance of detected photoelectrons from not only the substrate but also the coating [113–115,117]. For instance, at high incident X-ray angles the photoelectrons ejected from a thin film on a substrate will come primarily from the substrate, whereas at low incident angles the photoelectrons ejected will come primarily from the coating. Applying such procedures, it is possible to calculate the thickness of the coating based on a compilation database of mean free paths at a number of sample-detector angles. All such methods rely, however, on a number of assumptions [113–118]. For instance, the assumption that the coating is uniform is not very practical, when industrial reality indicates that it is very much dependent on the application system and procedures.

Wagner et al. [119] applied XPS to the analysis of 31 silicon-oxygen compounds and discovered that the difference between the energies of the O $1s$ and Si $2p$ photoelectron lines in inorganic silicon compounds was almost invariant, between 429.0 and 429.6 eV. However, for silicon polymers, significantly larger line differences arise, between 429.8 and 430.1 eV. The study remained almost an academic curiosity until the significance of the research to analyzing silanes and siloxanes adsorbed onto glass fiber surfaces became apparent at the University of South Australia [120]. This difference between the energies of the O $1s$ and Si $2p$ photoelectron lines can be used to characterize the surfaces of any silicon-oxygen materials as resembling either that of a siloxane or that of an inorganic silicate. The degree of silane or siloxane coating on a glass surface can be assessed using a method that is less affected by variances in the adventitious carbon levels present on all surfaces than previous XPS methods [113]. Although XPS has already been extensively applied to the study of silane adsorption onto many substrates, including E-glass, its potential for analyzing glass sizing agent adsorption is still in its infancy.

Transmission or attenuated total reflectance (ATR) FTIR spectroscopy on glass surfaces is difficult because of the strong incipient scattering of glass [121]. DRIFT has been used successfully in qualitative analysis of treated powder substrates. However, it extends only to semiquantitative measurements [122], as it relies on the scattering properties of the sample or substrate [123]. Characteristic FTIR vibrations of coatings on glass below 1600 cm^{-1} are difficult to identify using DRIFT, because of the strong Si—O and B—O absorbances of E-glass fibers in this region [124,125]. An overlayer of KBr is reported to alleviate this problem [123], but it is not that easy to reproduce this technique, and it is better if the adsorbing material has characteristic vibrations that can be monitored away from this region. Unlike silica, E-glass fibers exhibit only a very broad Si—OH vibration [124], so it is difficult to monitor the interactions between the coupling agent and the different surface silanols. DRIFT has, however, been employed in the region 2500–3100 cm^{-1}, primarily to confirm the extent of functionalized siloxane adsorption to glass fiber surfaces [3,4]. The only peak due to the glass substrate in this region is the first overtone of the B—O stretching vibration at 2671 cm^{-1} [126].

Many studies have used DRIFT to analyze fibers [122,127–130]; however, the direct measurement of silane adsorption onto glass fibers by DRIFT has only recently been developed by our group [3,4], and so its potential for analyzing the adsorption of glass sizing agents is also still in its infancy. The technique requires precise configuration of the FTIR spectrometer and modification of a commercially available DRIFT assembly. Our objective throughout the development of the technique was to identify spectroscopically submonolayer adsorption directly from E-glass fiber surfaces. Several spectroscopic options were pursued at the outset (including the KBr overlayer [123], procedure). Technology transfer has now seen the DRIFT technique implemented at the Åbo Akademi University and Ahlström Glassfibre in Finland. Along the way, we have learned that while some objectives may be difficult, worthwhile achievements are attained only by systematic pursuit of clearly defined scientific goals.

VII. SPECIFIC EXAMPLES—ONE SILANE AND ONE SILOXANE

Our objectives have led us into new ways to gather and treat both XPS and DRIFT data from silanes and siloxanes adsorbed onto E-glass fibers. A specific example will now help to show the effectiveness of this approach. Recently, we have investigated several commercial silane coupling agents, such as vinyl-tris(2-methoxyethoxy) silane, applied to E-glass fibers. Vinyl-tris(2-methoxyethoxy) silane (VTMES) was applied to E-glass fibers from dry ethanol as outlined in Table 3. Sample VTMES 1 was washed only with ethanol and then methylene

Table 3 Application Procedures for Silanes and Hydroxyl-Terminated PDMS

Silane or siloxane	Concentration (%)	Solvent	Posttreatment—sequential organic solvent wash
VTMES 1	4.98	Ethanol	Ethanol; methylene chloride (2 × 75 mL of each)
VTMES 2	3.78	Ethanol	Methylene chloride; toluene; hexane; acetone; methylene chloride (100 mL of each)
VTMES 3	5.57	Ethanol	Methylene chloride; toluene; hexane; acetone; methylene chloride (100 mL of each)
Hydroxy 1	3.67	Toluene	Methylene chloride (200 mL)
Hydroxy 2	3.67	Toluene	Methylene chloride; toluene (200 mL of each)
Hydroxy 3	3.67	Toluene	Methylene chloride; toluene; hexane (200 mL of each)
Hydroxy 4	3.67	Toluene	Methylene chloride; toluene; hexane; then acetone (200 mL of each)
Hydroxy 5	3.67	Toluene	Methylene chloride; toluene; hexane; acetone; methylene chloride (200 mL of each)

chloride, but samples VTMES 2 and 3 were subjected to a series of organic solvents: sequentially methylene chloride, then toluene, hexane, acetone, and finally methylene chloride again. E-glass fibers from Ahlström Glassfibre (see composition in Table 2) were used, with an average diameter of 10.5 ± 2.0 µm. The three samples were analyzed by a range of techniques (see later), including XPS and DRIFT.

The difference in the O 1s and Si 2p line energies for these samples are shown in Fig. 5. The XPS results indicate that the surface of the treated glass fiber sample VTMES 1 resembles that of a siloxane (429.9 eV), but the surfaces of the other two samples resemble unsized E-glass (429.3 eV). After argon ion etching, the surface of VTMES 1 also resembles that of E-glass, as the siloxane coating is effectively removed by the sputtering process.

The DRIFT spectra of VTMES 1, 2, and 3 are shown in Fig. 6. Such spectra support the XPS results, because substantial C—H vibrations are present for the sample VTMES 1, but such vibrations are barely seen for samples VTMES 2 and 3. The DRIFT spectra suggest the silane coupling agent (in samples VTMES 2 and 3) is almost removed after the extensive organic solvent washing procedure employed. As the silane coupling agent hydrolyzes and condenses to a siloxane on the glass surface during application, the adsorbing siloxane material becomes more hydrophobic and less soluble in ethanol than its silane counterpart and so adheres to the surface rather than remaining in the ethanol. Such adsorbed siloxanes, formed through the condensation of hydrolyzed silanes, physisorb to glass surfaces. Physisorbed siloxane oligomers are effectively removed when the

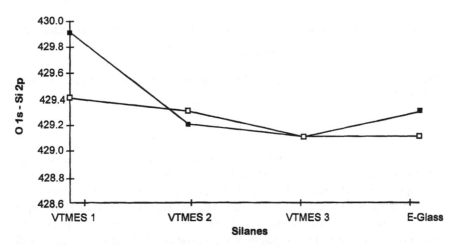

Figure 5 Difference between O 1s and Si 2p line energies for silane on E-glass fibers. (■) O—Si; (□) O—Si after etch.

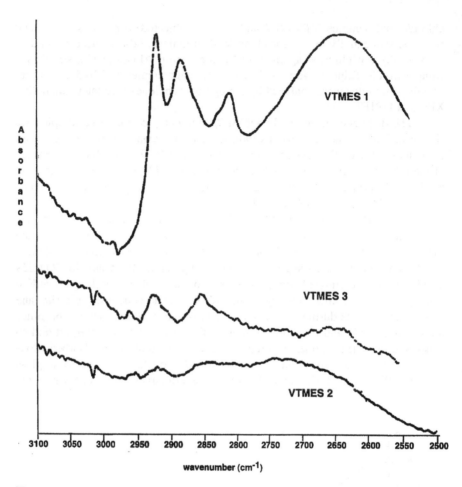

Figure 6 DRIFT spectra of vinyl-tris(2-methoxyethoxy) silane on E-glass fibers in the region for C—H IR vibrations.

treated glass surfaces are washed with suitable solvents, i.e., solvents (such as toluene and hexane) whose Hildebrand solubility parameters are closer to that of the physisorbed siloxane. Washing the treated fibers with less suitable solvents, such as ethanol and methylene chloride (i.e., a large difference in Hildebrand solubility parameters between the solvent and the physisorbed siloxane), results in less effective removal of the physisorbed siloxanes. Physisorbed siloxanes will chemisorb (the formation of siloxane bonds to the glass surface—see Fig. 3) on

Table 4 Hildebrand Solubility Parameters

Polymer or solvent	Solubility parameter $(\text{cal cm}^{-3})^{1/2}$
Poly(dimethylsiloxane)	7.3
n-Hexane	7.3
Toluene	8.9
Methylene chloride	9.7
Acetone	10.0
Ethanol	12.7

Source: Ref. 131.

oven drying the treated glass fibers at 100°C. Chemisorbed siloxanes or silanes will not, however, be removed even after extensive washing in a wide range of organic solvents.

Such results can be readily correlated with the different Hildebrand solubility parameters for the solvents used in washing the fiber samples (see Table 4). In our experience, wet acetone is the most effective solvent for removing physisorbed silanes. Siloxane polymers are another matter. The smaller the difference in Hildebrand solubility parameters between the siloxane and the polymer, the more soluble that polymer will be in the solvent [131]. The hypothesis can be further evaluated by testing the effects of the washing procedure (i.e., the sequential rinsing with methylene chloride, then toluene, hexane, acetone, and finally methylene chloride again) upon a siloxane having two terminal hydroxy groups that can interact with the glass surface.

There are a number of factors that affect the adsorption of polymers onto surfaces [75,132–134]. These include

The type of solvent
The polymer molecular weight and polydispersity
The concentration
The time allowed for adsorption
The number of reactive groups per molecule
The functionality of reactive groups on the polymer
The temperature
For aqueous systems, the pH
The type of posttreatment or drying conditions
The nature of the substrate itself

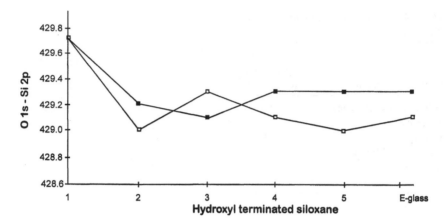

Figure 7 Differences between O 1s and Si 2p line energies for hydroxyl-terminated siloxane on E-glass fiber. (■) O—Si; (□) O—Si after etch.

Adsorption of functionalized siloxanes[135] or other polymers [136] onto glass surfaces can result in patchwise adsorption. Chemical and physical heterogeneities on the surface can be responsible for patchwise adsorption, which is thought to proceed in a multistep growth process. Each step occurs with smaller and smaller probability but results in bigger and bigger patches.

Hydroxyl-terminated polydimethylsiloxane (PDMS) was applied to E-glass fibers in toluene, and samples were removed during each stage of the washing procedure (see Table 3). The siloxanes used were not water soluble. The differences in the XPS O 1s and Si 2p line energies for these samples are shown in Fig. 7. The surface of the sample (hydroxy 1) washed only with methylene chloride approaches that of a siloxane polymer (429.7 eV), whereas that washed subsequently with toluene (hydroxy 2) is clearly similar to unsized E-glass. There is no change in the surface of hydroxy 1 even after a 5-min argon ion etch, indicating that there must be more siloxane present on the surface than for the silane VTMES 1. The surfaces of the other samples (hydroxy 3, 4, and 5), subjected to further washing, not surprisingly resemble that of the unsized E-glass.

The FTIR spectrum of the neat silicone oil and the DRIFT spectra of hydroxy 1 and E-glass fiber are shown in Fig. 8a. The DRIFT spectra of the samples (hydroxy 1, 2, 3, 4, and 5) sequentially removed during the washing procedure are shown in Fig. 8b. All the DRIFT spectra support the XPS data. Sample hydroxy 1 has strong C—H vibrations resembling those of the neat silicone oil, whereas after the toluene wash (second washing step), all subsequent samples have weak C—H vibrations, which do not resemble those of the neat siloxane. The initial wash in methylene chloride did not effectively remove the physisorbed siloxane, but toluene, with a closely matched Hildebrand solubility parameter, did.

Functionalized siloxanes adsorb to E-glass fibers, but only from organic solvents. Physisorbed siloxanes, like physisorbed silanes, are removed if washed extensively with a range of organic solvents. Toluene and hexane remove most physisorbed siloxanes quickly and effectively. Although some of the siloxane remains on the surface, the degree of coverage varies and is dependent on the functional groups pendant on the siloxane backbone. However, if the siloxane-modified surfaces are not rinsed or washed with appropriate organic solvents, substantial amounts of siloxane can desorb and migrate to the composite surface during processing of the final product. Unwashed siloxane-treated E-glass surfaces resemble poly(dimethylsiloxane) surfaces. Siloxanes bearing appropriate functional groups may eventually offer all of the advantages of silane coupling agents, but with the greater control and reproducibility inherent in the adsorption of defined polymeric reagents on surfaces.

It was just such considerations that led the Polymer Science Group to develop several basic siloxane microemulsions. One such microemulsion successfully completed pilot trials by a glass fiber manufacturer. The key to formulating a polymeric alternative to silane coupling agents arises out of the interfacial bonding shown in Fig 3, where silane hydrolysis after application to a hydroxylated surface causes siloxane formation. On this basis alone, suitably modified siloxanes should also be able to function as coupling agent alternatives.

In summary, then, XPS and DRIFT have proved powerful analytical techniques for characterizing the adsorption processes of siloxane polymers, bearing diverse functional groups, onto E-glass fibers [137]. In particular, the O $1s$ and Si $2p$ line energy difference is a useful means of assessing the degree of surface coverage of a siloxane quickly and effectively. It is possible to assess the strength of the attachment by examining the modified surface after it has been washed with a series of organic solvents, particularly using solvents with Hildebrand solubility parameters similar to that of the adsorbed polymer [138].

Silanes do not chemisorb immediately, because no silane remains if extensive washing is applied within 15 min after silane application (the same is not true for all functionalized siloxanes). Each silane will chemisorb with time (the time differs from silane to silane), so that extensive washing will no longer remove all the adsorbed silane. From aqueous solutions, it is likely that hydrolyzed silanes that have started to condense into oligomeric siloxanes physisorb first from solution. In aqueous solution both oligomeric siloxanes and hydrolyzed silanes physisorb rapidly onto glass surfaces, but removing all physisorbed species from the surface will require washing with a series of organic solvents, as the monomeric and polymeric species adsorbed have different solubility parameters. Such initial observations permit research to be directed now toward understand-

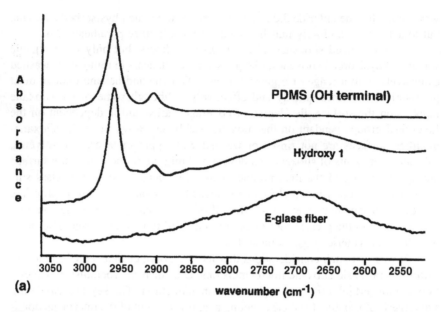

Figure 8 (a) IR spectrum of neat hydroxyl-terminated PDMS oil, DRIFT spectrum of hydroxyl-terminated PDMS (hydroxy 1) on E-glass fibers, and DRIFT spectrum of unsized E-glass fibers in the region for C—H IR vibrations. (b) DRIFT spectra of hydroxyl-terminated PDMS on E-glass fibers in the region for C—H IR vibrations.

ing how glass sizing formulations, in which both polymeric and monomeric species abound, adsorb.

We have now developed a wide variety of spectroscopic techniques, including DRIFT, XPS, AFM, Raman, NMR, streaming potential, high-resolution thermogravimetric analysis (TGA), secondary ion mass spectrometry (SIMS), and very recently time-of flight (TOF) SIMS, to probe silane-substrate interfacial regions (Fig. 9 illustrates our *initial* instrumental targets for examining silane–glass fiber substrates). An understanding of complex phenomena demands data from a broad spectrum of analytical techniques. New objectives in understanding the complex adsorption behavior of sizing formulations have been set, and the future research targets will stretch the range of experimental techniques already developed (see Fig. 9) into new areas.

VIII. GLASS SIZING AGENTS

Silane coupling agents are but one of the many ingredients in commercial sizings that are applied to glass fibers. They play an important role in interfacial adhesion

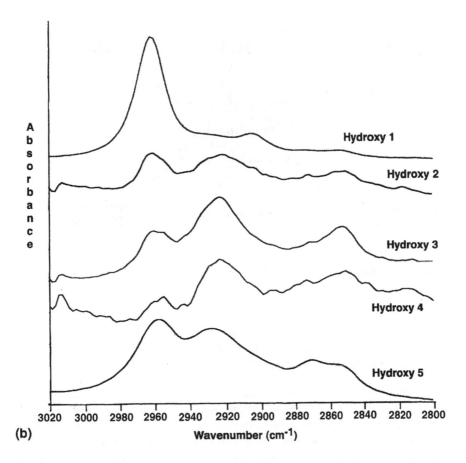

(b)

in many composite systems, but the mechanisms by which they improve product properties are still not well understood. Sizes and subsequent surface treatments that use silanes are often developed entirely empirically as a consequence. The structure of the coupling agent, the variables associated with silane deposition, and the respective concentrations of a plethora of additives are all adjusted in large matrix experiments, using end-product properties as criteria for improved or diminished performance.

In investigating complex phenomena, the key question is, what exactly do we wish to learn? The very reason that so many different approaches to coupling agents exist in the literature is that industry and university researchers both have different answers to this one key question. In dealing with sizing agents on glass fibers three key issues arise, and the answers to the following interrelated questions will determine the development of this research area:

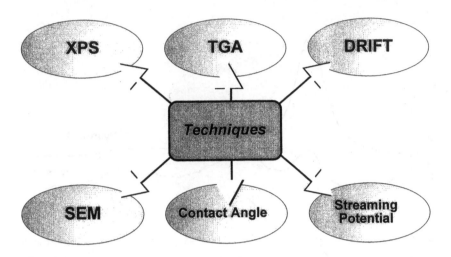

Figure 9   Analysis techniques for E-glass fibers.

    1.   Is the size going to adsorb onto the fiber?
    2.   Is the size going to desorb from the fiber?
    3.   Will the size accomplish its intended task?

Industry is most interested in the answer to the third question (as it should be). The first two questions do help scientists to understand and evaluate the third question, but if the perfect sizing formulation could be found without any prior knowledge, industry would not be too concerned if it never learned why that particular sizing formulation was successful. As any research field matures, it does become necessary to understand the reasons and processes that constitute successful materials (sizing agents), as performance levels are often sought or extended beyond initial or existing limits. The answers to the "largely academic" first two questions then become much more important. Understanding the best way to use alkoxy silanes in sizing formulations has developed empirically. As their use has increased, understanding the processes associated with their use as adhesion promoters has not only become useful but has now proved essential. The previous lack of suitable analytical techniques during the 1970s and early 1980s has passed, as surface-specific instrumental techniques have advanced at a tremendous pace in the past 10 years. Today, detailed studies of surface interactions are available through a host of advanced surface analytical techniques, and there is no longer any reason why surfaces such as silica should be used as simplified models of any inorganic oxide surface, such as glass. The resources are there that, combined with a little ingenuity, will allow the surface in question to be studied directly as changes occur on that surface.

REFERENCES

1. EP Plueddemann. Silane Coupling Agents. New York: Plenum, 1982.
2. FD Osterholtz, ER Pohl. In: KL Mittal, ed. Silanes and Other Coupling Agents. Utrecht, The Netherlands: VSP, 1992, p 119.
3. PS Arora, JG Matisons A Provatas, RStC Smart. Langmuir 11:2009, 1995.
4. LG Britcher, DC Kehoe, JG Matisons, RSC Smart, AG Swincer. Langmuir 9:1609, 1993.
5. DJ Vaughan, RC Peek Jr. Proc 29th Annual Technical Conference, SPI Reinforced Plastic Composites Institute, Sect 13-D, 1974, pp 1–4.
6. DW Dwight, FM Fowkes, DA Cole, MJ Kulp, PJ Sabat, L Salvati Jr, TC Huang, J Adhesion Sci Technol 4:619, 1990.
7. FM Fowkes, DW Dwight, DA Cole, TC Huang. J Non-Cryst Solids 120:47, 1990.
8. H Ishida. In: H Ishida, G Kumar, eds. Molecular Characterisation of Composite Interfaces. New York: Plenum, 1985, p 34.
9. EP Plueddemann. In: H Ishida, ed. Interfaces in Polymer, Ceramic and Metal Matrix Composites. New York: Elsevier, 1988, p 17.
10. JG Matisons, AE Jokinen, JB Rosenholm. J Colloid Interface Sci 194:263, 1997.
11. JG Matisons, AE Jokinen, JB Rosenholm. Silicones in Coating II, Paint Research Association, Orlando, 1998, paper 9, pp 1–14.
12. JG Matisons, AE Jokinen, JB Rosenholm. J. Mater. Sci. Lett., 17:149, 1997.
13. DJ Ondrus, FJ Boerio. J Colloid Interface Sci 124:349, 1988.
14. FJ Boerio, RG Dillingham. In: KL Mittal, ed. Adhesive Joints: Formation, Characteristics, and Testing. New York: Plenum, 1984, p 541.
15. K Esumi, K Meguro. Bull Chem Soc Jpn 56:331, 1983.
16. PR Moses, LM Weir, JC Lennox, HO Finklea, JR Lenhard, RW Murray. Anal Chem 50:576, 1978.
17. DJ Hook, TG Vargo, JA Gardella, KS Litwiler, FV Bright. Langmuir 7:142, 1991.
18. MR Rosen, ED Goddard. 34th Annual Technical Conference, SPI Reinforced Plastic Composites Institute, Sect 19-E, 1979, p 1.
19. MG Bader, JE Bailey, I Bell. In: W Kriegel, H Palmour, eds. Ceramics in Severe Environments. Vol 5. New York: Plenum, 1972.
20. AG Atkins. J Mater Sci 10:819, 1975.
21. JO Outwater. J Adhesion 2:242, 1970.
22. DL Angst, GW Simmons. Langmuir 7:2236, 1991.
23. JM Park, RV Subramanian. J Adhesion Sci Technol 5:459, 1991.
24. CG Pantano, TN Wittberg. Surf Interface Anal 15:498, 1990.
25. B Arkles In: R Anderson, GL Larson, C Smith, eds. Hüls Silicon Compound Register and Review. 5th ed. Piscataway, NJ, Hüls America: 1991, p 59.
26. D Wang, FR Jones, P Denison. J Adhesion Sci Technol 6:79, 1992.

27. KW Allen. J Adhesion Sci Technol 6:23, 1992.
28. EK Drown, H Al Moussawi, LT Drzal. J Adhesion Sci Technol 5:865, 1991.
29. CP Tripp, ML Hair. Langmuir 8:1120, 1991.
30. GE Maciel, DW Sindorf. J Am Chem Soc 102:7606, 1980.
31. DW Sindorf, GE Maciel. J Am Chem Soc 103:4263, 1981.
32. DW Sindorf, GE Maciel. J Am Chem Soc 105:1487, 1983.
33. DW Sindorf, GE Maciel. J Am Chem Soc 105:3767, 1983.
34. DJ Ondrus, FJ Boerio, KJ Grannen. J Adhesion 29:27, 1989.
35. JE Sandoval, JJ Pesek. Anal Chem 61:2067, 1989.
36. KMR Kallury, UJ Krull, M Thompson. Anal Chem 60:169, 1988.
37. CG Pantano, LA Carman, S Warner. In: KL Mittal, ed. Silanes and Other Coupling Agents. Utrecht: VSP, 1992, p 229.
38. H Wang, JM Harris. J Am Chem Soc 116:5754, 1994.
39. FD Blum, W Meesiri, H-J Kang, JE Gambogi. In: KL Mittal, ed. Silanes and Other Coupling Agents. Utrecht: VSP, 1992, p 181.
40. B Arkles. Chemtech 7:766, 1977.
41. MJ Owen. In: JM Ziegler, FWG Fearon, eds. Silicon-Based Polymer Science: A Comprehensive Resource. Adv Chem Ser 224. Washington, DC: American Chemical Society, 1990, p 705.
42. DE Leyden, JB Atwater. In: KL Mittal, ed. Silanes and Other Coupling Agents. Utrecht: VSP, 1992, p 43.
43. N Shirai, K Moriya, Y Kawazoe. Tetrahedron 42:2211, 1986.
44. FM Fowkes. J Adhesion Sci Technol 1:7, 1987.
45. E Osmont, HP Schreiber. In: DR Lloyd, TC Ward, HP Schreiber, eds. Inverse Gas Chromatography. ACS Symp Ser 391. Washington, DC: American Chemical Society, 1989, pp 230–247.
46. AC Tiburcio, JA Manson. J Appl Polym Sci 42:427, 1991.
47. AC Tiburcio, JA Manson. J Adhesion Sci Technol 4:653, 1990.
48. K Tsutsumi, T Ohsuga. Colloid Polym Sci 236:38, 1990.
49. R Doremus. Glass Science. New York: Wiley, 1973.
50. L Holland. The Properties of Glass Surfaces. London: Chapman & Hall, 1964.
51. D Rosington. In: L Pye, H Stevens, W La Course, eds. Introduction to Glass Science. New York: Plenum, 1972, p 101.
52. A Kruger. In: JD Nowotny, LC Four, eds. Surface and Near Surface Chemistry of Oxide Materials. Materials Science Monographs, Vol 47. Amsterdam: Elsevier, 1988, Chap 9.
53. W Weyl. Glass Ind 26:12, 1945.
54. W Weyl. Glass Ind 28:231, 1947.
55. W Weyl. J Am Ceram Soc 32:367, 1949.
56. L Holland. Glass Ind 44:268, 1963.
57. H Williams, W Weyl. Glass Ind 26:275, 1945.
58. S Kister. J Am Ceram Soc 45:59, 1962.
59. F Ernsberger. Phys Chem 21:146, 1980.
60. H Ishida. In: KL Mittal, ed. Adhesion Aspects of Polymeric Coatings. New York: Plenum, 1983, p 45.
61. D Haaland. Appl Spectrosc 40:1152, 1986.

62. V Davydov, A Kiselev, L Zhuravlev. Trans Faraday Soc 60:2254, 1964.
63. M Hair. Infrared Spectroscopy in Surface Chemistry. New York: Marcel Dekker, 1967.
64. L Little. Infrared Spectra of Adsorbed Species. London: Academic Press, 1966.
65. CG Armistead, A Tyler, F Hambleton, S Mitchell, J Hockey. J Phys Chem 73: 3947, 1969.
66. T Horr. PhD thesis, Ian Wark Research Institute, University of South Australia, 1992.
67. A Rastogi, J Rynd, W Stassen. Proc. 31st Annual Technical Conference, SPI Reinforced. Plastic Composites Institute, Sect 6-B, 1976, p 1.
68. M Hair. In: G Goldfinger, ed. Clean Surfaces: Their Preparation and Characterization for Interfacial Studies. New York: Marcel Dekker, 1970, pp 269–284.
69. M Mohai, L Bertóti, M Révész. Surf Interface Anal 15:364, 1990.
70. E Shafrin, W Zisman. J Am Ceram Soc 50:478, 1967.
71. A Paul. J Mater Sci. 12:2246, 1977.
72. H Scholze, D Helmreich, I Berkardjiev. Glastech Ber 48:237, 1975.
73. WH Casey, B Bunker. Rev Mineral 23:397, 1990.
74. J-C Petit, J-C Dran, MG Della. Nature 334:621, 1990.
75. SM Cohen, T Cosgrove, B Vincent. Adv Colloid Interface Sci 24:143, 1986.
76. G Tesoro, Y Wu. In: KL Mittal, ed. Silanes and Other Coupling Agents. Utrecht: VSP, 1992, p 215.
77. FD Osterholtz, ER Pohl. In: H Ishida, G Kumar, eds. Molecular Characterisation of Composite Interfaces. New York: Plenum, 1985, p 157.
78. FD Osterholtz, ER Pohl. In: DE Leyden, WT Collins, eds. Chemically Modified Surfaces in Science and Technology. London: Gordon & Breach, 1987, p 545.
79. FD Osterholtz, ER Pohl. In: DE Leyden, ed. Silanes, Surfaces and Interfaces. New York: Gordon & Breach, 1986, p 485.
80. C-C Lin, JD Basil. Mater Res Soc Symp Proc 73:585, 1986.
81. S Savard, LP Blanchard, J Leonard, RE Prud'homme. Polym Composites 5:242, 1984.
82. F Devreux, JP Boilot, F Chaput. Phys Rev A 41:6901, 1990.
83. N Nishiyama, K Horie, T Asakura. J Appl Polym Sci 34:1619, 1987.
84. ER Pohl, CS Blackwell. In: H Ishida, ed. Controlled Interphases in Composite Materials. Amsterdam: Elsevier, 1990, p 37.
85. KJ McNeil, JA DiCapri, DA Walsh, RF Pratt. J Am Chem Soc 102: 1859, 1980.
86. AA Humffray, JJ Ryan. J Chem Soc B 1138, 1969.
87. JR Chipperfield, GE Gould. J Chem Soc Perkin Trans II 1324, 1974.
88. E Akerman. Acta Chem Scand 10:298, 1956.
89. E Akerman. Acta Chem Scand 11:373, 1957.
90. C Eaborn, R Eidenschink, DRM Walton. J Chem Soc Chem Commun 388, 1975.
91. ER Pohl. Proc. 38th Annual Technical Conference, SPI Reinforced Plastic Composites Institute, Sect 4-B, 1983, p 1.
92. H Slebocka-Tilk, RS Brown. J Org Chem 50:4638, 1985.
93. KA Smith. J Org Chem 51:3827, 1986.
94. DC Kehoe. Silicone Modification of E-glass Fiber Surfaces. PhD thesis, University of South Australia, 1996.

95. FD Blum W Meesiri, H-J Kang, JE Gambogi. In KL Mittal, ed. Silanes and Other Coupling Agents. Utrecht: VSP, 1992, p 181.
96. ME McGovern, KMR Kallury, M Thompson. Langmuir 10:3067, 1994.
97. R Wong. J Adhesion 4:171, 1972.
98. SV Matveev. Biosensors Bioelectron 9:333, 1994.
99. LV Phillips, DM Hercules. In: DE Leyden, ed. Silanes, Surfaces, and Interfaces. New York: Gordon Breach, 1986, p 235.
100. AS Piers, CH Rochester. J Chem Soc Faraday Trans 90:1253, 1995.
101. B Arkles, J Steinmetz, J Zazyczny, P Mehta. J Adhesion Sci Technol 6:193, 1992.
102. DR Bennett, JG Matisons, AKO Netting, RstC Smart, AG Swincer. Polym In 27: 147, 1992.
103. P Auroy, L Auvray, L Léger. J Colloid Interface Sci 150:187, 1992.
104. T Cosgrove, CA Prestidge, B Vincent. J Chem Soc Faraday Trans 86:1377, 1990.
105. W Noll. The Chemistry and Technology of Silicones. 2nd ed. New York: Academic Press, 1979, pp 443, 470.
106. VM Litvinov, BD Lavrukhin, AA Zhdanov. Polym Sci USSR Ser A 27:2474, 1985.
107. VM Litvinov, BD Lavrukhin, AA Zhdanov. Polym Sci USSR Ser A 27: 2482, 1985.
108. IL Dubchak, AI Pertsin, AA Zhdanov. Polym Sci USSR Ser A 27:2340, 1985.
109. A Viallat, JP Cohen-Addad, A Pouchelon. Polymer 27:843, 1986.
110. SK Duplock, JG Matisons, AG Swincer, RFO Warren. J Inorg Organomet Polym 1:361, 1991.
111. LG Britcher, DC Kehoe, JG Matisons, AG Swincer. Macromolecules 28:3110, 1995.
112. JG Matisons, A Provatas. Macromolecules 27:3397, 1994.
113. D Briggs, MP Seah, eds. Practical Analysis by Auger and X-ray Photoelectron Spectroscopy. New York: Wiley, 1983.
114. TL Barr. Crit Rev Anal Chem Pt 1:113, 1991.
115. OAJ Baschenko. Electron Spectrosc Relat Phenom 57:297, 1991.
116. L-S Johansson. Surf Interface Anal 17:663, 1991.
117. JE Fulghum. Surf Interface Anal 20:161, 1993.
118. K-I Nishimori, K Tanaka. J Vac Sci Technol A8:3300, 1990.
119. CD Wagner, DE Passoja, HF Hillery, TG Kinisky, HA Six, WT Jansen, JA Taylor. J Vac Sci Technol 21:933, 1982.
120. LG Britcher. PhD thesis, Ian Wark Research Institute, University of South Australia, 1997.
121. H Ishida, JL Koenig. Polym Eng Sci 18:128, 1978.
122. PJ Brimmer, PR Griffiths. Anal Chem 58:2179, 1986.
123. SR Culler, MT McKenzie, LJ Fina, H Ishida, JL Koenig. Appl Spectrosc 38:791, 1984.
124. E Lipp, AL Smith. In: AL Smith, ed. The Analytical Chemistry of Silicones. New York: Wiley, 1991, pp 49, 325–333.
125. RA Condrate, Sr. In: LD Pye, HJ Stevens, WC LaCourse, eds. Introduction to Glass Science. New York: Plenum, 1972, p 101.
126. D Haaland. Appl Spectrosc 40:1152, 1986.
127. M Fuller, P Griffiths. Appl Spectrosc. 34:533, 1980.

128. O Ahmed, O Gallei. Appl Spectrosc 28:430, 1974.
129. H Hecht. Anal Chem 48:775, 1976.
130. M Fuller, P Griffiths. Anal Chem 50:1906, 1978.
131. EA Grulke in Polymer Handbook, 3rd ed., J Bandrup, EH Immergut (eds.), Wiley Interscience, 1989.
132. M Kawaguchi, A Takahashi. Adv Colloid Interface Sci 37:219, 1992.
133. A Hariharan, S Kumar, T Russell. Macromolecules 23:3584, 1990.
134. TGM Van de Ven. In: Kinetic Aspects of Polymer and Polyelectrolyte Adsorption on Surfaces. Report MR 266. Pulp & Paper Research Institute, Canada, October 1993.
135. WJR Botter, S Ferreira, F Galembeck. J Adhesion Sci Technol 6:791, 1992.
136. HM Fagerholm, C Lindsjö, JB Rosenholm, K Rökman. Colloids Surfaces 69:79, 1992.
137. A Chilkoti, BD Ratner. Surface Interface Anal 17:567, 1991.
138. CW Chu, DP Kirby, PD Murphy. J Adhesion Sci Technol 7:417, 1993.

Index

ISBN 0-8247-0239-5